THE URBAN TRANSPORTATION OF IRRADIATED FUEL

THE URBAN TRANSPORTATION OF IRRADIATED FUEL

Edited by

John Surrey

Head of the Energy Policy Programme
Science Policy Research Unit, University of Sussex

Foreword by

Ken Livingstone

Leader, Greater London Council

MACMILLAN PRESS
LONDON

First published 1984 by
THE MACMILLAN PRESS LTD
London and Basingstoke
Companies and representatives
throughout the world

Typeset by
Wessex Typesetters Ltd
Frome, Somerset

Printed in Great Britain by
Camelot Press Ltd
Southampton

British Library Cataloguing in Publication Data

The urban transportation of irradiated
fuel.
1. Nuclear fuels – Transportation –
Safety measures – Congresses
I. Surrey, John
363.1′79 TK9360
ISBN 0–333–36938–6
ISBN 0–333–36939–4 Pbk

Contents

Foreword by Ken Livingstone, Leader of the Greater London Council

The GLC's concern about the transportation of spent nuclear fuel through its densely populated areas dates back to 1980, when members of all parties expressed their concern about the lack of information on safety questions. The Conference held in London in April 1983, of which this book is a record, was therefore a culmination of the Council's concern and a desire to address the issues, which transcended party boundaries.

I am grateful to those eminent speakers who took part, and the conference organisers, who endeavoured to provide the public with a wide spectrum of views on the subject, both scientific and psychological. Thanks are also due to the many people who attended and contributed to the debate, which formed such an important feature of this Conference. The questions and fears which they expressed are included in this account of the proceedings, as they rightly should, thus fulfilling one of the aims of the conference to be a medium of two-way information flow.

In November 1983, six months after our Conference highlighted, amongst other potential dangers, the very real deficiencies of tests on flasks, and in particular the use of scale models or computer modelling to simulate an accident, the CEGB have announced that they will now stage full-scale tests. They are also withdrawing from service a faulty design of Magnox flask. At the same time the Inspector presiding over the Inquiry into the proposal to build a pressurised water reactor at Sizewell announced the appointment of an additional assessor to advise on matters concerning the transport of irradiated nuclear fuel.

The legitimate fears of Londoners cannot be swept aside, and therefore I hope that this book will serve to better inform both the regulatory authorities of the nature of these fears, and the population of London, and indeed of other cities, of the way in which some of these problems are addressed. The final aim must be to remove any perceived potential threat from our cities, and I hope this book is a significant step in that direction.

Acknowledgements

The Greater London Council would like to thank the following for their assistance in arranging the conference and, subsequently, this book.

Conference organisers	Technica Ltd, 11 John Street, London WC1.
Conference programme manager	Patricia King-Smith
Technical director	David Slater
Technical coordinator	Bethan Morgan
Conference ticket manager	Judith Page
Exhibition organiser	Pamela Glick
Printing/artwork	Frank Spencer
Secretarial services	Krystina Noble-Campbell

Council officers
Rick Kelly
John Beatson
Valerie Howell
Gerry Noakes
Carol White
The staff and management of the Connaught Rooms

Conference Book

Editor	John Surrey
Managing editors	Patricia King-Smith Bethan Morgan

Notes on the Contributors

Robert Barker is now a part-time consultant on transportation matters, having recently retired from the US NRC and IAEA where he had been in charge of the Agency's international transport safety programme since 1979. He has been directly involved with the transportation of radioactive material since the early 1950s, compiling the first Los Alamos Monitors Handbook in 1953. From 1954 he worked for both the IAEA and US AEC and was personally involved in the development of both US and International Standards. In the 1970s he prepared several environmental impact statements on transportation to and from nuclear power reactors. He is an engineer and health physicist by training.

Marcel J. Berthier has been employed in the French Ministry of the Interior since 1977, where he has been concerned with peacetime nuclear matters. Within this Ministry, the Civil Security directorate has been responsible for the setting up of arrangements in the departments for the storage and transport of nuclear and combustible materials. Marcel Berthier has engaged in work for international organisations including the International Atomic Energy Agency (IAEA). He has made on-site studies of the solutions appertaining to these problems in many countries. His work currently involves the encouragement and coordination in France of the use of radiological response equipment, training of fire officers and keeping responsible 'Securité Civile' departments informed. He divides his time equally between the transport of radioactive materials and control of their efficient routing.

Jean-Claude Charrault was born in 1932. He graduated from the Ecole Centrale des Arts et Manufacture de Paris in 1957 and in 1958 joined the French Commissariat à l'Energie Atomique. In 1960 he became Directorate-General for Research of EURATOM and since 1971 has worked for the Commission of the European Communities as Head of Division (Nuclear Policy).

Roger Clarke has worked actively in the radiological protection field for 14 years, firstly as a research scientist carrying out work on the environmental and biological effects of radiation. In 1978 he joined the National Radiological Protection Board to head its Assessments Department and in 1983 became Board Secretary. Dr Clarke has published widely in the open literature on the topics of health effects of radiation, assessment of the consequences of routine and accidental releases of radioactive material and emergency planning. He is a consultant to the United Nations in the field of effects of atomic radiation, acts as an independent Expert for the Commission of the European Communities and is a member of a number of international committees concerned with radiological protection.

Richard Clutterbuck has a Cambridge MA in Mechanical Sciences and a London PhD in Politics. He served in the Royal Engineers for 35 years, retiring as a major-general in 1972 and is now Reader in Political Conflict at Exeter University. He participated as an engineer in the nuclear trials on Christmas Island in 1958. He is a director of Control Risks Ltd which specialises in international security against terrorism. He has written nine books, of which the most recent are *Kidnap and Ransom* (1978), *Britain in Agony* (1980), and *The Media and Potential Violence* (1983).

Tony Cox is a Director of Technica Ltd, and has an overall responsibility for the company's scientific and engineering work in addition to leading specific projects in the areas of hazard-analysis and risk-assessment. He is internationally known for his work on the subject of vapour-cloud dispersion. His main current interest is in the development of hazard-analysis techniques for more general and practical application. He graduated from Cambridge with first-class honours in Mechanical Sciences and then carried out research into air pollution in the Department of Mathematics, Imperial College, London, for which he was awarded a PhD. He is a Chartered Engineer, Member of the Institute of Energy, and Fellow of the Royal Meteorological Society.

Commander Cree joined the Metropolitan Police in 1958 and has served in various parts of London. From May 1980 to October 1983 he was in charge of a number of branches at New Scotland Yard including that which dealt with planning for major incidents and the coordination of policing for public-order events. During his period at New Scotland Yard he was involved in the planning for policing the Royal Wedding in 1981, the visits of the Pope and the President of the USA in 1982. He left New Scotland Yard on 3 October 1983 to take command of a district based at Southwark.

Brian Evason graduated in Civil Engineering from Southampton University in 1969. He joined Taylor Woodrow Construction Ltd and worked in various civil engineering capacities until 1976, including work on feasibility and design studies for concrete oil-platforms for the North Sea and elsewhere. He then transferred to a department specialising in problems outside the common experiences of the profession, particularly structural dynamics. His special interest is the response to impact of concrete structures, in developing a fuller understanding of the behaviour and finding reliable methods of prediction. He presented a paper on the subject to the 6th SMIRT Conference (Paris 1981).

Lord Flowers was appointed Rector of The Imperial College of Science and Technology in 1973 following six years as Chairman of the Science Research Council. He is Chairman of the Committee of Vice Chancellors and Principals. He is a theoretical physicist who spent much of his early career at UKAEA Harwell. He is probably best known as past Chairman of Computer Board for Universities and Research Councils (1966–70), Royal Commission on Environmental Pollution (1973–6), Standing Commission on Energy and the Environment (1978–81) and the University of London Working Party on the Future of Medical and Dental Teaching Resources (1979–80). He was President of the European Science Foundation from 1974–80. He is a Fellow of the Royal Society, holder of numerous honorary degrees and fellowships and an Officier de la Légion d'Honneur. He is a Member of the House of Lords Select Committee on Science and Technology.

David Hall has been the Director of the Town and Country Planning Association since 1967. He is a chartered town planner and has held senior posts in English local government. He is Chairman of the Council for Urban Studies Centres and a Board Member of the Habitat International Council. He was visiting professor to the Centre for Metropolitan Planning and Research, Baltimore, in 1975 and consultant to the Quebec government in 1978. He has lectured and broadcast extensively in Britain and overseas, and is the author of numerous articles on planning matters. He has a specialist interest in the decision-making process for large projects such as nuclear power stations.

Peter Jacques is the Secretary of the TUC's Social Insurance and Industrial Welfare Committee and head of the TUC's Social Insurance and Industrial Welfare Department. He is a TUC member of the Health and Safety Commission, a member of the Industrial Injuries Advisory Council, and a member of the Social Security Advisory Committee. He was also a member of the Royal Commission on the National Health Service. Within the TUC he is responsible for social security and related collective bargaining on issues such as occupational pensions, NHS spending and industrial relations, health and safety at work, legal services and environmental policy.

Robert M. Jefferson is manager of the Transportation Technology Center at Sandia National Laboratories in Albuquerque, New Mexico, USA, whose activities are devoted to research and development addressing the transportation of radioactive materials. This Department of Energy-funded activity is responsible for technical evaluation of shipping systems, development of new materials and technology, testing, risk assessment, evaluation of institutional considerations, and information collection and dissemination. In addition to his full-time duties at Sandia, Robert Jefferson was an adjunct professor of nuclear engineering at the University of New Mexico for 16 years. Additional professional background includes work on radiation effects, reactor design, and operational responsibility for a range of reactors, accelerators and other radiation-producing devices.

Professor Terence R. Lee has been Head of the Department of Psychology at the University of Surrey since 1971 and was pro-Vice Chancellor at the same University from 1977 to 1981. He studied psychology at Magdalene College, Cambridge, gained a first degree in 1949 and a PhD in 1954 and he then held posts at the universities of Exeter, St Andrews and Dundee. His publications include *The Perception of Risk* (Royal Society, 1982), *The Public's Perceptions of Risk and The Question of Irrationality* (Royal Society, 1981), *Psychology and the Environment* (Methuen, 1976), and *Psychology and the Built Environment* (with D. V. Canter, Architectural Press, 1974).

Charles E. MacDonald is Chief of the Transportation Certification Branch, US Nuclear Regulatory Commission, Washington DC. He worked for the US Merchant Marine Academy in 1956 and was employed by the Combustion Engineering Inc. for mechanical design of pressurized water reactors and radiological safety from 1956 to 1963, with 2 years unrelated experience in the US Navy. From 1963 he was employed by the US Atomic Energy Commission and the US Nuclear Regulatory Commission in the regulation of the nuclear industry. Since 1972 he has been responsible for the review and approval of applications for the packaging of radioactive material. He is active in the American National Standards Institute and American Society of Mechanical Engineers for the development of standards for the safety of transport of radioactive materials.

Gordon MacKerron is a Fellow of the Energy Programme at the Science Policy Research Unit, University of Sussex. With a background in economics and project evaluation, he has worked since 1978 on issues of energy policy and economics, specialising on the UK and international experience of electricity use and nuclear power. He has published a number of papers in this area and was an adviser to the Monopolies Commission in its investigation of the CEGB (1980/1). He also appeared as a witness at the Sizewell Inquiry for the Electricity Consumers' Council.

Harbans L. Malhotra is an International Fire Protection Consultant; he is Chairman of the Committee on Fire Resistance Tests for both the BSI and ISO, and Chairman of the RILEN Coordination Group, and UK Representative on the General Council. He retired in 1983 from the Fire Research Station after thirty-two years as their Senior Principal Scientific Officer and head of the Building and Structures Division. He has specialised in research into the properties of materials, particularly of concrete, at high temperatures and into the behaviour of structures in fire. He is a member of the Concrete Society and the Institute of Structural Engineers Committee on Fire.

Fred Millar, PhD, is Director of the Nuclear and Hazardous Materials Transportation Project of the Environmental Policy Institute. He has testified before the US Congress and numerous state and local legislative bodies, has helped to initiate two national level lawsuits against the US Nuclear Regulatory Commission (NRC) on nuclear waste transport and has led the lobbying effort that forced the NRC to reveal the rail and highway routes for spent fuel shipments. His research has provided citizens' groups all over the US with information on safety problems and legal issues. He assisted state attorneys-general in New York and Ohio in their lawsuits against the federal attempt to override local and state laws regulating nuclear transportation. Prior to joining EPI he was a professor of sociology and a consultant to the Ohio Public Interest Campaign.

Raymund O'Sullivan joined the Department of Transport in 1969 and since 1981 has held the post of Transport Radiological Adviser to the Secretary of State, and Head of the Radioactive Materials Transport Division.

David W. Pearce is Professor of Political Economy at University College London. He is the author or editor of eighteen books including *Cost-Benefit Analysis* (Macmillan), *Environmental Economics* (Longman), *Social Projects Appraisal* (with C. A. Nash) (Macmillan), and *Decision-Making for Energy Futures: a Case Study of the Windscale Inquiry* (Macmillan). He has produced some 100 articles in learned journals, is on the editorial board of six journals, has acted as a consultant to OECD, Commission of the European Communities, ECE (Geneva), the World Bank and the Governments of Egypt and Thailand. He is a member of the National Radiological Protection Board. Married with two children, David Pearce lives in Bedford.

Dr Marvin Resnikoff is an international expert in nuclear fuel reprocessing and waste management. He is currently staff scientist and co-director of the Sierra Club Radioactive Waste Campaign. As Project Director at the Council on Economic Priorities he authored the study 'The Next Nuclear Gamble' which details the hazard of transporting nuclear fuel and outlines safer options. He has served as a consultant to the New York Attorney-General on transportation of irradiated fuel and to the Illinois Attorney-General on the expansion of spent fuel tools. He was also part of an international team of experts reviewing plans to locate a reprocessing and waste disposal operation in West Germany. He is a graduate of the University of Michigan with a PhD in theoretical physics.

Morris Rosen is Director of Division of Nuclear Safety, Safety Standards Program, Advisory Services, Radiological Protection. His most relevant past experience includes years at the IAEA, one year as IAEA expert in Korea and 8 years as Chief Accident Analyst for the US Nuclear Regulatory Commission. He worked for 5 years at General Electric, in design of nuclear systems, and 5 years Construction Engineering in design of nuclear power plants.

Leonard Solon is Director of the New York City Health Department Bureau for Radiation Control, and associate professor in the Department of Environmental Medicine of the New York University Medical Centre. He is a co-patentee of the laser photocauteriser employed in the treatment of ocular disorders such as detached retina, and co-inventor of a method of powering laser devices using nuclear energy sources. He is certified as a health physicist by the American Board of Health Physics. He was previously Assistant Chief and then Chief of the Radiation Branch of the United States Atomic Energy Commission Health and Safety Laboratory. He was also a member of the Technical Consultants Panel of the Atomic Energy Commission, Division of Military Application.

John Surrey is Head of the Energy Group at the Science Policy Research Unit (SPRU, University of Sussex), Adviser to the House of Commons Select Committee on Energy, and member of the editorial boards of *Energy Policy, Energy Economics* and the *Energy Journal*. Before joining SPRU in 1969 he held posts in the Department of Economic Affairs, the CEGB Planning Department, and the British Transport Commission. He has written extensively on energy policy and nuclear power, including public opposition. His books include *Energy Policy: Strategies for Uncertainty* (1977, with P. L. Cook), *The European Power Plant Industry* (1981, with W. B. Walker), and *Energy and the Environment: Democratic Decisionmaking* (1978).

Des Wilson is Chairman of Friends of the Earth and of CLEAR, the Campaign for Lead-Free Air, he is also on the Board of SHELTER, the national campaign for the homeless.

Brian Wynne is currently leader of a project on institutional settings and environmental policies at the International Institute for Applied Systems Analysis, Laxenburg, Austria. He normally teaches science and technology policy at Lancaster University, and is the author of several publications on risk assessment and sociology of science, including a recent book, *Rationality and Ritual: The Windscale Inquiry and Nuclear Decisions in Britain*.

1 Aims and Context

JOHN SURREY

The Urban Transportation of Irradiated Fuel might seem an esoteric if not prosaic title for a book. So what is irradiated fuel and why is transporting it through urban areas sufficiently problematic to make it worth writing a book on the subject?

Irradiated fuel is the uranium (metal or oxide) that has been inside a nuclear reactor releasing its energy as heat which is converted to steam to generate electricity. In the process of releasing energy, the fission reaction transforms some of the uranium and produces radioactive waste. So that the fission or chain reaction can continue it is necessary to replace the partly spent fuel at regular intervals with fresh fuel. After being removed from the reactor the irradiated fuel is put into temporary storage at the nuclear plant to allow it to 'cool', that is, for some of its radioactivity to decay sufficiently that it can be loaded into special containers, known as 'flasks', and despatched to a reprocessing plant where the remaining uranium can be separated from the plutonium and 'high-level' radioactive waste; the recovered uranium can then be refabricated into fresh fuel elements and the high-level waste stored indefinitely in one form or another. Alternatively, the spent fuel can be left for a longer period in temporary storage at the nuclear plant site – in cooling ponds or in dry storage flasks – and eventually shipped away for permanent storage without reprocessing and without recovering the remaining uranium and the plutonium. In both cases the irradiated fuel sooner or later has to be transported from the nuclear power station and it is this stage of the nuclear fuel cycle which is the subject of this book.

Here it is appropriate to point out that although the transport of irradiated fuel has occasionally arisen as an issue in the nuclear controversies which have erupted in the US, Western Europe and Japan over the past decade, it has not surfaced as one of the central issues and it has not until recently had widespread prominence or attention. The main issues have centred on the safety of the reactor itself – especially the emergency core cooling system; the separation during reprocessing of plutonium and its possible diversion for nuclear weapons purposes or its potential use in future fast breeder reactors; the problems of longer-term storage or disposal of highly radioactive waste; and the risks of sabotage and 'nuclear blackmail' at various points in the fuel cycle.

Just as an onion has many layers, so too have the controversies over nuclear power – strip away one layer of concern and another appears. This has meant that the lines of battle keep shifting from one concern to another and the alignment of forces also changes as different groups and individuals come into the arena on particular issues. This fluidity has made it difficult both to allay some of the more exaggerated or misconceived anxieties and to identify the truly irreconcilable differences between the various sets of protagonists and opponents. Both 'sides' of the controversies represent coalitions which to some extent change according to the particular concern at issue.

It is in this broader context of the nuclear controversies that the transport of irradiated fuel through urban areas has surfaced as yet another layer of concern in recent years. Once more, new participants have been brought into the arena. They include local politicians, public health officials and environmentalists in the urban areas through which irradiated fuel is transported, and trades union representatives of the railwaymen involved in transporting the flasks and the firemen who would have to attend any accident. On this particular issue these new interest groups come together with other groups which have long opposed nuclear power, but mainly on other grounds. For the latter, it is probably true to say that irradiated fuel transport is a side issue in that they are forthrightly opposed on other grounds and their opposition would not weaken if there were some satisfactory resolution of the irradiated fuel transport problem. On the other hand, to the extent that their concern is based on local or highly specific concerns, it may be reasonable to suppose that some of the new interest groups would not oppose nuclear power if their anxieties over the modes and routes of transport were allayed; that is to say, their objections might subside if nuclear fuel flasks did not pass through their own locality or if they were satisfied that existing transport methods and emergency arrangements were foolproof.

Compared with other nuclear-related risks, irradiated fuel transport has one important distinctive feature. Whilst many people have strong views on nuclear power, irrespective of where they live, the only people who are directly affected by nuclear power installations are those who happen to live in the vicinity – this being defined as the area at risk in the very low probability event of a serious accident, e.g. a core meltdown leading to a large radiation release. In most cases major nuclear power installations are sited some way from heavily populated areas and the numbers directly affected are a very small proportion of the total population. To the (limited) extent that a nuclear power station increases local employment and incomes, some local people feel that they actually benefit from its existence; the same may be true if it contributes to local tax revenues (thereby reducing the general tax burden in the area) or if the electric utility provides 'compensation' in the form of lower electricity prices for local customers or contributes to the development of the local infrastructure, by paying for the building of a library, school, hospital or road.

However, the transport of irradiated fuel is different in that most railways and most roads go through built-up areas for part of their length. This is particularly true in small, heavily populated countries as in western Europe and Japan. In this sense irradiated fuel transport brings many more people into the area which is seen to be 'at risk'. Moreover, very few, if any, of those who thus feel themselves to be 'at risk' derive any direct benefits, which are not available to the rest of the population. For them the possibility that their electricity bills may be (marginally?) lower because nuclear power contributes part of total electricity supply is at best an extremely indirect benefit, if indeed it is recognised at all. Thus they are conscious only of what they perceive as an extra and unnecessary risk – and one which they find great difficulty in placing into the context of other risks which they face in the course of their daily lives.

The argument that irradiated fuel should be re-routed away from built-up areas carries strong intuitive appeal. Indeed, were it possible to re-route it so that no one would feel 'at risk', it would be the obvious thing to do. As we shall see, cost hardly enters into it since transporting nuclear fuel amounts to a tiny proportion of the total costs of nuclear-generated electricity. But in many instances this argument is tantamount to a plea to shift the risk to others who live near an alternative route.

It is this consideration which makes it extremely difficult for national governments to respond positively to requests to re-route this traffic. To do so would be to create a precedent which would be taken up by those living near the alternative routes. Thus national authorities fear that acceding to re-routing requests would be to grant a virtual veto on the movement of irradiated fuel on all routes. National authorities therefore tend to emphasise the intrinsic safety of the heavy steel flasks, the stringency of the regulations and the precautions taken, the efficiency of the emergency arrangements, and the low level of risk to the public compared with other sources of risk which are commonly accepted.

THE GLC CONFERENCE

This book is an outcome of an international conference sponsored by the Greater London Council and which was held over three days in April 1983 in the Connaught Rooms in London. For some time previously there had been mounting concern in parts of London through which consignments of irradiated fuel were passing by rail – both from the east and from the south via Willesden Junction, en route to the reprocessing plant at Sellafield (formerly Windscale) in Cumbria. Around 600 people attended the conference and heard presentations from twenty-six speakers and many engaged in lively discussion – a good deal of which is reflected in the final chapter.

The object of the Conference was defined by Sir Ashley Bramall, Chairman of the GLC, who opened the proceedings:

> Nuclear power is the black magic of our day, surrounded – for most of us – by an incalculable amount of mystery and a corresponding amount of fear. In this Conference we are trying to bring facts to bear on the many different dimensions that there are in this problem, when you add to the problem raised by nuclear energy itself the complications of irradiated fuel transport through the largest urban area in Western Europe. There are many different perceptions of these problems.
>
> Perhaps the biggest divide of all is between the group, represented in this room, of those who have important knowledge on the various aspects of the subject to be discussed, and the rest of us – prominently represented by myself – who know nothing about them at all. Again, there are the different perceptions of those who are professionally or by employment reasons interested and connected with the various aspects of the problem, and those who, although certainly not uninterested, are from that point of view, disinterested.

The various perceptions have got to be brought together by public authorities like our own, in a programme of practical action.

As the Authority mainly responsible for this vast urban area, we are very grateful to all those who have come today to pool their different perceptions and their different aspects of knowledge. We have succeeded in bringing together people with a wide and deep knowledge of the problems with which we are involved, from many different disciplines, and important international contributions. Perceptions of these matters are very difficult in different countries and the experiences of different countries can teach us a great deal.

I think it is an important duty of the GLC as the Authority responsible for this vast urban area, to bring these problems out into the open and to try to get the greatest possible consensus and the biggest application of knowledge to them, so that we can discuss them and, one hopes, act on them from a basis of information and knowledge.

Sir Ashley's introduction emphasises three strands which featured repeatedly in the Conference – the need to bridge the divide between the experts' assessment of the 'objective' risks and the public's perceptions of the same risks; the need to learn from international experience and different approaches to the problem; and the need to see all its dimensions.

Since the GLC Conference was held three months into the public inquiry on the proposed Sizewell B pressurised water reactor, at a time when it was already evident that the inquiry would be the biggest (in terms of volume of evidence) and longest on record, it is reasonable to ask about the relationship between the two events and, in particular, whether the Conference added anything significant to the mountain of evidence accumulating before the inquiry inspector. By the time the reader reaches the end of the book, he will have formed his own opinion as to the value of the Conference. But at this stage it may help if I inject my personal view that the Conference served an entirely different need from that of the inquiry.

For one thing the Conference obtained contributions from 'experts' who, having no direct interest in its outcome, presented no evidence to the inquiry. More important perhaps, by allowing some 600 people to participate, it provided an important opportunity for learning and attempting to 'bridge the divide' highlighted by Sir Ashley Bramall. Public inquiries serve an important but essentially different purpose from that which Sir Ashley had in mind. They allow opposing views on controversial collective decisions to be cross-examined through the medium of expert witnesses and legal counsel and they enable an independent umpire to reach an impartial verdict on the balance of argument – even though, in Britain at least, that verdict is not binding upon the Government. The quasi-judicial procedures of public inquiries are meant to establish wider legitimacy for whatever final decision emerges; that is to say, part of its purpose is to maintain public confidence in the decision-making process. As in courts of law, justice must not only be done but also be seen to be done.

Although the terms of reference of the Sizewell inquiry, and the way they have been interpreted by the inspector, are broader than for previous inquiries, public inquiries are essentially concerned with specific projects. Site-specific considerations therefore tend to weigh heavier in the final balance than wider policy considerations. An enlightened inspector may take evidence on contextual issues that he judges to be relevant, but his verdict must centre on the question of whether a particular project at a particular location should proceed. Even then it may be difficult to arrive at a 'right' decision on the basis of the evidence presented.

Indeed this is illustrated by the stance adopted at the Sizewell inquiry by the National Electricity Consumers' Council, the watchdog for some 19 million electricity consumers. On the basis of the voluminous pre-inquiry evidence the Council could not judge whether investment in Sizewell would lead to a lower or higher price of electricity for current or future consumers. Consequently its own evidence has been highly sceptical and the Council has reserved its own judgement until the inquiry has run its course. Since the consumers' watchdog itself represents a huge constituency – virtually every household in the land – its eventual judgement will be highly germane to the degree of political legitimacy achieved by the inspector's verdict.

It is easy to see that the transport of irradiated fuel is a type of problem that does not fit at all comfortably in the public inquiry mould – even though the Sizewell inquiry will cover those aspects of the problem that are relevant to the proposed Sizewell PWR. The issue is much broader than the precise arrangements for future spent fuel from Sizewell (which, in any case, is unlikely to be reprocessed until well into the next century). The question of whether Sizewell traffic should go through London is intimately connected with the safety and acceptability of existing nuclear fuel traffic from other sources, both via London and other actual or potential routes.

Risks are at the heart of the concern and it was clear from the outset that the Conference would be dominated by the following central issues of public concern:

(1) The behaviour of the spent fuel and the amount of radionucleides that might be released to the environment as a result of major accidents and the radiological consequences for public health.

(2) The integrity under accident or sabotage conditions of the steel flasks that carry the irradiated fuel, including the question of whether scale model tests and mathematical modelling can satisfactorily predict the structural effects of mechanical shock and extreme fire conditions (given the complex geometry and materials problems, especially concerning the integrity of the seals); and whether (as in Germany and Japan), full-scale tests are judged to be necessary in order to demonstrate to the public that the flasks are indeed safe.

(3) The precautions for safeguarding the flasks while they are in transit, including the need to ensure that they are not involved in accidents with other hazardous materials and protected from terrorist attacks.

(4) The effectiveness of emergency-response arrangements, including the training of firemen, police and ambulance drivers and the definition and coordination of responsibilities between the numerous organisations concerned.

(5) Whether on-site dry storage, although reducing the need for interim storage and hence the early risks to the public, might create additional long-term risks.

(6) Whether shipments imported for reprocessing (in Britain and France) conform strictly with national regulations.

It is clear that all of these are technical questions that require expert assessments. However, as Sir Ashley Bramall pointed out, questions of technology policy must be put into a wider framework – public decision-making has to allow for social, economic and political considerations, as well as purely technical ones. The organisation of the Conference and the contents of this book recognise that need. However, before coming to the structure of the book, it is worth providing a brief historical and international overview of nuclear power as background to the meat of the book and context to the specific issue of irradiated fuel transport.

For those who may wish to skip over it, in the following section I argue that (1) economics explain much of the almost universal slump in nuclear ordering (outside France) since 1974; (2) learning processes – both as they affect design, construction and operation and as they affect safety regulations – have turned out to be much longer and more complex than previously thought; (3) retaining a viable nuclear power option for the longer-term future will be more difficult to the extent that demand for nuclear plants remains low, nuclear design and management teams contract and age and know-how begins to decay; (4) few reactor suppliers can count on obtaining a series of export orders and whether the suppliers can survive on a profitable basis depends on whether they can do something to reduce the costs of nuclear power and to reduce their own development, design and manufacturing costs, and (5) with the exception of France, because of its large nuclear programme since 1973, traffic in spent nuclear fuel is generally at a much lower level than expected 10 years ago and, indeed, in the US there is virtually none at all.

THE INTERNATIONAL SITUATION OF NUCLEAR POWER

Explaining the rise and fall

By contrast with the previous 10 years, which was a period of prototype reactor development in the US, Britain, Canada and France, 1966–74 was a period of vigorous commercialisation. In retrospect, it can also be seen as one of unbridled optimism regarding demand growth, the step jumps in reactor size taken to achieve the cost targets, and the solutions that were needed to the technical and political problems of fuel reprocessing and storing highly radioactive waste.

Up to 1973 the nuclear industry's forecasts universally indicated that nuclear power would be cheaper than electricity generated from oil at the world market price of around $2 a barrel (compared with around $28 a barrel in mid-1983). The expectation of cheap nuclear energy and high profits from the sale of nuclear plant and services encouraged numerous countries to enter the nuclear business in the period 1966–73. Nearly all chose to build light water reactors under licence to the US industry which had developed their designs from the US submarine programme.

The quadrupling of the oil price in 1973–4, coupled with widespread forecasts of huge 'energy gaps', prompted numerous governments to announce big programmes of nuclear expansion. Apart from France, those plans were universally frustrated. Since 1975 the general experience has been one of cancellations, cost-escalation and a prolonged slump in ordering. Few industries have undergone such a reversal so early in their development. After 8 lean years, the international outlook is still very depressed for the nuclear plant industry.

However, the depletion of oil and gas, the power of the Middle East producers (despite the current glut) and fears over supply security explain the continued interest of governments in retaining industrial nuclear capabilities as a long-term energy option. Similarly, the industry's belief that market prospects will revive in the 1990s – due to an upturn in electricity demand and the need to replace fossil fuel plants installed in the 1950s and 1960s – explains why nuclear plant suppliers are trying to hang on in the business.

In several countries vigorous commercialisation sparked off equally vigorous public opposition, but the main causes of the collapse of market prospects were economic. The rising oil price reduced world economic activity and that in turn reduced electricity demand and the need for new base-load generating plant. Much existing generating capacity was oil-fired and the short-term scope for switching to other fuels was small. Higher oil prices therefore raised electricity-generating costs, both directly as regards oil-fired plant and indirectly by allowing more price headroom for coal. Higher fuel costs fed through into higher prices and further reduced demand.

At the same time, nuclear plants were taking much longer to build and capital costs – especially capitalised interest charges over the long construction periods – were rising sharply. In some countries the capital cost escalation was accentuated by higher interest rates and expensive modifications (often after construction was finished) necessitated by more stringent regulatory requirements. To some extent the regulatory changes may have been a response to public anxieties, but in the main they were due to a learning process on the part of the regulatory authorities as design and operating experience accumulated and as thinking on safety standards matured.

In the US and several other countries, utilities were increasingly squeezed for investment finance due to rising operating and construction costs, high costs of borrowing and difficulty in winning approval from rate commissions to raise prices sufficiently to finance large development programmes which would at best benefit the next generation of consumers. Reduced profitability lessened the ability of investor-owned utilities to borrow on capital markets. Even where utilities were publicly owned, the uncertainties over future construction periods and costs, future regulatory requirements and over nuclear plant operating performance made the utilities even more hesitant. For the industrialising countries of the Third World, balance of payments problems and rising indebtedness also served to choke off demand for nuclear plant.

This is an unashamedly economic interpretation; I think the collapse of the market would have occurred had there been no public opposition and no Three Mile Island accident. That, of course, does not imply that if market prospects revive public attitudes would be unimportant.

Nuclear capacity and operating performance

If we simply take the current stock of nuclear plant installed worldwide, which is around 140 GW (gigawatts), and add to it the plant still under construction and firmly planned, nuclear generating capacity in the non-communist countries would rise to around 330 GW in 1990. This is an upper estimate because it assumes that no further nuclear plant will be retired, no current orders cancelled and no delays to current construction schedules.

TABLE 1.1 *Nuclear plant capacity (GWe, rounded)*

Where installed	*PWRs 1981*	*PWRs 1990*	*BWRs 1981*	*BWRs 1990*	*PHWRs 1981*	*PHWRs 1990*	*GCRs 1981*	*GCRs 1990*
North America	36	87	18	40	5	15		
Japan	7	11	8	18				
W. Europe	22	92	9	24			11	17
Elsewhere	1	15	2	5	1	4		
Total:	66	205	37	87	6	19	11	17

Table 1.1 shows the expected build up, which is chiefly due to the orders placed before the collapse in demand. Light water reactors (PWRs and BWRs), which originated in the US, predominate. The only other current design is CANDU, the Canadian PHWR design. Gas-cooled reactors were developed in Britain and France; two Advanced Gas-cooled Reactors are being built in Britain to keep the

industry alive; but gas-cooled technology has had no international appeal for nearly 20 years (i.e. since the very low-cost estimates for PWRs and BWRs began to appear). Two-thirds of the build-up consists of PWRs, the design originally developed by Westinghouse in the US.

German ordering may at last be beginning to pick up, but the nuclear plant industry generally remains very depressed. The only exception has been France, where the previous government pushed through a massive programme to ensure that base-load supply would be wholly nuclear by 1985 and to reduce dependence on imported oil. Indeed, the prospect is that so many PWRs will have been installed by about 1987 that some will be working at low load factors or producing mainly for export to adjacent countries with higher-cost electricity. The apparent success of the French nuclear programme signals an early reduction in domestic ordering, so that the French nuclear industry, too, faces a crisis unless it can obtain export orders.

The first consequence of the general collapse of nuclear ordering is that, outside France, nuclear power's contribution to energy supply remains small in relation to the desired transition away from oil (see Table 1.2). Whether or not nuclear power could ever have been the main and appropriate vehicle for reducing dependence on oil and natural gas (given the non-substitutability of many of the demands for hydrocarbons, e.g. road transport and much process heat), oil dependence is still a fact of life for energy policymakers.

TABLE 1.2 *The contribution of nuclear power (% of primary energy)*

	Nuclear power		*Oil*	
	1980	*1990*	*1980*	*1990*
North America	4	9	43	33
Europe	3	8	51	41
Pacific (mainly Japan)	4	10	60	48
IEA Total:	3	9	47	38

SOURCE International Energy Agency, 1982.

Another direct consequence is that, despite the large amount of re-design and engineering work necessitated by changing regulatory requirements – which has kept design and project management activities ticking over – the nuclear plant suppliers are rapidly running out of work. Servicing the increasing stock of reactors and their nuclear fuel requirements may be lucrative, but it brings no work to the expensive facilities required to build major components and sub-systems which are specific to nuclear plants; nor do they provide the experience necessary for nuclear design and construction teams if they are to stay on a learning curve.

Several nuclear plant vendors have rationalised internally to reduce overheads. AEG in Germany withdrew several years ago. General Electric in the US has not been actively seeking new BWR orders for some years. All suppliers, including now the French, are chasing export orders to tide them over.

The key question is to what extent the long dearth of orders will affect the industry's efficiency in designing and building nuclear plants, should demand eventually pick up. Without orders, will it be possible to retain key technical staff and to update know-how in relation to new technical information and changing regulatory requirements? Will the firms that make major components and sub-systems (e.g. PWR pressure vessels, steam generators and circulating pumps) switch their engineering and fabrication resources to other products rather than face continuing underemployment? This is particularly important for a contracting industry where the reactor vendor depends on a supporting infrastructure of specialist sub-contractors. If demand stays flat for another 10 years, would it be cheaper at a future date to re-establish the production structure than to subsidise the retention of existing resources over the intervening period? Will the industry undertake the development work necessary to be able to introduce improved thermal reactor designs in the 1990s?

If future demand is the industry's main worry, the second is the operating performance of some existing reactors, for that will affect utilities' attitudes. From an economic viewpoint the distinctive feature of nuclear plants is that their running costs are low but their capital costs are high. It follows that they should work at full capacity all the time they are available. Nuclear power economics are very sensitive to the load factor which is achieved in relation to the design capacity. The higher the load factor, the lower the cost per kWh; and vice versa. This axiom was reflected in economic appraisals on which the investments were made; they always assumed that nuclear plants would achieve lifetime load factors of at least 70 per cent in relation to design capacity.

Table 1.3 shows that operating performance has in fact been highly variable. The average load factors show CANDU's as out-performing all other classes and light water reactors (especially PWRs) in western

TABLE 1.3 *Operating performance of commercial nuclear plant, 1979–81 (mean load factor (%)[a] (no. of reactor-years in period))*

	PWRs	*BWRs*	*PHWRs*	*GCRs*
USA	55.2 (128)	60.9 (68)		
Canada			84.5 (24)	
Japan	50.6 (25)	62.3 (31)		70.2 (3)
W. Europe	67.4 (63)	54.8 (42)		51.0 (85)
France	66.7 (21)			51.4 (15)
Germany	69.6 (17)	33.2 (15)		
UK				49.4 (64)
Italy	10.5 (3)	0 (3)		51.4 (3)
Sweden	52.0 (4)	69.3 (16)		
Switzerland	82.5 (8)	89.1 (3)		

NOTE
(*a*) Load factors relate to design capacity and one weighted by reactor size.

Europe out-performing similar reactors in the US. Having been very low from 1975 to 1978 the load factors of light water reactors in Japan were, by 1981, beginning to match the US performance (which itself is low in relation to performance in western Europe).

Behind these averages lies an intriguing picture of highly variable performance within each class of reactor, except CANDU. The Canadian reactors have reached peak performance in under 2 years' operation and, partly because problems have been anticipated and rectified before becoming serious, their subsequent operation has been relatively trouble-free. Performance among PWRs and BWRs varies widely and independently of age, size and technical characteristics.

For PWRs of Westinghouse design, the early two-loop models with a design rating of around 600 MW have generally achieved 70 per cent load factors after 2 years' operation. But for the three- and four-loop models (around 900 MW and 1200 MW respectively) the average load factor has so far been around 60 per cent and there is little sign of maturation. Steam generator tube leaks have been a recurring problem, generally setting in after the third year and sometimes causing long shutdowns at great cost in terms of replacement power (and great capital cost, too, where plugging and sleeving have not been appropriate and it has been necessary to install new steam generators). Despite also being subject to persistent steam-generator problems, the two-loop model has achieved good performance. By inference, the poor performance of the current three- and four-loop models cannot be wholly due to steam-generator problems and, to the extent they are difficult to pinpoint, these 'unexplained' problems may be difficult to avoid in designing similar reactors in future.

For BWRs the picture is even less clear; the average load factor masks great variability between similar plants and the early onset of equipment failures masks any positive maturation effect. Unlike PWRs, few BWRs can be identified as reliable performers. The most widespread problem causing non-availability is pipe cracking duc to the use of inappropriate steel for feedwater pipework. Despite the superior performance of German PWRs over US PWRs in every size class, German BWRs have had an unhappier history on the whole than US BWRs.

This leads me to draw two conclusions. The first, from US experience, is that US utilities installing light water reactors should be prepared to keep a larger margin of idle plant on the system as insurance against the breakdown of nuclear plant or, if the utility has strong interconnections, for the same reason it should be prepared to buy (high-cost) power generated by other utilities. The second stems from successful European experience with PWRs. This shows what can be achieved with meticulous attention to design details, materials, quality control, intelligent operation to get the most out of the plant, and preventive maintenance undertaken as far as possible during annual shutdowns for refuelling. Both conclusions challenge the notion that the stage has been reached where buying reactors 'off the shelf', as it were, would guarantee good performance.

Disappearing uranium security

The compound growth rates used in forecasting the demand for nuclear power in the early 1970s implied that world uranium resources would be depleted rapidly. They powerfully supported the arguments for expanding reprocessing capacity (to recover partly spent fuel) and to build fast breeder reactors (essentially substituting plutonium and using the remaining uranium fifty to sixty times more efficiently than in existing thermal reactors). This line of argument found its strongest appeal as the uranium price rose dramatically in the mid-1970s – the spot price rose from $6 per pound in 1972 to $43 in 1978.

Rather than signalling diminishing returns in uranium mining, however, the price increase was due to the combination of several short-run factors. The increase in civil demand expectations came suddenly after a long period of rundown of military demand after the US weapons' stockpile was built up in the 1950s. By the late-1960s only the lowest-cost mines were still producing and exploration was miniscule. As civil demand rose, the US Atomic Energy Commission – which controlled the bulk of western enrichment capacity – intervened to regulate the market. It required utility customers to contract much further ahead for enrichment services. As the utilities were forced to seek long-term contracts this, too, added to the short-term pressure of demand.

As the uranium price increased, Westinghouse declared that it was unable to honour its guarantee to provide fixed-price uranium to the utilities to which it had sold reactors. They, too, were forced to seek further large quantities at the time when the price was rising sharply and the costs were indicated by the multi-billion-dollar lawsuits filed by the utilities against Westinghouse and by Westinghouse against an alleged cartel of leading uranium producers.

By 1979–80 the situation was easing. This was due to the collapse in reactor ordering and public inquiry rulings which cleared the way to the opening up of major low-cost uranium resources in Australia and western Canada. By 1982 the spot price was down to $25 per pound – in real terms not much more than 10 years earlier.

Dealing with partly irradiated fuel from light water reactors and with high-level radioactive waste has presented much more severe problems than foreseen in the early 1970s. Pending politically acceptable technical solutions, large quantities of 'spent fuel' are accumulating in temporary storage at nuclear plant sites. Although it could be argued that the mounting stock of partly spent fuel represents a long-run insurance against diminishing returns in uranium mining, that has to be balanced against the facts that estimated reprocessing costs have been rising steeply and that generally accepted solutions for indefinite high-level waste storage have yet to be found.

The receding fear of uranium scarcity has removed any economic justification for building fast breeder reactors until they are economic compared with thermal reactors. Having so far spent £2200m (1982 prices) on FBR R & D, the British Government and the CEGB do not now expect any commercial demand for FBRs over the next 25 years. In France, with the 1300 MW Superphenix FBR nearing completion, it is questionable whether either the French Government or Electricité de France will be prepared to finance further FBRs; this is, of course, against the background of an electricity system which is already largely nuclear and of relatively low-cost PWRs. In other countries FBR plans also appear to be at a virtual standstill as the new technical and cost information is being evaluated.

Implications of international experience

1. The role of utilities The evidence strongly suggests that 'buying reactors off-the-shelf' cannot guarantee low construction costs and good operating performance. Close coordination is required between the utility client, architect-engineer and plant supplier at all stages from initial planning to commissioning, operation, maintenance and repair. A thorough understanding is required of 'design intent' (i.e. all aspects of the technology and tolerances, etc., not just access to the licensor's technical codes and operating manuals). Ontario Hydro and Electricité de France cooperate closely with their respective plant suppliers and regulatory authorities and perform the architect-engineer function themselves. In Canada and France these large, technically proficient utilities, intimately involved in all the technical decisions, have provided the conditions of success. In Germany Siemens retains full responsibility for supporting R & D design and project management; this is a different model but can be appropriate where the reactor vendor is technically very strong and the utilities fragmented and technically rather weak. Both examples contrast with the position in the US where neither the well-developed capabilities of the vendors nor the large independent firms of architect-engineers have been able to compensate for a fragmented utility industry which, on the whole, is technically passive and at arms length from the regulatory authority.

2. Unit size, learning and standardisation It would appear that rather too much and too early emphasis was placed on the cost advantages of very large reactors. The scale economies have not generally been achieved to the degree that was expected, whilst complexity and unanticipated difficulties have increased with step jumps in design. Even Electricité de France, with its massive commitment, has admitted that its 1300 MW units now show no cost advantage over its 900 MW units. This may not always be the case, but the evidence (especially with CANDUs in Canada and PWRs in Germany) points to the importance of learning from prototypes, understanding 'design intent' and intricate problem-solving before commercialisation and scaling-up. Standardisation, which has so far been attempted only in France – with results which, so far, are impressive – is risky while the technology and regulatory requirements are still evolving. How and when to standardise are tricky decisions.

3. R & D and the industrial infrastructure German success with PWRs illustrates the value of undertaking supporting R & D over many years to improve reliability, maintaining close control over project management, and entrusting the manufacture of the major components and sub-systems to highly proficient subcontractors. Due to the use of superior alloys and water-chemistry techniques, German PWRs have avoided the problem of steam-generator tube leaks which persist with PWRs of Westinghouse design. By contrast, the story with German BWRs was quite different, with the result that German BWRs have performed worse than US BWRs. In the US, France, Germany and Britain, nuclear R & D institutions account for most of the nuclear R & D expenditure and are themselves very dependent upon political support for the continuation of fast reactor work. Their priorities seem to favour long-range technologies which, by definition, have uncertain economic pay-offs. However, the clients' needs appear to require the resolution of current and near-term problems of reliability. This apparent mismatch is being reflected in the clients themselves taking responsibility for much more of the latter type of work involving current reactor systems. This is true of the CEGB and of US utilities which have combined through the Electric Power Research Institute to cope with problems of reliability.

4. Regulatory activity The rise of regulatory activity seems to have stemmed more from learning by the regulators than from public disquiet over real and imagined hazards. It has thus been part of the general process of technological maturation. Solutions are urgently needed for reprocessing and waste management problems; in some countries they are a precondition for further nuclear ordering. Particularly important questions for the UK, if it adopts the PWR, would be whether the British regulatory authority could resist pressures to adopt the increasingly rigorous standards being set in some other countries and whether this would result in open-ended cost escalation. In a sense, in opting for an international reactor system, one is also opting for international safety regulations which, as we have noted, are fluid.

5. State support In all countries with civil nuclear programmes the state has played a leading role in promoting the technology and subsidising R & D on a scale accorded to no other 'infant' industry. Whereas in the US and Germany, the industry tends to blame its predicament on 'failure of political leadership', the problems clearly call for more than the type of political support which can be provided in most western democracies. Again, France is the exception. The nuclear commitment in France has involved the use of Presidential authority; a legal and administrative system which provides no avenues for dissent to be upheld; lack of effective Parliamentary scrutiny; the Government's willingness to finance the nuclear programme with massive foreign borrowing (with future exchange-rate risks); the over-arching responsibility of the Minister for Industry for all aspects of nuclear development and safety; and the far-reaching, Government-inspired restructuring of the entire French power plant industry. No other liberal democracy has offered such scope for autarky in peacetime and it would be foolish to imagine that other western countries could (or should) copy the unequivocal, centralised commitment and the technocratic style of decision-making which is characteristic of the French approach to industrial policy in general and nuclear power in particular.

6. *Nuclear power prospects* World nuclear plant supply capacity greatly exceeds demand and this situation is likely to persist throughout the 1980s. There is vigorous competition for the few export possibilities which come up and in some cases orders have been taken with insufficient safeguard agreements or restraint in the transfer of sensitive technology – with obvious risks for nuclear proliferation. However, few if any reactor suppliers can count on obtaining a series of export orders and whether the suppliers can survive on a profitable basis depends on whether they can overcome the twin problems of capital cost escalation and variable operating performance and whether, through rationalisation, they can reduce the costs of development, design and manufacture of major components and sub-systems. This will be especially difficult as long as there are very few home orders and home markets remain closed to foreign competition. Utilities and governments are increasingly likely to resist having to shoulder a mounting bill in order to protect an industry with limited prospects.

7. *Irradiated fuel traffic* With the exception of France, because of its large nuclear power programme since 1973, traffic in irradiated fuel is generally at a much lower level than expected 10 years ago and,

indeed, in the US there is virtually none at all. The main reasons are the virtual collapse of reactor ordering since 1974, combined with unresolved problems at the back end of the fuel cycle – particularly the lack of major commercial reprocessing plants outside France (La Hague) and Britain (Sellafield) and the non-acceptance in some countries of proposed methods of indefinite storage of the 'high-level' radioactive waste returned from the reprocessing plant. As a result, increasing quantities of irradiated fuel are accumulating at nuclear power stations, particularly in the US, and this has led to increasing interest in on-site dry storage.

GUIDE TO THE BOOK

A book such as this, consisting of contributions by many authors and with important linkages between many of the chapters, could no doubt be structured in a variety of different ways. For that reason, I have resisted the temptation to divide the text formally into separate sections, for that might have implied a rigidity of classification greater than the material warrants. Instead, I have attempted to group the chapters in such a way as to produce a reasonable degree of coherence to give the reader a conspective view of each topic or aspect of the subject.

The first grouping comprises Chapter 2 by Lord Flowers, Chapter 3 by Des Wilson and Chapter 4 by David Pearce. Each author provides a framework for clarifying the issues as he sees them. The main contrast is between the position of Lord Flowers, who argues that what happens at other parts of the nuclear fuel cycle is of little relevance to the question of whether it is safe to transport nuclear fuel through urban areas, and that of Des Wilson, representing Friends of the Earth, for whom irradiated fuel transport is just another unacceptable risk of an unsafe and uneconomic technology. However, both seem to agree that irradiated fuel transport would not be a central issue in a debate about the future of nuclear power.

The second group comprises Chapters 5 to 13, which provide some international perspectives. Morris Rosen describes the role of the International Atomic Energy Agency in establishing international guidelines for the safe transport of radioactive materials. Jean-Claude Charrault outlines the role of the European Community in relation to nuclear power and how it sees it reducing the Community's dependence on oil imports. The facts pertaining to irradiated fuel transport in Britain are contained in the chapters by the CEGB, Raymund O'Sullivan and Commander Cree. Chapter 10 by Colonel Berthier gives similar information for France, the only country with a large nuclear power programme and a rapid build-up of irradiated fuel transport – from all parts of France (and overseas) to the reprocessing plant at La Hague on the Normandy coast. Chapters 11 to 13 present three different views of the situation and problems in the US, where virtually no irradiated fuel transport is taking place due to a lack of commercial reprocessing plant, where there are transport restrictions in numerous states, and where there seems to be fundamental conflict between the responsibilities of local authorities that place primary emphasis on public health and the US Department of Transport, which is responsible for ensuring the transport of radioactive materials, and where the limitations of the railways mean that irradiated fuel has to be transported by road.

The third grouping consists of Chapters 14 to 20, all of which represent 'expert' assessments of the various risks involved and the means of safeguarding against them, including the chances of accidents leading to radiation release and their potential consequences, the integrity and testing of nuclear fuel flasks and fire tests.

The fourth grouping (Chapters 21 to 26) brings together a variety of broadly non-technical aspects, including public perceptions and psychology, the economic aspect, the security aspect, implications for both local decision-making and the trades unions.

Finally, the 'Open Forum' (Chapter 27) brings together the various viewpoints expressed and the questions raised from the floor throughout the Conference. This chapter is meant both to complement the information and arguments in the previous chapters and to reflect a wide range of attitude-forming concerns.

2 Issues and Non-Issues

LORD FLOWERS

We are embarking on a three-day examination of the nuclear power programme. We shall be studying it from a particular point of view – irradiated nuclear fuel arising, perhaps, in Sizewell, transported through urban areas such as London, on its way to the reprocessing plant in Cumbria. This is an important aspect, but one that has received comparatively little public attention so far. The point at which irradiated fuel arises is a nuclear power station, and there has been a great deal of discussion about that. The Sizewell inquiry taking place at present is examining the proposal to introduce a reactor system into Britain – the pressurised water reactor or PWR – which is commonplace elsewhere, but new to us. A great deal of information is available about the design of such plants and about their operation. Much of it is being paraded before us and critically examined at Sizewell. It is not our task to conduct an independent inquiry into the PWR. If anyone has views about that, let him express them before the Sizewell inspector. Similarly the point of treatment of the waste is the plant in Cumbria, and that was thoroughly investigated in the Windscale inquiry a few years ago.

Today, we are not concerned with the point of arising, or the point of treatment, but with what happens in between, especially if the fuel passes through urban areas. It is no concern of ours whether the new Sizewell power station uses a PWR or some other kind of reactor. To a sufficiently close approximation for our purpose, the amount of radioactivity possessed by the irradiated fuel – the extent to which it is inherently dangerous – is determined by the power output of the station and not by the type of reactor used to produce that power. In any case the cost of safe transport of irradiated fuel is a very small proportion of the whole, so the minor differences which arise from one reactor system to another could not possibly decide the choice of system. Important though it is to the nuclear industry, the choice of reactor system is, for us at this conference, a non-issue.

Of course, whether there is a new power station or not at Sizewell, or anywhere else for that matter, is at first sight an issue for us because if there are no nuclear power stations there is no irradiated fuel to transport. This is recognised by the fact that in this conference there are a number of talks about the extent of the nuclear power programme, even about whether we need one at all or whether we could instead rely on alternative sources of energy.

I said that 'at first sight' it is an issue for us. On further reflection, I believe we shall, most of us, come to accept that the argument is fallacious. I have already hinted at the reason. The costs of the transport arrangements are comparatively small and, as we shall see, the risks associated with transport are also small. In any case, containment flasks can be modified and routes can be changed. Transport arrangements, though deserving of careful examination in detail cannot of themselves be a determinant of the nuclear power programme. If we want to have a conference on whether we should have a nuclear power programme at all, by all means let us do so, but the urban transport of irradiated fuel would not play much of a part in such a conference – it would be a very minor consideration. But surely, you may be thinking, if we double the size of the nuclear power programme, shall we not also double the amount of irradiated fuel being transported and therefore double the risk? The answer is that we shall certainly double the amount of fuel eventually being transported, but the risk can be made as small as we like, because there is nothing particularly difficult or costly about it compared with other components of the nuclear power system.

Therefore, the extent to which we rely on nuclear power will not be determined by transport considerations – it is for us at this conference a non-issue. Now, if some of your invited speakers now decide to go home they will be quite right! More likely they will disagree, but then let them at least explain why what they have to say is relevant to the matter before us.

Before I move on to the job in hand, let me deal briefly with one other non-issue. There are those who would have us declare a nuclear-free zone in, let us say, Harringay, much as there are those that would welcome home rule for Wigan. It may be that to ban the deployment of nuclear weapons within certain areas of the world would be a useful step towards making nuclear warfare less likely. That is what is meant by a nuclear-free zone: one containing no nuclear weapons. As far as I know, London contains no nuclear weapons and is never likely to do so, unless someone drops one on us. Some people, however, think that because some area contains no nuclear weapons, it should also contain no nuclear installations of any kind – no nuclear power stations and no transport facilities for nuclear fuel. Therein lies a massive misapprehension – that nuclear weapons and nuclear power inevitably go hand in hand. One understands the source of the misapprehension readily enough. The first public manifestations of nuclear power were the tragedies of Hiroshima and Nagasaki which brought to an end the second world war. Moreover, it was out of the wartime weapons programme that the civil nuclear power programme was born. That is the origin of the fear that many people have of all things nuclear. Some people imagine that if an accident befell a nuclear power station it would explode like an atomic bomb. It would not. Some people, no doubt, think the same of the transport of irradiated nuclear fuel. We shall come to that shortly.

In this country, nuclear weapons manufacture and civil nuclear power are kept well apart. Our power stations are subject to international inspection to ensure that they stay well apart. The highly specialised materials and facilities required to manufacture weapons simply could not be hidden away on the site of a power station. Important though they may be in another context, therefore, nuclear-free zones are, for this conference, a non-issue.

It seems that nobody has been asked to say what irradiated nuclear fuel consists of or how it arises in the first place; so let me spend a few minutes on that. A nuclear reactor burns uranium to produce heat, which is used, as in any power station, to raise steam which drives the turbines which generate the electricity. We are concerned with the first stage only – the burning of uranium; the other stages are much the same as one finds in a fossil-fuelled station. Unlike a coal-fired power station, however, one cannot simply shovel the uranium into the furnace; or, like gas or oil, inject it into the burners. Uranium burning is much more sophisticated. Uranium, usually in the form of oxide pellets, is fabricated into extremely complicated, highly engineered fuel assemblies consisting of long rods sheathed in a suitable alloy encased in a tube with coolant passing through at high pressure. The coolant is there, of course, to take the heat away. In the PWR this coolant is water; in the British-designed reactors it is carbon dioxide gas. The nature of the coolant does not much matter for our purpose today.

Large numbers of fuel assemblies are inserted into channels in the reactor and there is a great deal of complicated equipment for supplying the coolant and monitoring such things as its temperature and pressure in every fuel channel. In the course of the nuclear reaction, some of the uranium in the fuel assemblies is destroyed, accompanied by a release of energy, most of which is transferred directly as heat to the coolant. As with coal, however, when uranium is burnt an ash is formed and this ash is called nuclear waste. It is highly radioactive – that is to say it emits a large amount of energetic radiation, much of which can only be absorbed by considerable thicknesses of solid materials. In the course of absorption, the energy of the radiation is also converted into heat. It is this radiation that is potentially harmful because it disrupts the cells out of which all living things are made. In large amounts it is lethal; administered in a controlled manner, however, as in radiotherapy, it can be highly beneficial. The latter is no concern of ours today.

Only a very small proportion of the uranium in the fuel assembly is burnt in the reactor. A further very small proportion is converted into another nuclear fuel called plutonium. Irradiated nuclear fuel therefore consists mainly of pellets of unburned uranium oxide within which are imprisoned some highly radioactive waste and a very small amount of plutonium. The fuel must be removed from the reactor at regular intervals because the waste would otherwise prevent the reactor from working properly. It is replaced by fresh fuel and the irradiated fuel is stored until it is ready to be taken away to be processed. In this processing, the plutonium is extracted for future use, the waste is removed for further treatment, and the remaining uranium can eventually be fabricated into fuel assemblies once again.

Radioactivity dies away in time. When a fuel assembly is removed from the reactor, the radioactive heat still being generated in the fuel decreases by a factor of ten in about half a day, by another factor of ten in about six months, and by a further factor of ten in about 3 years. Some of it will remain for thousands of years, but by that time the heat release will be negligible for practical purposes although its potential biological effect will still be significant. As long as large amounts of heat are being generated (due to radioactivity) in the fuel assembly, it is necessary to continue cooling it. The irradiated fuel assembly is therefore kept on site for some months, or years, under engineered cooling until the heat production has fallen sufficiently to allow its transport to the reprocessing plant to take place conveniently.

Although now sufficiently cool for the purpose, the fuel is still extremely radioactive. It is still potentially lethal to human beings and injurious to the environment. It must therefore be carried in flasks which contain it safely against accidents in transport, which absorb the radiation still being emitted, and which provide a measure of cooling against the residual heat still being generated by that radiation. Such a flask is no mean thing; if designed to carry a few tons of irradiated fuel, the flask will weigh, typically, 100 tonnes. It is a massively engineered steel structure. It is not at all fragile, nor is anybody going to walk off with it on the spur of the moment. There is no need for me to describe these flasks in detail because others will do so. Let me quote just one descriptive passage from the Central Electricity Generating Board (CEGB) evidence to the Sizewell inquiry:

> All PWR fuel flasks are of cylindrical construction. To contain and shield the radioactive material the flask is constructed from massive steel. . . . The lid is securely attached to the flask body by many large diameter, high tensile bolts and is sealed by compressible elastomeric 'O' rings. The flask contains no moving parts which can deteriorate during transport. There are no instruments which might give faulty readings, and there are no power supplies which could fail. There are valve penetrations through the flask wall in order to fill the flask with water or nitrogen, and also to vent it during preparation for transport. The lid and base sections of the flask, the fore and aft sections during transport, are protected by shock absorbers currently fabricated from balsa wood encased in steel. In all, the flask is massive but relatively simple and the resultant transport process is straightforward.

These flasks are designed to be safe in normal operation. That is to say, the radiation which escapes from them is well below that allowed for workers and public alike, well below the natural background to which we are all exposed. They are also designed to survive severe accidents such as head-on collisions between two trains, or falling off a bridge or being immersed in fire or under water, without impairment to the containment of radioactivity, and they have been subjected to a realistic testing programme.

The British design of flask costs about £400,000 and a special rail vehicle to transport it, another £400,000. The cost of a round trip from Sizewell to Sellafield in Cumbria and back is about £7,000 per flask and the annual maintenance costs are about £10,000. These sums are peanuts compared with the costs of the major nuclear facilities.

The prime responsibility for flask design and for safety rests with the

CEGB. According to their Sizewell evidence, two Acts of Parliament have particular relevance to the transport of radioactive materials: the Radioactive Substances Act, which gives powers to the Secretary of State for Transport to make any regulations he sees fit; and the Nuclear Installations Act, which imposes on the licensee absolute liability for personal injury or damage attributable to nuclear matter in the course of carriage in the United Kingdom. The Department of Transport certify the flasks as fit for use and they do so according to regulations drawn up by the International Atomic Energy Agency for the Safe Transport of Radioactive Materials. The roles of the Department of Transport and the IAEA will be described to you, and there will also be a great deal of information about the design and testing of the fuel flasks and of the transport vehicles, together with the experience of some other countries.

Let me complete this part of my remarks by saying that the transport of irradiated fuel is not something new. Since 1969, over 900 tonnes of light water reactor fuel have been transported in the United Kingdom in over 400 flask journeys and the current rate is about 90 flask movements per year. There has been only one derailment on running track reported so far and that involved an empty flask being returned to a power station from Cumbria.

Thus the transport of irradiated fuel is well tested and takes place in flasks and on vehicles and according to procedures which are approved and regulated both nationally and internationally. It must be said, however, that with flask movements representing less than 0.004 per cent of rail freight (in terms of laden wagon-miles), it is not surprising that accidents involving flasks have been so rare. As our new nuclear capacity comes into maturity, with local irradiated fuel storage saturated, more movements will take place and perhaps more derailments are then to be expected. But that is not to say that radiation accidents will then ensue because the flasks are designed to be safe even in extreme accident conditions.

There is no doubt that great care has been devoted to all these matters. I have not, myself, found grounds to complain that all is not well. The Sizewell inquiry can be expected to report on the particular arrangements proposed for Sizewell. However, the Government also have two independent committees of eminent experts which stand ready, in principle, to advise on whether such arrangements are satisfactory in all respects and whether the tests cover all eventualities. The first is the Radioactive Waste Management Advisory Committee, under the chairmanship of Sir Denys Wilkinson. This body stands ready to be consulted about all aspects of waste management in the normal course of events. It has so far confined itself mainly to questions arising in the processing of irradiated fuel and the safe storage and eventual disposal of the waste which remains after processing. It reports to the Secretary of State for the Environment. The second is the Advisory Committee on the Safety of Nuclear Installations, which concerns itself with the design of installations so that they will be safe in accident conditions. It is chaired by Sir Morris Sugden and reports to the Health and Safety Commission which is responsible for the Nuclear Installations Inspectorate.

It is not immediately clear to which of these two bodies the Government should turn for advice about the transport of irradiated nuclear fuel. Transport is an unavoidable part of waste management; on the other hand, a flask can be regarded as a nuclear installation. Similarly, the public requires assurance that transport arrangements and the handling of flasks are satisfactory in the ordinary course of events, but they would also like to be assured that the flasks are adequately designed to withstand all reasonable accident conditions without breach of containment. I do not wish to involve myself in a trade dispute between these two committees. What concerns me is that neither appears to have been consulted about the safety of irradiated fuel in transport. It may even be the case that their offer of help has been spurned by the Government. It seems to me that this simply is not good enough. If the Department of Transport were to approach the two chairmen, both of them Fellows of the Royal Society, those distinguished and sensible people would be able very readily to combine forces in order to determine whether any improvements should be suggested. I believe this Conference would be reassured to know that an independent expert assessment was being undertaken. I am sorry to have to say that to the best of my knowledge it is not, and I would urge the Government to think again. It may be that the Government feel that in this matter they are not in need of independent advice; but we can tell them that the public are certainly in need of independent reassurance, and that is just as important.

Almost all the transport of irradiated fuel takes place by rail. At Sizewell, for example, only the first mile, from the power station to the railhead at Sizewell Halt, will be by road; the rest of the journey to Sellafield will be by rail. The vehicles used by British Rail are of modern design; they have very low centres of gravity for additional stability, and they incorporate bogie units which are not prone to derailment. According to the Sizewell evidence, all flasks are moved on dedicated trains, so that, for example, no inflammable material or explosives are carried on the same train. The vehicles are ordered so that in the event of a collision there is protection from impact by the locomotive in front and the brake-van behind. The vehicles have automatic braking and they are maintained to a high standard and, because of their weight, they are restricted to a maximum speed of 60 miles per hour.

It must be emphasised that British Rail have a long history of the safe carriage of potentially dangerous materials and it is their custom to rehearse emergency procedures at frequent intervals. During 1981, for example, incidents involving the transport of irradiated fuel were simulated seven times in different parts of the country, with the local CEGB power station staff in attendance. It is the view of the Department of Transport (given at the Windscale inquiry) and of British Rail and the CEGB that the inherent safeguards provided by the design of the flasks and the nature of the transport arrangements make unnecessary the carrying of special escorts or the choice of special routes. Personally, I agree with this judgement. According to the CEGB, the annual probability of there being a train accident which might result in serious damage to a fuel flask in the area of Greater London is about one in ten million. This probability is so small that it is difficult to grasp except, perhaps, by analogy. It is about the same as the annual probability of my being killed by lightning, another eventuality of which people are more scared than the facts warrant.

So the chance of a serious accident is very small. Nevertheless, what might the consequences be? The CEGB claims that in no credible accident would the flask be breached so that even if it were severely damaged no radioactivity could be released; the flask is so designed as the international regulations require. But suppose they are out of luck and a breach occurs or the lid seal is disrupted – what then? Almost all the radioactivity is imprisoned atom by atom in the solid fuel element material, from which a small portion can be released only if the fuel heats up sufficiently due to its own radioactivity. This might happen if almost all the cooling water were to drain out of the breach; but even so, many hours would elapse before the fuel was hot enough for any serious release of radioactivity. During this period there would be plenty of time for corrective action to protect the crew, the public and the environment under the well-established Flask Emergency Plan administered by the CEGB, and required by the International Atomic Energy Agency.

I have to say that, with all the precautions that are taken, the chances of an accident with serious consequences during the transport of irradiated fuel are much less than in the transport of many other industrial materials. If it were not for the sheer boredom, I would not hesitate to accompany a nuclear flask on its travels day after day. That, however, is not quite the point. We are dealing with an emotive issue and when we are emotional we pay scant regard to objective assessments. There are even those who like to play upon the emotions of others achieving some fell purpose of their own. It is a free country; we have the right to speak as we wish. What is important for all of us, however, is to recognise that in many industrial activities, but especially in the nuclear industry, we are dealing with risk at two different levels. We are dealing, on the one hand, with objective estimates of risk – with the principles of safe design, the assessment of malfunction and the provision of appropriate precautions; and on the other hand we are dealing with the public perception of risk which may be very different. The first is the business of the engineer, the second of the politician. It is the duty of the engineer to press upon us the need to be objective; to point out perhaps that it will cost us dear if we are not. It is the duty of the politician to try to assess and influence the state of public opinion. In the end, it is the politician, as the representative of the people, who will determine whether the engineer shall have his way and on what terms. It is in order to hear what the engineers have to say that the politicians of County Hall have summoned this gathering, and I salute them for it. It is a pity that some of the engineers (the CEGB) have another engagement. The date was not of their choosing and they have provided us with information both written and visual. Nevertheless, it is a pity they are not here.

The CEGB, for their part, have made valiant attempts to come to terms with the realities of public perceptions. For them, however, there is a problem of credibility. If they provide a remedy for a problem they believe does not exist, do they soothe the public misconception, or do they compound it? Those Londoners who fear the transport of irradiated fuel through London might be comforted to know that it was to be redirected through somebody else's back-yard; or it might convince them that they were right to be alarmed in the first place. Let me remind you of the old story of the man who wanted to buy an absolutely reliable motor car. On being shown a Rolls-Royce he observed that it had a starting handle. (I said it was an old story.) He enquired whether this meant that the starter motor sometimes did not work. 'Oh no, sir,' said the salesman, 'it is absolutely reliable, the starting handle is provided just in case.' There ensued the sort of argument we shall probably have at some point today. Later Rolls-Royce discontinued the starting handle, recognising perhaps that superfluous precautions could mislead the customer.

I think we have to accept that the problem of credibility is a serious one. The scientist in me says that the CEGB and British Rail are right: the additional risk of routing irradiated fuel through the densely populated area of London is negligible, compared with other industrial risks that we readily accept every day. The politician in me, and I have a bit of that too, says that it is the public perception of risk that has to be allayed, not the objective risk. In these circumstances there is room for compromise, on both sides, with experience being the judge in the long run.

The last topic I want to touch on is terrorism. What are the prospects that a determined group of well-trained terrorists might attack or seize a nuclear flask in transport; and if they did, what might they hope to do with it? For obvious reasons this is a difficult subject to discuss in public. We do not want to provide useful hints for would-be terrorists. Nevertheless some general points can legitimately, and safely, be made.

Let us first remember that a nuclear flask weighs about 100 tonnes. It is not easily transferred from one vehicle to another unless suitable and very conspicuous cranes are at hand. Neither road vehicles of the size required nor trains are easily stolen in this country, and if they were they would still be difficult to hide. It is a simple matter to arrange to be in constant radio contact with the flask and its vehicle, so that one knows immediately if and where its journey has been unexpectedly interrupted. Nevertheless, it is possible to conceive of unlikely, very unlikely, circumstances in which the flask could be stolen and its whereabouts kept secret for some hours. It is really an exaggeration to suppose that it could be much longer than a few hours. So – what next?

The most important point to make is that irradiated fuel is not a nuclear explosive – not any sort of explosive – and there is no way that a terrorist group could convert it into explosive material no matter how skilled they were unless they had access to large-scale industrial facilities constructed for the purpose to which they could transport the flask undetected. Both this and the conversion of the material so obtained into a bomb would be exceedingly complex and dangerous operations which certainly could not be completed quickly. There would be plenty of time for the security services to intervene. More likely, having successfully seized a nuclear flask, the terrorist group would try to mount a blackmail attempt with it. In other words it would become a straightforward hijack operation in which they would threaten to blow up the flask unless their demands were satisfied. Given the state of public understanding of matters nuclear, that would cause great concern wherever the hijack attempt took place; if it were in a highly populated area such as London, there would undoubtedly be a hysterical reaction which would be, by far, the most serious outcome.

So people are panicking, but what further damage could the terrorist group attempt? The worst thing they could do would be to carry out their threat to blow up the flask. I would be surprised if someone does not put before you how many deaths would be caused if the contents of the flask were uniformly distributed over a large area of London. I do not intend to do so, because that is precisely what would *not* happen. Every competent scientist knows that a state of uniformity is extremely difficult to achieve in practice, however hard you try. Remember that the flask is a massive steel structure constructed to resist impacts. If the terrorists possessed a portable stand-off weapon, that would not suffice to breach it. If they possessed a sufficient quantity of high explosives, however, and expert knowledge of how to use it, they might be able to cause a breach equivalent to about a one-inch diameter hole. The flask would thus have suffered damage no worse than we have discussed before under the heading of accidents in transport. Some local discharge of radioactive material would follow but there would be no general release unless the cooling water in the flask was almost completely lost. Even then, the self-heating of the fuel, without which no substantial release is possible, and even then it would be only a small portion of the total, would take many hours, during which emergency action would certainly be taken with overwhelming force. It would be the Iranian Embassy all over again. It would provide much excitement for several days, together with a clean-up problem necessitating some temporary local evacuation of the kind that sometimes arises with an industrial accident.

I suspect it would cause considerably less disruption than the withdrawal of labour from our public utilities – electricity for instance, or, in more recent memory, water.

Thus the worst consequence of terrorist action – in the unlikely event that they succeeded in seizing a nuclear flask in the first place – would be due to hysteria among the public. That hysteria would result from public perception that the risk involved in terrorist action is much greater than it really is. It is by playing on that misconception that the terrorist would hope to achieve his ends. By contrast, it is by trying to understand what the risks really are that we learn to defeat the terrorist. Throughout this talk, I have tried to distinguish between the objective assessment and the public perception of risk. That they are not the same must be the fault of both sides. The public are to blame to the extent that they make no attempt to quantify their fears. They do not put industrial risks into the context of what they are prepared to accept in the ordinary way of life. They do not always balance risk against benefit. They confuse issues with non-issues. In this, overwhelmingly, they are encouraged by the media and by television, because it's only bad news that makes a good story. Industry, for their part, are to blame to the extent that in trying to explain their objective assessments, they sometimes fail to grapple with the realities of public fears. It is irrational to ignore public fear, no matter how deep the misapprehension on which that fear is based. It is also to court frustration because, in the end, it is the public that will decide the future of the nuclear industry. We have to try to bridge the credibility gap from both sides. I hope that is what this conference is about.

3 Nuclear Power – Is it the Only Way?

DES WILSON

Representing Friends of the Earth, I was specifically invited to address the wider question: Do we need nuclear power anyway? Clearly, if the answer to that question is 'no' and if we act accordingly, then the problems raised by this conference will no longer exist. While I do not suggest it is automatically appropriate to tackle the side-effects of a policy by abandoning the policy altogether, when in every other respect the policy is highly questionable, then the case for abandoning it to eliminate side-effects becomes much greater. That, in a nutshell, is our point: that we should not need to worry about the urban transportation of irradiated fuel because we do not need the fuel anyway. We are wrongly transporting the wrong product in order to implement the wrong policy for the wrong reasons.

Before coming to our critique of nuclear energy and our alternatives, there are a number of broad points I should make. First, while this issue should be considered in the context of the energy needs of the nation and the options for meeting those needs, it is not only an energy issue, it is also now an economic issue, an industrial issue, a public safety and security issue, a civil liberties issue, an environmental issue, and in so much as it affects generations to come, it is a moral issue too.

Nuclear energy is an issue striking at the heart of the way we are governed. Nuclear reactors are the most expensive high-technology product that anyone has ever tried to buy or sell on a commercial basis. The colossal sums of money required to research and develop any type of reactor are beyond imagination. Yet decisions taken in this country to proceed with the first and second nuclear programmes, based on the Magnox and the AGR respectively, were taken behind closed doors in

Whitehall, without Parliamentary debate and with no public involvement whatsoever. That decisions of such economic magnitude should have been taken by so few, with virtually no consultative process, is indicative of the constraints which nuclear policy making tends to place on our well-tried democratic processes and institutions. Leaving safety and other considerations aside, the sheer magnitude of the financial commitment required to undertake a nuclear power programme is such that the decision to pre-empt a large portion of the nation's resources cannot be left in the hands of the nuclear scientists and engineers.

While technical and scientific input is obviously essential, the final decision on whether nuclear power should be used to solve our energy problems is one for the people. While technicians and scientists devised the nuclear bomb, no one has ever suggested they should be responsible for its employment. The same applies to nuclear power. It is therefore crucial that the whole issue is openly debated, involving as many individuals and organisations as possible, and that the final decisions are taken from one point of view only – what is best for the community as a whole in the widest sense and in the longer-term.

It is just such a debate that Friends of the Earth in this country has been urging for nearly a decade. It might have been hoped that the Sizewell PWR inquiry would have provided a forum for a wide-ranging national debate, but for a variety of reasons we have been disappointed. I will return to that matter a little later.

I have said that nuclear power is an issue that relates to democracy itself, partly because the expenditure involved is immense and must be diverted from other areas of need. But, and perhaps more fundamentally, nuclear energy is an issue about which there is no social consensus. If national opinion polls are to be believed, over the past few years more people have consistently opposed the expansion of nuclear power than supported it. The issue is deeply divisive, and becoming more so. Governments facing such division can either seek to promote a new consensus, through authentic public education and debate, or they can attempt to muzzle the opposition while themselves retreating into the secret recesses of Whitehall. I am afraid that the present government has chosen the latter course. It fears – and I quote from a leaked Cabinet document – 'that a broad-ranging public inquiry would arouse prolonged technical debate', and it is convinced that it will 'make more rapid progress towards its objective by a low profile approach which avoided putting the government into a position of confrontation with protestors'. When the 'protesters' may conceivably include more than half of the population, this is demonstrably an anti-democratic approach to nuclear decision-making.

The public does not easily accept nuclear power. There is reason to believe that the more people learn about nuclear reactors and their waste products, the less they accept nuclear power. Ministers must face reality when deciding whether or not to expand the role of nuclear power, and reality consists not only of energy demand projections, predicted fossil-fuel shortages, oil prices, and industry's needs. People's views of nuclear energy are equally real and valid, and of considerable importance. Public acceptance of nuclear power is essential for the success of the nuclear industry and Ministers who do not realise this act at their peril.

The position adopted by those responsible for promoting an extension of the nuclear power programme is that negative views on the subject are promoted by irresponsible scaremongers. This position is at variance with the facts. It must be recognised that the significant resistance to a continued expansion of nuclear power has developed and persisted in a situation where the balance of resources in informing the public is tilted overwhelmingly towards the proponents of nuclear power. Not only is a programme of nuclear power being foisted on to a public which manifestly doesn't want it, but the same public is being used to pay for extravagant campaigns to have its opinion reversed. Cries of 'scandal' have rightly followed a recent attempt by Ministers to employ public relations consultants in their campaign in opposition to CND, but in a more covert form, this has been going on for years in the case of the civil nuclear power debate.

Current plans are to build ten more nuclear reactors in Britain, beginning with the PWR at Sizewell. It was the PWR that was the centre of a major emergency at Three Mile Island at Harrisburg, Pennsylvania. This threatened the possible release of radioactivity and over half a million residents nearby had to be alerted for evacuation. The record shows a nuclear catastrophe was only narrowly averted. Nuclear advocates have attempted to argue that the accident demonstrated how effective were the reactor safeguards. However, the sheer number of consecutive failures which occurred during that accident show how difficult it is to ensure safety in a PWR. It is noteworthy that partly as a result of the near-disaster at Three Mile Island, a recent US study has estimated that the probability of such an accident may be up to 100 times higher than earlier so-called 'authoritative' risk studies suggested. The event has been traumatic for local residents, both at the time and since. The best present estimate of the cost of the clean-up programme is put at over $1000 million and replacement electricity will cost several times more. It is suggested that the design chosen for the Sizewell B plant is different, but in fact fundamental weaknesses remain.

Then there is the issue of radioactive waste – the nuclear millstone. All nuclear stations generate lethal high-level nuclear waste and this has to be isolated from human beings for hundreds of years. If the Romans had used nuclear power we would still be guarding their waste. As custodians of this land for our children and their children, we have to seriously consider the legacy we will leave them in this respect.

At the Sizewell inquiry later this year Friends of the Earth will present the case, and call witnesses from a number of different countries, that the PWR has not been demonstrated to be safe – that it is an inherently hazardous technology. We will, of course, publish our case in full at that time. Because of the other areas I wish to cover I will therefore resist the temptation to expand on our concern about safety at much greater length.

Friends of the Earth is participating in the Sizewell inquiry because we believe that the technical case for the PWR is unsound. At the Windscale inquiry – the last major nuclear inquiry – Friends of the Earth argued that, on both economic and technical grounds, the proposed thermal oxide reprocessing plant was premature, and that it made more sense to store spent fuel than reprocess it. There is now little question that these arguments were correct. I am not being unduly immodest in saying this, since I was not involved with Friends of the Earth at the time. What I wish to point out is that the promoters of nuclear energy, try as they might, do not have a total monopoly on expertise. The public can shop around for their information on nuclear power, and if they do so they will be better informed.

But, as I have said before, the question of whether we should continue with nuclear power is not exhausted by purely technical argument, but involves moral issues as well. I would like to illustrate this, albeit briefly, by turning to the question of nuclear proliferation – a question which is as important as any facing the world today.

Nuclear trade, in common with nuclear reactors, has its risks. It has become increasingly difficult to disguise the fact that civil and military nuclear enterprises are mutually enhancing. The Nuclear Non-Proliferation Treaty (NPT) which was intended to sever this link, actually promotes the use of nuclear energy while being marginally effective in deterring those countries which are determined to put it to less benign uses. The NPT is also rightly perceived as a case of 'one law for the rich and another for the poor', since those countries keenest to sell reactors abroad also tend to be those possessing the nuclear weapons which are denied the buyer. When reactor orders are scarce, as they are today, the already inadequate safeguards of the NPT are likely to be overlooked.

It is quite clear that several countries that have either acquired or are in the process of acquiring, nuclear weapons have done so through ostensibly civil channels. The terms of the Non-Proliferation Treaty, and the simple facts of nuclear technology, make proliferation an almost unavoidable consequence of civil nuclear trade. The proliferation problem does not, at present, appear to have a solution. Even so, Britain is by no means doing all it could to minimise the military consequences of its arms trade in nuclear hardware. The National Nuclear Corporation, for instance, is currently trying to sell a Magnox reactor, a British design which also happens to be a very efficient producer of plutonium (the nuclear explosive) to Chile. This sale is being negotiated in the full knowledge that Chile has not yet signed the Non-Proliferation Treaty. Asked by the *New Statesman* to comment on his 'welcome' to Chile as a customer, despite repressions, torture and continued political arrests in that country, Dr Ned Franklin, Managing Director of NNC replied:

> We should wish to be satisfied that the present government was tolerable to the people and the recent history of a country is relevant to this issue. On the other hand, democracy is not the norm in human affairs and we cannot restrict ourselves on moral grounds from dealing with non-democratic governments if they are judged to be stable and acceptable to their people.

He did not offer a view on what level of oppression the directors of NNC held to be 'stable and acceptable.'

The problem of proliferation is acute today, and, as the use of nuclear power increases, it will undoubtedly get worse. Britain's nuclear industry, like those of other countries, is kept alive on a diet of orders from captive markets and vague hopes of export opportunities on almost any terms. In a competitive energy economy it would almost certainly be dead. Continued support for this industry is something which may, in the short term, satisfy the powerful lobby it represents, but may well be something we will come to regret. For this reason we are hopeful that the Sizewell inquiry will concentrate not just on the narrowly technical areas of safety, economics, etc., but consider no less important the wider questions as well.

The Greater London Council should be thanked for organising this conference. I have to say, however, that in other respects there is every justification for deep concern about the way the debate on nuclear energy is being conducted in this country. In particular, we have a demonstrably unbalanced public inquiry into the introduction of the pressurised water reactor at Sizewell. This is not just the view of Friends of the Earth and other objectors, but even of the Central Electricity Generating Board. That organisation has acknowledged that it is unfair that objectors are not financed with public funds. Few who have spent a few hours at the Sizewell inquiry will need further argument on this point. On the one hand, you have the ranks of expensive legal minds and CEGB experts; on the other, small groups of objectors who have to raise their own funds by every means possible, who are only able to attend spasmodically, and only able to present their case within strict financial limits. The Secretary of State for Energy, Nigel Lawson, has obstinately refused to allow any public funding of objectors, yet the Government and CEGB case is entirely financed by public money. As the inquiry drags on and becomes longer and longer so the finances become even more loaded on the Government and CEGB side.

For this reason, the inquiry is unbalanced and unfair and we have warned the Secretary of State that while its conclusions may well be satisfactory to him, we reserve our right to reject its predictable findings on the grounds that it was financially rigged to achieve his objective of forcing upon the British people a form of energy provision that is unnecessary, economically insupportable and unsafe. He has undermined the integrity of the inquiry before it even began, despite every attempt to warn him of this by the inspector, the CEGB, ourselves, and the media. He must understand that there can be no satisfactory outcome for him, no matter what the inspector decides. Nor has Mr Lawson even waited for the end of the inquiry. Within its first week he made a major speech in public making it clear that he was committed to an expansion of nuclear energy.

The CEGB itself does not escape censure. Its arrogant treatment of this conference speaks for itself. This is organised by the largest local authority in Britain and for the CEGB contribution to be a short film and a paper, freeing them of any necessity to answer questions or take part in any debate, is an outrage. Any suggestion that its participation would prejudice the inquiry is nonsense. First, since the inquiry began, the CEGB representatives have appeared in the media debating the issues raised. Second, objectors, including Friends of the Earth, do not find participation in this conference prejudicial to the inquiry. Their arrogant treatment of this conference is unacceptable. They are not here because they are contemptuous of the democratic process and they are not here because they wish to avoid answering embarrassing questions. It is our view that if they do not have the courage and integrity to come to this conference, their film should have been rejected, and they should be presumed to have no worthwhile input to make.

Friends of the Earth have no desire to see the decisions on nuclear energy taken emotively. We accept that this is a crucial issue that should be considered responsibly and seriously. On the other hand, we know from the tremendous response to our appeal for funds to support our case at the inquiry, that there is widespread public concern on this issue, and the Minister and the CEGB must clearly understand that there is no way that they can foist their desires on the people of this country unless they are able to persuade the people that their policy makes sense. The loading of the inquiry, and the treatment by the CEGB of occasions such as this, speak volumes of their present attitude.

Before moving to wider energy policy let me make a brief point about Friends of the Earth. We exist simply as an expression of the concern of an increasing number of people that if we as occupants of a planet forever devour and squander its resources, forever exploit its every advantage to satisfy our immediate appetites, forever pollute its air, soil and water, forever assume that all other living species exist only to serve our needs and wants, then one day that planet will become uninhabitable. We are concerned about the environmental crisis already with us – the extensive pollution, the loss of species at an unprecedented level, the worldwide problems of loss of cultivated land, the destruction of tropical rain forests, the desecration of places of beauty and habitats of wildlife and the reckless waste of resources, notably energy resources. In particular we are concerned as custodians of the planet for our children and their children. It is they who will enjoy the benefits or suffer the disadvantages of our activities today.

Friends of the Earth is not a negative force. We do not oppose technology and fresh initiatives as a matter of principle. Our thinking is not Luddite. We are not 'anti' any action so much as 'pro' alternatives. For instance we oppose transport domination by the motor car, but only within a strategy of being *for* environmentally clean and socially advantageous public transport. Likewise, we oppose plans to further develop Britain's nuclear energy programme, but in the context of being for alternative sources of energy that we believe are more beneficial in the longer term. We are not anti-industry – we are pro socially responsible industry.

We have not chosen to oppose nuclear energy as a matter of principle; we oppose it because after involvement in the issue and much research over many years we believe it is unsafe, creates extremely serious problems of waste management and disposal, is vulnerable to exploitation by negative political forces and urban terrorists, is uneconomic, and leaves an unacceptable legacy for the generations to come. We oppose it above all, however, because we support a completely satisfactory alternative.

I had planned to say that we were at the crossroads in terms of energy policy, but I fear you would be entitled to ask 'What energy policy?' We have no energy policy in Britain. We have a policy for electricity, a policy for coal, a policy for gas, and a policy for oil, but no overall energy policy. Each industry vies and struggles with one another for customers in the High Street and subsidies from the Exchequer. No utility has any incentive to promote conservation of energy with any vigour; rather the incentive is to sell more appliances and more fuel. With Ministers, in our view incorrectly, and the CEGB, in our view incorrectly, predicting greater energy use and preparing to meet that projected use by producing more energy, there is no drive towards conservation. If the gas producers were able to convert all their customers to North Sea gas, why cannot the electricity producers convert their customers into low-energy consumers by insulating each and every loft and draught-stripping each and every window? It has been reliably suggested that the return on investment of such basic insulation work would be three years. The most vulnerable in our society, the old, who form a growing percentage of people in this country, the sick, the young and the unemployed, are those who spend most time in the home and who require the greatest supply of domestic energy and who can least afford to pay the costs. Our lack of energy policy turns social need upside down.

Energy policy cannot be divorced from wider national priorities – the investment of public money, the meeting of social needs, the elimination of unemployment. The housing stock in this country is deteriorating dramatically; colder and damper houses lead to an increase in ill health. Friends of the Earth has argued for years, and does so with greater urgency now, that a national home insulation programme paid for by Government would produce smaller fuel bills, warmer and thus healthier homes, reduce the demand for electricity generation and above all, create many more jobs in every community in the land. It would be the first step down the road to a sustainable and progressive energy future.

Such a step has been taken in other countries. The Swedish referendum of 1980 urged the phasing out of nuclear power; by the beginning of the next century Sweden is likely to derive 60–70 per cent of its energy supply from sources which are only marginally, if at all, in use today. The path which Sweden is following involves a high degree of energy efficiency coupled with energy supply from renewable energy sources, with emphasis on the solar option.

The recent history of energy conservation policy in this country is quite another story. The present Government has abdicated to the market place its responsibility for leadership and direction. The Central Electricity Generating Board, with its statutory obligation to provide secure and economical supplies of electricity, shows no interest in saving energy. It concentrates instead on supplying electricity which we then waste through inadequate thermal insulation and inefficient appliances and industrial processes. As our competitive position vis-à-vis countries like West Germany and Japan is eroded by our inefficient use of energy, we are spending £2½ million per week on research into the fast-reactor and fusion, technologies which on optimistic forecasts will not provide commercial energy supplies in less than 30 years, if ever. In contrast to these high-risk and futuristic technologies, the cost-effectiveness of conservation has been detailed in numerous British reports and publications, the first being for the International Institute for Environment and Development's 'Low Energy Strategy for the UK'. The results of this work have been reinforced by more recent studies such as 'An Energy Strategy for the UK – Opening the Solar Option' by Earth Resources Research for Friends of the Earth. The Association for the Conservation of Energy (ACE) has just produced a report showing that the introduction of a serious programme of energy conservation in buildings would have, besides the obvious social benefits, the effect of creating 155,000 new jobs. These jobs would be unskilled or semi-skilled and could largely be created in deprived inner city areas. Such a programme could, according to the ACE report, offer an internal rate of return of 13–16 per cent.

The Select Committee on Energy in its report on energy conservation in buildings concluded that cost-effective measures with pay-back periods of 5 years or less could produce savings in the area of 30 per cent of delivered energy consumption in both domestic and non-domestic buildings, using existing technology. The major obstacle which the Committee saw to a programme of energy efficiency which would reach these targets was 'the fragmentation of responsibility amongst government bodies and above all the lack of political will at the heart of government which smothers the efforts of the Department of Energy's Conservation Division.' The fact that the Government has cut the total conservation budget appeared to the Select Committee to reflect 'the Government's fundamental disinterest in conservation, especially where public expenditure is involved.' The Committee went on to conclude, 'There are now many conservation measures that are so much more cost-effective than most energy supply investment, that the caveats expressed by the Department of Energy seem quibbles. Such reservations do not represent reasons for not stimulating highly cost-effective energy conservation investment.'

In this country the approach to energy policy-making is asymmetric. We have for so long emphasised the necessity to increase the supply of energy that we have neglected to examine the economic advantages of reducing demand. The view that the economic health of a nation is predicated on ever-increasing consumption of energy is rapidly becoming an anachronism, one which we can no longer afford. The bias towards a supply-side view of the energy problem is incorporated in our institutional framework. The Select Committee expressed its reservations on this point when it stated that

> We are thus concerned at the misallocation of resources which results from investment in supply being appraised at the required rate of return (5 per cent), while many conservation projects are required to satisfy much higher rates of return. We consider it a testimony to the irrationality of present energy policy that investment in additional supply capacity by the coal, electricity, gas, and oil industries, is, in practice, assessed by different criteria than those applied to conservation by local authorities and other public sector bodies responsible for buildings.

How ironic that a Conservative Government should be so disinterested in conservation. How ironic that so-called radical organisations like Friends of the Earth should be the ultimate conservatives. Yet this seems to be the case. The Government does not just passively ignore energy conservation, it actively suppresses information which supports the kind of case Friends of the Earth and other groups have been making for years. A Whitehall report written by Sir Derek

Rayner, who had been appointed to assess the efficiency of Government departments, was withheld from the public because it pointed to the Government's complete lack of a coherent conservation policy. A follow-up report put forth conclusions so unpalatable to the Government that it was suppressed on the specific instructions of Energy Minister Nigel Lawson. This second report set out to answer the criticisms which had earlier been made by the Select Committee, that Whitehall had no clear idea of whether investing about £1300 million in a nuclear power station was as cost-effective as spending a similar amount to promote energy conservation.

One of the unpalatable conclusions was that

> There are significant and continuing national benefits to be gained from increased conservation investment. If the nation is to receive the maximum benefit from the investment resources that it spends, then greater emphasis needs to be placed on energy conservation. Far from minimising the role of government, this study suggested that money now devoted to research into new sources of energy could, in the short term, more advantageously be used to promote conservation.

It further estimated that on the basis of present economic investment information 'the potential that is available in the commercial and industrial sectors may amount to some £5000 million (saving 11 per cent of present delivered energy consumption) and might, in the absence of other obstacles, be achievable in five years.'

It is now fairly clear that the chief obstacle to improved energy efficiency in Britain is the ideology of the present Government. In place of the reports to which I have been referring, Mr Lawson published a third report, the drafting of which was done under his supervision. This official study, which has been published and given as evidence to the Sizewell inquiry, is sceptical about the value of any further role for the Government in energy conservation. It states that, because of all the uncertainties, 'It is difficult to find an economic justification for direct Government involvement in fuel choices made by consumers.'

Lawson appears to be in a crowd of one here, since no other major report supports his conclusion. His ruthless attempts to force through his own preferences, irrespective of the alternatives and of the advice he gets, is rapidly assuming the proportions of a national scandal. What we have here is an appalling muddling of priorities. Huge sums are spent on research into high-prestige nuclear technologies in the hope that the investment may begin to pay off in 30 years. Meanwhile, less costly measures which could save as much as one-third of the energy presently consumed in Britain within 5 years are dismissed for no logical reason.

Friends of the Earth calls for a moratorium on the nuclear programme for at least a decade, and for the measures argued for in the ERR, ACE, Select Committee and Rayner reports to immediately be put into effect by an Energy Conservation Agency. An Energy Conservation Agency would counterbalance established energy supply industries and coordinate a national conservation effort. Above all, it would be responsible for planning and setting up a major national homes insulation programme, administering the injection of finance through local government in the forms of loans or grants to tenants and owner-occupiers alike, and supporting local authority or voluntary organisation work forces as well as private industry to feed into the programme. It should be responsible for developing higher standards of thermal insulation for new buildings and regulating their introduction. It should also be responsible for establishing energy-efficiency targets for all energy uses and mounting major publicity and education campaigns.

Indeed, if this programme is to be taken seriously, local authorities, working with this agency, would ultimately have to be responsible for seeing that every single house within their area, and thus every single house in the country, is properly insulated. One cannot overemphasise that the money spent on this would pay for itself. Furthermore, this would be a labour-intensive exercise that would create employment and pour money currently being spent on unemployment back into the economy in a far more constructive way.

In addition, the moratorium should be used as a time for extensive and fully public national debate on Britain's energy policy, and in particular on the energy sources to be adopted in the future. The debate should involve both producers and consumers of energy and should encompass not only technical and economic aspects of energy generation but the social and ethic aspects as well. The debate about energy policy is as much a debate about the kind of society we want to create as it is about technical choice.

Two decades of muddle-headed priorities have led to the PWR proposal for Sizewell. Approximately £10 billion have been poured into the nuclear option and yet nuclear electricity today contributes less than 2 per cent of our total delivered energy. Many people confuse electricity with energy. Electricity is but a small proportion of our total energy use – some 12 per cent. Of that segment, nuclear electricity represents between 13–16 per cent. There is such an excess of electricity generating capacity in England (about 30 per cent) and in Scotland (about 70 per cent) that even on the coldest night of the year if all the nuclear power stations were to be switched off we would hardly notice.

The CEGB case is that they are preparing for demand to come. They claim that coal-fired power stations are coming to the end of their useful lives and cannot be renovated and cannot be replaced. They argue that we will run out of coal – in 300 years or so – and that we will need an alternative source of fuel. How ironic, too, that it is the environmentalists who are always described as the prophets of doom and yet it is the CEGB who seek to justify their unnecessary proposals by pessimistic projections. However, let's at least acknowledge their desire to plan ahead for we would like to do so too.

We believe the answer to Britain's energy problems, and a lot of other problems as well, is a two-fold approach: first, the programme of conservation to which I have referred; and second, the urgent exploration and then the development of alternative renewable sources of energy supply. The cost of even the short-term nuclear programme would more than finance the comprehensive implementation of both these approaches.

Let us now look at the exciting prospects offered by renewable energy resources. Once more, the attitude of the Government to the potential of this solution has been astonishing. Whilst £220 million of research money will be pumped into the nuclear option this year, the research and development budget for renewable energy has been reduced from £14 million to £11 million. This represents a virtual dismissal of the whole concept of renewable energy technology. As the Friends of the Earth book *Solar Prospects* written by Mike Flood has demonstrated, other countries are already achieving results and meeting energy needs from the sun, the wind and the waves.

We propose district heating – combined heat and power. Every year 20 per cent of the UK's primary energy is wasted in reject heat from power stations and it is projected that this waste will increase by 40–60 per cent over the next 20 years. More than 40 per cent of our end-use energy need is low-grade heat for space and water heating. If fuels were converted directly into heat to meet these needs, with electricity as a by-product, current and projected waste could be enormously reduced. Combined heat and power on a district basis could meet the needs of 33 per cent of households by the year 2000 and of 80 per cent or more by the year 2025 – almost all the UK's urban households.

Combined heat and power is a reliable technology. Several European countries already meet a considerable proportion of their space and water heating requirements from combined heat and power stations, both in industry and as a result of district heating systems. District heating systems supply a quarter of the total heating load in Denmark, 20 per cent in Sweden, and 40 per cent in the Soviet Union. The UK is falling far behind other countries in the provision of heat from district heating – combined heat and power schemes – and yet a programme of eight to ten such stations would be far cheaper than a programme of PWRs of similar capacity.

Solar energy can combine with and eventually replace combined heat and power in supplying much of Britain's space and water heating requirements. District heating networks could be designed to facilitate the changeover. Flat-plate collectors for solar water heating have been in widespread use all around the world and there are currently 5000 installed in the UK. Flat-plate collectors with short-term storage could supply more than half the annual hot-water requirements of most homes. More efficient collectors could supply virtually all of the hot-water requirements if used in conjunction with inter-seasonal heat storage.

Friends of the Earth projects that by the year 2025 solar heating systems could meet 20 per cent of energy needs. This isn't some kind of 'Star Wars' thinking; today there are estimated to be well over 3 million flat-plate collectors in operation around the world. The technology exists and is proven. The book *Solar Prospects* illustrates the extent that Britain has fallen behind the rest of the world. For instance, building firms in Saskatchewan, Canada, are constructing houses that require no external heat source and rely on the sun to provide space and winter heating, and Saskatchewan winters are harsher than in the north of Scotland. Japan aims to equip 20 per cent of its houses and 25,000 offices and factories with solar water heaters by 1990. That is in addition to the 3 million solar heaters it already has. Denmark aims to meet 10 per cent of its electricity demand from the wind by 1990. Plants and waste can be converted into alcohol substitutes for petrol. All cars now running in Brazil are using blends of petrol and alcohol.

Mike Flood, in *Solar Prospects*, points out some of the reasons why renewable technologies are more attractive than most conventional energy technologies:

They can be matched in scale to the need, and can deliver energy of the quality that is required for a specific task, thus reducing the need to use premium fuels or electricity to provide low grade forms of energy such as hot water (which can be supplied in many other ways).

They can often be built on, or close to, the site where the energy is required; this minimises transmission costs.

They can be produced in large numbers and introduced quickly, unlike large power stations which have long lead times, often of 10 years or more. Rapid planning and construction lowers unit cost and allows planners to respond quickly to changing patterns of demand.

The diversity of systems available also increases flexibility and security of supply. In contrast, over-dependence on imported fuels makes a country more vulnerable to political pressures from producer nations and multinationals. Generic faults in power plant, serious breakdowns, industrial action or simply bad weather can jeopardise the supply of electricity.

Let me also quote what Sir Martin Ryle, Nobel Prize winner in Physics and Fellow of Trinity College, Cambridge, wrote recently. He said that

> The Government-proposed nuclear programme would be only an insignificant contribution to the energy which will be lost by oil/gas depletion and it is extremely urgent to put in hand other programmes which can make an effective contribution. The immediate attention to saving energy now wasted, the installation of district heating schemes in urban areas and a modest introduction of renewable energy sources could, by the end of the century, replace most of the non-transport uses of oil without increasing our present coal production.

Sir Martin went on to say that 'The obsession with nuclear-based electricity as the main source of our future energy supplies has led to inadequate development of these alternative programmes.' He concluded by saying that

> If nuclear power were to be abandoned it would be seen by some – especially those who have given their working lives to its development – as a retrograde step. But this is not a new problem in engineering; the extremely challenging aerodynamic, structural control and

> instrumentation problems which were solved in the development of Concorde, have produced a magnificent piece of engineering, but one which appears to have no future in the present world. We must put all our energies into solving the difficult problems, in many disciplines, which are involved in renewable sources . . . .

To sum up, then, the whole energy policy – or lack of an energy policy – in this country is open to criticism. Forecasts of demand and costs are highly defective. Nuclear power is about the worst conceivable option on safety grounds, economic grounds, and social grounds. If we are not to squander a fortune of public money in dangerous and unnecessary technology, we need to reappraise forecasts of demand and costs and explore far more economical and sensible alternatives. The nuclear option fails to meet our energy problems and needs. A wide range of energy-efficient fossil-fuel and renewable energy technologies which are available are more reliable, cheaper, quicker to implement, and less harmful environmentally than nuclear power.

The nuclear power merchants are not taking us forward but holding us back. The case for a major energy conservation programme on energy, economic and employment grounds is overwhelming. The case for renewable energy is irresistible. We should reject the imposition of nuclear power and instead move forward imaginatively to grasp the opportunities that a rational and sustaining energy policy can create for this country and peoples all over the world. We have a unique opportunity with the resources provided by North Sea oil to choose and finance and implement the kind of solutions that generations to come will recognise as bold, imaginative, but above all correct. They will recognise those measures were the correct ones for them. They will be able to look back at you and me and our generation and say – they really were friends of the earth.

4 Some Simplifications for Decision-Making

DAVID W. PEARCE

Transporting nuclear fuel, spent fuel rods and nuclear waste is an integral part of the nuclear fuel cycle. To object outright to this aspect of the fuel cycle is to reject nuclear power as an energy source in the UK's energy future. That is a position some will find wholly consistent and for them the transportation issue is simply one more aspect of the risks of nuclear power. To others, nuclear power is an essential feature of energy policy, so that the degrees of freedom in debating the transportation issue really relates to: (1) the extent to which its associated risks should dictate the choice of a particular *type* of fuel cycle, and (2) the mechanisms for minimising that risk. We can then look at both standpoints and ask what they imply, if anything, for the structure of decision-making in the UK.

CHOOSING BETWEEN THE OPTIONS

While there is any number of variations on the precise nature of a nuclear fuel cycle, we can safely limit ourselves to just two for current purposes. These are:

(1) the 'throwaway' cycle in which, once used, spent fuel is disposed of to some storage system; and
(2) the 'recycling' cycle in which spent fuel is transported to a reprocessing centre for recovery of uranium and plutonium which can then be returned to the fuel cycle or stored, or if of military grade (in respect of plutonium), diverted out of the civilian power system.

In the United Kingdom we have a mixture of both cycles with Magnox fuel being reprocessed at Windscale, now renamed Sellafield, and AGR fuel destined for reprocessing at the THORP plant at Sellafield and which was the subject of the 1977 Windscale inquiry. For various reasons, a great deal of spent fuel is stored under water at Sellafield.

If concern over the transportation of irradiated fuel is so great, one would expect one option within the throwaway cycle to be considered, namely, storage for very long periods at the nuclear station site. This would remove the problem of transportation. The obvious cost is that we would have to multiply the kinds of facilities at Sellafield by however many power stations in the UK; that would present formidable problems of monitoring and control. Arguably, the risks of geographically dispersed long-term storage of spent fuel are greater than the risks of centralised storage. Moreover, such a policy would preclude the reprocessing option. Seen in a cost-benefit framework, we would have to weigh up the costs of increased control in the geographical dispersion option together with the costs of forgoing the reprocessing option, against the benefits of dispersion which would accrue as reduced risks from transportation and, some may argue, the benefits of not concentrating spent fuel on one site which could then be the focus of terrorist attention. In speaking of cost-benefit, I am not suggesting a fully monetised cost-benefit analysis. I am suggesting that the framework of *thinking* in cost-benefit terms is the correct one (Pearce 1979; Pearce, Edwards and Beuret 1979). In turn, the issue is then what is the proper institutional framework for debating these costs and benefits.

In all this it is as well to remember that concentrating on irradiated fuel is somewhat misleading. Even if spent fuel is confined to power station sites for long periods, there remains the transportation of the fuel itself from fabrication plant to power station, and the problem of irradiated waste for deep-sea disposal or, perhaps eventually, disposal to deep formations underground.

REJECTING NUCLEAR POWER

If nuclear power is rejected *per se*, the debate over the transport of irradiated fuel would seem somewhat beside the point. In essence, all that can be argued from such a standpoint is that the transportation risk is one more adverse feature of nuclear power along with the risks of routine radiation, the chance of an accident, the civil liberties issue, the proliferation problem and so on. Two comments seem in order. First, if this is the view of those who object in a wholesale manner to the land transportation of irradiated fuel it is intellectually honest to point it out rather than to focus attention on one aspect of a fuel cycle in a political context where one knows that to object to that aspect is to preclude virtually all the options for using nuclear power. It is conceivable, but I suggest very unlikely, that someone holds the view that the transportation risk is the dominant adverse feature of nuclear power. Second, the experience of the last few years has surely demonstrated that the institutional context for debate over the virtues of nuclear power is not the local public inquiry. I shall return to the central issue in a moment.

THE RE-ROUTING ARGUMENT

The straightforward suggestion so far is that one either accepts nuclear power as part of energy policy and seeks to minimise the risks associated with it, as with any source of energy, with due regard for the impact of risk reduction on the economics of energy supply, or one rejects nuclear power on a wholesale basis, in which case the transportation issue is not likely to be a dominant and pervasive aspect of that rejection. This dichotomy is important, because the pros and cons of nuclear power are not suited to the public inquiry system, whereas the methods of minimising risks from spent fuel transport initially appear to be.

To illustrate this latter aspect, consider the view, widely held, that however convenient it is to use transportation systems that run through urban complexes, irradiated fuel should be routed through areas where the population at risk is minimised. Since such an option involves questions of land-use in the sense familiar to public inquiries, the discussion of such options appears suited to the local inquiry framework. Again, however, a number of comments are in order. First, if the routing involves a final destination such as Sellafield, and if the power station sources are as diversely located as Hunterston, Torness, Sizewell, Dungeness, Hinkley Point and Wylfa, the local authorities involved might be numerous, as might the local objectors placed at risk by the transportation system. The local public inquiry is then fairly evidently not the proper forum. The public inquiry commission, if it were ever introduced, might be.

Second, we shall have one more example of the phenomenon of

'geographical shift' whereby one local authority declares its antipathy to a route through its own territory in the hope of shifting the problem to other routes and other local authorities. If, on the other hand, the purpose of geographical shift is to encourage *all* authorities to engage in such activity, the objective becomes more clearly defined as outright opposition to nuclear power *per se* and our previous remarks apply.

Third, geographical shift implies that alternative routings are possible. Here the choice between alternatives could involve a complex weighing up of the limitations of travel by rail and the flexibility afforded by road, but each of which may have very different risk factors in terms of the chances of an accident.

Fourth, and in its favour, routing to minimise risk reflects a legitimate concern with both reducing the probability of accident and with the size of the consequences of an accident. (It remains a costly curiosity that much of the discussion about the risks of any activity, including using nuclear power to create electricity, continues to be carried out in terms of probabilities only. Risk embraces both these dimensions.)

THE PUBLIC INQUIRY AS FORUM

The public inquiry has now become a focus for debates over nuclear power, or, for that matter, the merits of motorways, forecasts of passenger travel by air and so on. The inquiry was never intended for this purpose. It remains an administrative procedure which presumes in favour of a development unless the objections are judged to be sufficiently in the public interest to outweigh that tacit presumption. Moreover, it affords the most vociferous and active objectors their role only by the sufferance of the system. Although the subjects it discusses have changed, its nature and role within the political system have not. It is none too surprising, then, to find discussions of the public 'need' for nuclear power, the apparent virtues of alternative energy sources, alternative lifetime, international dimensions of proliferation, discussions of human rights and so on all taking place within that forum but without the real possibility that those arguments could sway a government of the day in its policy towards nuclear power. This apparent contradiction which adds so much to the frustration of the third-party objectors is built in to the public inquiry system as we have wrongly modified it in recent years, and notably in the case of the Windscale inquiry. We are pretending that the system is a legitimate political forum when it cannot be because we cannot have an administrative procedure usurping the proper function of government or Parliament. The inquiry system as applied to such 'major' issues is a fiction, as those who have followed the procedures as a means of decision-making seem agreed (Pearce, Edwards and Beuret 1979; Wynne 1983).

Once this is accepted it is also fairly easy to see why the much debated public inquiry commission has never been invoked. For it, too, would effectively remove the procedure for debate and evaluation from the political arena and place it in a context of so-called expert debate between parties none of whom can legitimately claim to represent public opinion. No one elects the inquiry inspector, the developer or the objecting parties with the exception of the local authorities who may well be involved (but who themselves may have called the inquiry).

IS SPENT FUEL TRANSPORT A 'BIG ISSUE'?

What this very brief overview of the issues suggests is that 'little issues' should remain in the province of the public inquiry but that 'big issues' should not. I wish, however, that I could offer a criterion for distinguishing the big from the small. Windscale's reprocessing plant was a 'big' issue because of the national and international aspects it raised. As such it should not have been the subject of a local inquiry in the form that actually occurred. All that it did was to launch us into the unreal world of debates about principles that cannot be commented upon by the local inquiry. If bigness is to be defined according to the number of persons affected, THORP was a big issue, but the Belvoir coal mine was not. The Mossmorran LNG terminal could arguably affect hundreds of thousands in the unlikely event of a serious accident, as might the installations on Canvey Island. But it would be hard to say they were national issues. There is without question a spectrum of several dimensions of social cost ranging from the small and localised through to the large and localised, to the small and national and the large and national.

Where does the transport of irradiated fuel fit in? In principle, it is an issue which potentially affects millions but that requires us to think of multiple simultaneous or sequential accidents. In practice, we are talking of localised risks. As such, there is a prima facie case for raising the objections to such transportation in the local inquiry framework.

But, legally, no one has applied for permission to transport the spent fuel through a given area. The issue thus seems to be in no man's land in terms of the chances for public scrutiny. If and only if public concern warrants it, there is a case for an inquiry of sorts which, like a planning inquiry commission, could sort out the generic issues from the site-specific problems of defined routes.

REFERENCES

Pearce, D. W. (1979) 'Social cost-benefit analysis and nuclear futures', *Energy Economics*, 1, No. 2.

Pearce, D. W., Edwards, L. and Beuret, G. (1979) *Decision-Making for Energy Futures: a Case Study of the Windscale Inquiry*, London: Macmillan.

Wynne, B. (1983) *Rationality and Ritual: A Case Study of the Windscale Inquiry*, London: British Society for the History of Science.

5 The Role of the International Atomic Energy Agency

MORRIS ROSEN

ROLE OF THE IAEA

The International Atomic Energy Agency was established in Vienna, Austria, in July 1957. It is an international organization and, like the World Bank and the World Health Organization, it is a specialized agency in the United Nations system. Although autonomous, the Agency sends reports of its work to the General Assembly and other United Nations organizations. The Agency is authorized by its Statute, among other things:

> to establish or adopt . . . standards of safety for protection of health and minimization of danger to life and property . . . and to provide for the application of these standards to its own operations as well as to operations (supported) by the Agency or . . . at the request of . . . parties . . . or of a State . . . to activities in the field of atomic energy.

The Agency's regular budget this year is about US$100 million. That money, which is contributed by the 111 Member States, is divided among several programme areas:

(1) Technical Cooperation and Assistance, primarily for developing countries;
(2) safeguards; and
(3) a number of technical programmes on specific subjects including: Research and Isotopes, Nuclear Fuel Cycle, Nuclear Power and Nuclear Safety.

The 1650 persons employed by the Agency are recruited from more than seventy-eight Member States. Most of them work at the Agency's headquarters in Vienna, Austria. The Agency also operates research laboratories in Seibersdorf, just outside Vienna, Monaco and Trieste, and has field offices in Toronto, New York and Geneva. Many functions of the Agency are also largely carried out through the use of experts provided by Member States – as consultants, participants on technical committees and advisory groups.

The Agency provides guidance to Member States in the form of safety standards, advisory material, technical reports and advisory missions. Also exchange of information is provided in seminars, symposia and meetings arranged or sponsored by the Agency, and through INIS – the International Nuclear Information Service, on a computer programme. Training and technical assistance are made available in various areas as the need arises in the form of training

TABLE 5.1 *International regulations for transport of radioactive materials based on IAEA Regulations*

Mode		*Regulations*
Sea	(IMO)	International Maritime Dangerous Goods Code 1975
Inland Waterways	(ECE)	European Agreement for the International Carriage of Dangerous Goods by Inland Waterways (ADN) – in draft
Air	(IATA)	International Air Transport Association Restricted Articles Regulations 1977
	(ICAO)	Convention on International Civil Aviation (Chicago Convention) Annex – Standards and Recommended Practices
Rail	(CIM)	International Convention Concerning the Carriage of Goods by Rail
	(RID)	Annex I – International Regulations Concerning the Carriage of Dangerous Goods by Rail 1977
Rail (east bloc countries)		The Regulations currently applied by states participating in the Council for Mutual Economic Assistance (COMECON)
Road	(ECE)	European Agreement Concerning the International Carriage of Dangerous Goods by Road (ADR) 1976
Post	(UPU)	Detailed Regulations for Implementing the Universal Postal Convention 1964

courses, workshops, regional seminars and advisory missions. In the field of transport of radioactive materials, the Agency issues regulations (Regulations for the Safety Transport of Radioactive Materials, 1973 Revised Edition (as amended)), but does not enforce them. They are enforced either through adoption by Member States, as in the UK, or through conventions or other actions by international bodies (see Table 5.1).

Transport safety

The Agency's activities in Transport Safety, which are carried out by the Division of Nuclear Safety as part of a subprogramme of Radiological Safety, were initiated in the late 1950s. In view of the expanding use of radioactive materials for peaceful purposes at that time, the Agency undertook to develop harmonized safety rules for transportation of those materials on as wide a basis as possible and for all means of transport. Based on existing good practices and the few simple regulations already in effect, the Agency began a synthesis of such rules in 1958.

The first edition of the Agency's Regulations for the Safe Transport of Radioactive Materials, Safety Series No. 6, was published in 1961. In addition to being applied to the Agency's operations, they were 'recommended to Member States and to International Organizations concerned as a basis for national and international transport regulations'. Soon after, the Agency published 'Notes on certain aspects of the Regulations' in which ideas underlying the Regulations were explained as part of a continuing programme to communicate the bases for and adequacy of the measures taken to achieve safety in transport.

The Regulations were revised and updated in subsequent years to reflect experience in their application, new trends in radiation protection and changes in the methods and technology. The most recent version, entitled Regulations for the Safe Transport of Radioactive Material, 1973 Revised Edition (As Amended), Safety Series No. 6, was published in 1979. It is available in the four working languages of the Agency, English, French, Spanish and Russian. A comprehensive review of the Regulations is currently underway for a Revision to be published next year.

By 1969, the Agency's recommended Regulations had been adopted by, or used as the basis for, regulations of almost all of the international organizations concerned with transportation and by most of the Agency's Member States.[1] As a result, the Regulations are applicable to transportation of radioactive materials almost anywhere in the world.

This was accomplished without a specific convention or treaty. This, we believe, occurred because the Regulations developed by the Agency were practical, understandable and effective and were among the first comprehensive requirements for safety in the transport of radioactive materials in the world. In certain cases, however, the international organization applying the regulations is giving effect to a convention. For example, the standards of the International Civil Aviation Organization are issued under the Chicago Convention and International Regulations Concerning the Carriage of Dangerous Goods by Rail for Central Europe are issued under the International Convention Concerning the Carriage of Goods by Rail.

GENERAL OVERVIEW OF TRANSPORT OF IRRADIATED FUEL

Radioactive material is any material or combination of materials which spontaneously emits ionizing radiation. Because of their radiation-emitting properties and other physical and chemical properties, radioactive materials are widely used in medicine, agriculture, industry and research activities. Radioactive materials are also a subproduct from nuclear power generation in the form of irradiated or spent fuel. In such connections, radioactive materials are transported in many forms and a wide range of quantities.

Transport requires careful attention to a wide range of administrative, technical, procedural and operating problems. The transport environment is predictable but uncontrolled. During transport, packages are subjected to handling by unskilled workers who may have limited knowledge of radiation-protection practices. In normal transport, the packages of radioactive material pass through the public domain, often through urban areas, in relatively close proximity to members of the general public and give rise to a negligible radiological impact. Such shipments may be subject to accidents in transport with a wide range of severities, although the severe accidents usually have very low probabilities of occurrence.

Transportation of irradiated or spent fuel from a nuclear power plant is an important activity associated with the operation of that plant. As shipped, nuclear fuel is in solid physical form, encapsulated in an inert cladding and, in addition, is contained in a sealed container designed to dissipate the decay heat, to prevent nuclear criticality, to shield the radiation emitted from the fuel to the extent necessary and to prevent release of the radioactive contents even if involved in a severe transportation accident, including fire.

Since the 1940s, when production of radionuclides in nuclear reactors began, the number and size of shipments of radioactive materials have grown each year. It is now estimated that several million shipments are made each year. Most of the shipments are small quantities, shipped by air, for use in hospitals or doctors' offices. For example, from data recently reported to the Agency on numbers of shipments made in 1980, a total of 100,000 packages were shipped in France, of which 70,000 were radiopharmaceuticals and 211 irradiated fuel; in the Federal Republic of Germany, a total of 350,000 packages were shipped, 315,000 were radiopharmaceuticals and 66 irradiated fuel; and in Italy, for 1982, a total of 200,000, 198,000 were radiopharmaceuticals and 63 irradiated fuel. These data show that more than 400 packages of spent fuel were shipped in the UK over a distance of 217,000 kilometres, and in six countries in Europe, 434 packages of spent fuel were shipped over 372,000 kilometres over a twelve-month period. Most of these were carried by road for some distance, then by rail and a few travelled by sea. Also in the USA, over the past 25 years, more than 4500 packages of spent fuel have been shipped.

Accidents and incidents

Although many million shipments of radioactive materials have taken place in the world, only a few hundred accidents or incidents have been reported in transportation involving packages of radioactive materials. Only a fraction of those accidents involved any release of contents or increased radiation levels and none resulted in serious injury or death attributable to the radiation aspects.

Transportation accidents and their potential effect on shipping containers have been extensively studied. The results of calculations of the radiological impact from transportation of all types of radioactive material show exposures in normal transport to be small and the risk from accidents very small.[2] Similar calculations for transport of irradiated fuel in Italy,[3] UK,[4] Federal Republic of Germany[5] and Japan[6] show the exposures in normal transport to be small in comparison with the exposures in other of the fuel cycle activities, and a very small fraction of the exposure from natural background radiation. UNSCEAR[7] states that 'estimates of the normalized collective effective dose equivalent commitment associated with irradiated fuel transport is in the range 10^{-3} to 10^{-2} man Sv $[GW(e)a]^{-1}$,' and estimates the total for operations in the nuclear fuel cycle to be 5.7 man Sv $[GW(e)a]^{-1}$.

With respect to the potential risk from accidents involving spent fuel, UNSCEAR indicates 'it is difficult to add any meaningful component . . . from accidents since there is a probability distribution of a whole range of accidents . . . different meteorological conditions . . . and widely different population distributions . . .' and gives no estimated values. Japan[8] concluded that no release will occur because of controls exercised in transport and Italy[9] concluded the annual expected impact on man is very small.

Based on the available facts and figures, we conclude that the risk of a public catastrophe from the transport of irradiated fuel has been virtually eliminated by strict standards, engineering design safety, and operational care; and that the likelihood of death, injury or serious property damage from the nuclear aspects of irradiated fuel in transportation is orders of magnitude less than the likelihood of death, injury or serious property damage from more common hazards, such as automobile accidents, fires, etc., and is comparable to that found in other safe industries.

THE TRANSPORT REGULATIONS AND RELATED IAEA ACTIVITIES

Safe, practical and enforceable regulations are considered to be critically important to continued development and utilization of nuclear energy.

Standards for Radiation Protection

In 1962, the Agency published the Basic Safety Standards for Radiation Protection, Safety Series No. 9, based on the (at that time) latest recommendations of the International Commission on Radiological Protection (ICRP). A revision was published in 1967, also based on ICRP recommendations. These standards prescribed maximum permissible levels of exposure to radiation and fundamental operational principles and provide an appropriate regulatory basis for the protection of the health and safety of employees and the public.

In collaboration with the World Health Organization, the International Labour Organization and the Nuclear Energy Agency of the OECD, the Agency has recently published the revised Basic Safety Standards for Radiation Protection.[10] These standards, based on the more recent recommendations of ICRP,[11] require that exposure to man be controlled within the following system of dose limitations:

(1) *justification of the practice:* no practice involving exposure to ionizing radiation shall be adopted unless its introduction produces a net benefit;
(2) *optimization of protection:* all exposures shall be kept as low as reasonably achievable, economic and social factors being taken into account;
(3) *individual dose limitation:* the dose equivalent to individuals shall not exceed the limits recommended for the appropriate circumstances by the ICRP.

Justification and optimization are source-related requirements which apply to all activities and sources of exposure, including transportation of radioactive materials. The individual dose limitation is an individual-related requirement which applies to doses from all activities and sources of exposure received by members of the general public and workers, other than exposures from medical practices and natural background radiation not technologically enhanced by man.

Among the actions involved in implementing the system of dose limitations are the following:

(1) The justification of a practice or an activity is considered by decision-makers usually before it is initiated, although ongoing practices may be examined at any time. The benefits to be derived from that activity should exceed (*a*) the benefits of any available alternative and (*b*) the detriment (that is, the expectation of harm from the radiation exposure) associated with that activity.
(2) For optimization, the marginal efforts incurred for reducing the exposure and the resulting change in the benefit (that is the change in the detriment resulting from reducing the exposure) are properly balanced for each source of exposure, under the constraint that in any case, the individual exposure must not exceed the individual dose limit.

(3) The individual dose limit for workers is an annual effective dose equivalent of 50 mSv (5000 mrem) and for members of the public, annual effective dose equivalent limits of 1 mSv (100 mrem) and 5 mSv (500 mrem) apply according to the circumstances of exposure. Dose equivalent limits for particular organs and tissues are also recommended in order to prevent non-stochastic radiation effects.

(4) The individual's exposure from each radiation source must be limited to assure individuals exposed to several sources are unlikely to exceed the individual dose limit for all sources. For that purpose, in practice, an individual dose 'upper bound' is defined for each source, usually as a fraction of the individual dose limit. This 'upper bound' dose limit is in practice used as a constraint on the process of optimization of the radiation protection for that activity or source.

(5) The dose to an individual worker or member of the public is a function of the levels of radiation and radioactivity in the environment, the individual's use of the environment and his personal habits. Where necessary, the worker's exposure can be monitored and doses can be controlled. However, all members of the public cannot be monitored individually and the use of the environment and personal habits normally cannot be controlled. Therefore, the individual, as well as the collective, dose must be restricted by controlling the levels of radiation and radioactivity released to the environment.

It has been shown that, in practice, application of the system of dose limitations is likely to ensure that the average individual doses will not exceed a small fraction of the individual dose limits,[12] and that very few workers approach the dose limits.

Transport Regulations

Since 1964, the Agency's Regulations for the Safe Transport of Radioactive Material have required that the provisions in the Agency's Basic Safety Standards for Radiation Protection be met in the transport activity. The basic principles underlying the Agency's Transport Regulations[13] were established in such a way that optimization of radiation protection in transport activities was achieved although several quantitative analyses were not carried out. More than 30 years of experience has shown that the package radiation limits and control procedures in the Regulations, together with the transport index system, have generally been effective in keeping the transport radiological impact to both workers and the public to very low levels.

In the revision of the Agency's Transport Regulations, currently underway, the system of dose limitations is being followed to the extent possible. With respect to *justification*, it is generally agreed that the overall benefits of the uses of radioactive material provide ample justification for the practice of transportation. Further, in the revision, it has been determined that there is adequate justification for each of the changes being made. Although cost-benefit analyses were considered for some of the specific changes in the Regulations, *optimization* of radiation protection in specific transport activities continues to be the responsibility of the individual Member State. Recommendations as to the annual individual effective *dose equivalent limits* have been added in the Regulations.

On the other hand, the basic principles in the Regulations on which protection and safety are based for accident conditions in transport are:

(1) limiting the quanity of radioactive material that may be transported in a single Type A package;
(2) design and testing of Type B packaging so as to withstand environmental conditions usually encountered in accidents in transport without significant loss of contents or effectiveness of shielding; and
(3) accepting the low risk and harm expectation from radiation exposure to low specific activity material and items exempt from specific requirements in the Regulation due to the low toxicity and low radiation levels intrinsic to the material or item.

Implementation of the Transport Regulations

The Agency's Regulations for the Safe Transport of Radioactive Materials provide standards for ensuring safe movement of all types of radioactive material, including spent fuel, by all modes of transport. Primary reliance for safety in transport is placed on packaging. The standards are mainly directed towards shippers. The development of detailed regulations for carriers is left to national authorities and international organizations.

Although primary reliance is placed on the packaging, safety in transport is provided by observing a variety of regulatory requirements and good safety practices, and giving care and attention to all details by all parties involved. The responsibilities are shared. The plant operator, as the shipper, must assure the material is properly prepared for transport; the carrier must exercise due care in transportation of the packaged material; and the consignee, in accepting the consignment, is expected to determine if leakage or loss of material has occurred in transit. A basic concept applied in developing the Regulations is to maximize the shipper's contribution to overall safety, thereby reducing the required contribution from the carrier to a minimum. Furthermore, carrier personnel are not required to have expert knowledge of radiation protection; they need only limited training, and are required to use only simple mathematics and common sense.

In 1977, an expanded programme in transport safety was undertaken by the Agency to give increased emphasis to implementation and assurance of compliance with the standards and to further understanding and acceptance of the level of safety in transport.

The Nuclear Safety Standards and their impact on Transport Regulations

NUSS, the Agency's programme for Nuclear Safety Standards, was initiated in late 1974 to establish internationally accepted safety codes and guides for nuclear power plants. The standards cover in detail the areas of governmental regulatory organization, siting, design, operation and quality assurance. The preparation of this set of documents, based on information supplied by nations with nuclear power plant experience, has been almost completed. A common understanding for safety has been reached; agreement, not only on the necessity and the requirements for a nuclear regulatory organization to properly conduct a nuclear power programme, but also on items such as the specific information and investigations required for safe plant siting, and specific design guidance to enable a systematic approach to safe design of all principal plant components and structures. These standards, however, are not industrial codes, such as those of the American Society of Mechanical Engineers (ASME) or the Deutsches Institut für Normung (DIN) in which minutely detailed procedures are given on how to carry out design requirements.

Many of the general principles found in the NUSS standards and the specific requirements for quality assurance set out in those standards are also used in or being added to the transport safety standards.

The quality assurance practices and principles developed for nuclear power plants and set out in the NUSS standards are equally applicable to packages containing irradiated nuclear fuels. In fact the principles of QA are applicable to all packages of radioactive material although some reasonable degree of freedom must be given as to the extent to which the QA programme is developed for transport packages – the so-called 'graded approach'. That is, for simple designs of packagings, a simpler QA programme; for complex designs, a more elaborate QA programme may be required. The requirements for QA programmes in Member States and standards for such programmes were discussed at some length in a meeting of experts convened by the Agency in November 1982. A technical report dealing with QA for transport packaging is planned to be issued next year.

The safety standards are designed to make it very unlikely that an abnormal exposure condition would occur in the transport of spent nuclear fuel. The qualification tests for the design of such packages and the independent review of each design by the competent authority provide assurance that the design meets the standards; that is, is capable of withstanding accident conditions it is likely to encounter in transport. The requirements for quality assurance, including certain inspections before the first use of each container and certain inspections before each use of the same container, provide assurance that each package when shipped has been constructed and loaded in accordance with the design specifications and requirements in the regulations. This makes it very unlikely that a release of excessive radiation or radioactive material would occur in transport or that an incident or accident would occur as a result of human error in the packaging and preparation for transport.

Nuclear liability

Compensation for damage caused by a nuclear accident is a matter of which as yet, fortunately, the world community has little practical experience. Despite the existence of nuclear facilities of all kinds in many countries and the numerous shipments of nuclear material around the world, there has been so far no nuclear accident with serious radiological consequences that would require indemnification for nuclear damage. In this connection, the Nuclear Regulatory Commission determined the Three Mile Island accident did not constitute an 'extraordinary nuclear occurrence' that could have put in motion the special indemnification system available under the United States legislation (the Price-Anderson Act).

Nonetheless, in view of the potential magnitude of damage and injuries that might arise from a nuclear accident and in order to provide the public with the assurance of adequate financial protection against damage which might result from certain peaceful uses of nuclear energy, special provisions for nuclear liability have been developed. Thus, during the 1960s, several conventions were adopted under the auspices of the OECD/NEA and the IAEA respectively: the Paris Convention of 1960, the Brussels Supplementary Convention of 1963, the Vienna Convention of 1963 and the Convention of 1971 on maritime carriage of nuclear material.

The common features of these conventions, which are all in force, consist of the following principles which have been adopted by most national legislations:

(1) channelling of liability to the operator of a nuclear installation for nuclear damage caused by a nuclear accident in or connected with his installation;
(2) limitation of the operator's absolute and exclusive liability in amount and in time;
(3) obligation for the operator to secure insurance cover or other financial guarantee up to the established limit of his liability;
(4) State intervention and indemnification in the event of compensation claims exceeding the operator's financial security;
(5) unity of jurisdiction of courts dealing with claims for compensation for nuclear damage.

With particular regard to the Vienna Convention on Civil Liability for Nuclear Damage,[14] adopted on 21 May 1963 by an international conference convened by the Agency, its entry into force on 12 November 1977 constituted a breakthrough in the long process of striving for an effective and world-wide framework for regulating nuclear liability matters. Among the present ten Parties to the Convention,[15] two have nuclear power plants in operation (Argentina and Yugoslavia) and three others are in the process of implementing a nuclear power programme (Cuba, Egypt and Philippines). The Convention has much to offer as a foundation for harmonizing national law on the subject-matter, in particular as regards developing countries. This is illustrated by the fact that even those which are not parties to it, such as Brazil and Mexico, have enacted legislation based on its provisions.

Exchange of information and training

Since work in one country may be relevant to other countries as well, international cooperation can make an indispensable contribution towards assuring nuclear safety. The Agency has always been a forum for the exchange of technical information to pool the collective knowledge and experience of the international community.

The existing Agency activities for the collection of operating information is being extended to include abnormal occurrences at nuclear power plants. Data obtained on abnormal occurrences based on seven categories of severity will be reviewed periodically to extract the most significant data and to determine lessons that can be learned from their analysis. The Nuclear Energy Agency in Paris is conducting a similar programme for members of that agency. The IAEA initiative is based on its broader membership which includes the CMEA and developing countries, the latter needing access to safety-related operating experience.

Similarly, in the area of transport safety, information on operating experience has been collected – information on the national competent authorities responsible for implementing the transport regulations in each Member State is collected and an up-to-date list published annually.[16] Copies of transport package approval certificates issued by the competent authorities are submitted to the Agency and a listing of valid certificates, noting countries validating certificates issued by other countries, is published semi-annually.[17] Data on shipments of radioactive materials during 1980 have been collected and a report will be issued when analysis is completed. Information on Member States which have adopted the Agency's recommended Transport Regulations have been obtained and up-to-date information will be requested in the next few weeks. This programme is also being expanded to include reports of accidents and incidents in transport involving radioactive material and radiation exposures resulting from transport of radioactive material.

In the past year, the Agency sponsored a training course at Harwell for managers of transport safety programmes on implementing the transport regulations. Plans are being made to offer a similar course in 1984.

Emergency response and preparedness

An effective emergency-response planning and preparedness programme is an essential part of the support mechanism for any nuclear facility. The justification for advance planning and arrangements is based on the fact that nuclear accidents have occurred despite excellence of design and care in operation. Guidelines for radiation emergency response planning and preparedness, including the handling of accidents during transport, have been issued to serve as advisory material to Member States.[18] The Agency's Radiation Emergency Assistance Plan with standing arrangements for emergency assistance to Member States in the event of a radiation accident, initially established in 1959, is being upgraded into a more comprehensive Nuclear Accident Assistance Plan. An inventory of the sources and types of assistance available for coping with radiation emergencies, first established in 1963, was last revised in 1980.[19]

CLOSING REMARKS

The conclusion that radioactive materials, including spent fuel, are being transported safely (including transiting of urban areas) is supported by the past record. To maintain this record, the Agency plans to continue its work to establish and maintain adequate, up-to-date standards for transport safety and to increase its efforts to promote harmonization of requirements for safety in the transport of radioactive materials among the various national and international organizations. We will continue to participate in and encourage studies to assess the radiological impact of transport throughout the world, provide for exchange of information on research in the area of transport safety and, where possible, expand our programme for providing guidance and assistance in the implementation of and compliance with the safety standards.

NOTES

1. Nuclear Legislation, Analytical Study, Regulations Governing the Transport of Radioactive Materials, OECD/NEA (Paris 1980).
2. 'Final environmental statement on the transportation of radioactive material by air and other modes', NUREG-0170, US Nuclear Regulatory Commission, Washington DC (December 1977); 'Transportation of radioactive materials in Sweden: A risk study', by Ann-Margret Ericsson, prepared by KEMAKTA Konsult AB for the Swedish Nuclear Power Inspectorate under Contract No. B14/77 (September 1979).

3. Personal communication (23 February 1982), Silvana Piermattei, Comitato Nazionale per l'Energia Nucleare, Rome, Italy.
4. 'Individual and collective doses associated with the transport of irradiated Magnox fuel within UK', H. F. MacDonald and J. H. Mairs (December 1978), RD/B/N4440, Job No. XJ022.
5. 'Results obtained with the computer code INTERTRAN for a transport of spent fuel elements from nuclear power plant Untermeser to Forbach by rail (Normal Mode Transport)', F. Lange, A. Meltzer (Koeln, 13 December 1982).
6. 'Safety analysis on transportation of radioactive materials by truck in Japan', Central Research Institute of Electric Power Industry, T. Nagakura *et al*., PATRAM 80 (November 1980).
7. 'Ionizing radiation: Sources and biological effects', United Nations Scientific Committee on the Effects of Atomic Radiation, 1982, Report to the General Assembly, United Nations (New York, 1982).
8. See note 6.
9. See note 3.
10. Basic Safety Standards for Radiation Protection, Safety Series No. 9, (1982 Edition), IAEA (Vienna 1982). Jointly sponsored by IAEA, ILO, NEA(OECD), WHO.
11. International Commission on Radiological Protection, ICRP Publication 26, Pergamon Press (Oxford, 1977).
12. 'The application of ICRP recommendations: Advice to the expert group reviewing the White Paper Command 884 "The Control of Radioactive Wastes" ', F. Morley, National Radiological Protection Board, UK (August 1977).
13. Advisory Group on Radiation and Safety Principles in Transport, AG 225, Report by the Chairman, J. Sousselier, IAEA (Vienna, July 1979).
14. Reproduced in the IAEA Legal Series No. 4, Revised 1976 Edition, 'International conventions on civil liability for nuclear damage'.
15. As of September 1982, the Vienna Convention was in force among the following states: Argentina, Bolivia, Cameroon, Cuba, Egypt, Niger, Peru, Philippines, Trinidad and Tobago, and Yugoslavia.
16. National Competent Authorities List No. 14, IAEA (October 1982).
17. List of Competent Authority Approval Certificates, IAEA.
18. The most recent publications include: 'Planning for off-site response to radiation accidents in nuclear facilities', Safety Series No. 55 (1981); 'Emergency response planning for transport accidents involving radioactive materials', IAEA TECDOC-262 (1982).
19. Document IAEA TECDOC-237 (1980).

6 The European Community Context

JEAN-CLAUDE CHARRAULT

Ever since the entry into force, in 1958, of the two Treaties of Rome, which established simultaneously the European Economic Community and the European Atomic Energy Community (generally known as EURATOM, the Community has held the view that nuclear energy represents an essential resource with many applications which can contribute to the prosperity of its peoples.

From the start, therefore, the institutions of the Community, including notably the Commission, were called upon to ensure the creation of conditions in which all Community users of nuclear materials would receive regular supplies. Those institutions were required, amongst other things, to promote research; to establish standards for the protection of the health of the general public; and to ensure that nuclear materials were not diverted to purposes other than those for which they were intended.

Thus for more than a quarter of a century the Community's institutions have been involved in the development of the nuclear industry within the Community. They have consequently acquired an extensive knowledge of its problems, including those relating to the transportation of irradiated fuel.

Although, as I said, nuclear energy was already clearly seen as important to the Community 25 years ago, its importance has of course meanwhile greatly increased. This is because of the way in which our economies have developed, becoming much more energy intensive, and because of the changing pattern of our energy supplies. For these reasons the Community has, over the past several years, developed a comprehensive energy strategy. I believe that this energy strategy constitutes the backdrop to any serious discussion on the theme of this Conference.

By 1973, at the time of the first oil crisis, the Community had become dangerously dependent on external sources of supply of energy; 64 per cent of all our energy was imported and an overwhelming part of our energy imports consisted of oil, mainly from the Persian Gulf. The 1973 crisis brought these uncomfortable facts fair and square to the attention of the Community and its Member States. The lesson was repeated in 1980 and 1981, on the occasion of the disturbances of the oil market which followed upon the Iranian revolution and, subsequently, upon the outbreak of the Iraq/Iran war. And gradually, under the impact of these events, the Community, along with other countries of the industrialised world, has developed a comprehensive energy strategy.

The Community's aim is to reduce oil's share of total energy consumption to 40 per cent by 1990 – chiefly by a more rational use of energy and by increased supplies of energy from non-oil fuels. If Member States' forecasts for future demand and supply prove sound, then we should be able to reach this target. After allowing for the effect of a more rational use of energy, this would mean that, between 1980 and 1990:

(1) consumption of coal and other solid fuels would increase by a quarter, from 223 to 276 million tonnes of oil equivalent (mtoe);
(2) gas use would also increase by a quarter from 169 to 210 mtoe;
(3) new and renewable energies would increase from 14 to 27 mtoe;
(4) nuclear power would more than treble, from 43 to 137 mtoe;
(5) even so, oil consumption would fall only slightly: from 494 to 441 mtoe.

Good progress has in fact already been made in changing the pattern of demand. Oil's share in total energy consumption fell from 61 per cent in 1973 to 50 per cent in 1981; in the same period the share of nuclear and solid fuels in electricity generation rose from 50 per cent to 67 per cent.

To a large extent these changes have been due to the impact of increasing oil prices. For the time being, this trend has been reversed; the incentives to pursue investments in oil alternatives are now weaker. However, in the case of nuclear power, the situation by the end of the decade is reasonably clear. Because nuclear investments have long lead times, most power plants which will enter service before 1990 are already under construction.

Let us now turn to the relative roles of the Community institutions, governments of Member States, and the economic operators. The Community provides both a framework within which Member States can operate and a support for their operations. As regards the framework, the Euratom Treaty gave the Community the general task of creating the conditions necessary for the speedy establishment and growth of nuclear industries. But it left the task of actually setting up those industries to others; they include the electricity-generating enterprises, the nuclear fuel cycle industry, the nuclear construction industry and ancillary industries, for example the provision of transport services.

The characteristics of these different economic operators vary greatly in the different Member States. Some are wholly publicly owned; some are wholly privately owned; and some are in mixed ownership. But all have one thing in common: they are subject to a high degree of regulation and control by the public authorities. At Community level, they are subject to the Euratom Treaty. At national level, they are subject to domestic legislation and various forms of government intervention. Indeed, it is the government of each Member State which, in the end, decides whether or not nuclear energy can be exploited at all within its own frontiers, and if so, on what conditions. On this subject, the Council of Energy Ministers last year arrived at some specific and relevant conclusions.

Firstly, although several Member State Governments remain very hesitant about any use of nuclear energy in their own country, the Council nevertheless emphasise that, for the Community as a whole, nuclear and solid fuels will remain, at least until the end of the century, the two most substantial alternatives to oil. This means that – since the common energy strategy provides for equivalence of effort by Member States in restructuring the pattern of energy supply – those countries which do not opt for nuclear power will have to go much further than the others in the use of coal. This, for example, is already happening in some Member States, especially Denmark and Ireland. Second, the Council of Ministers acknowledge that, though there may be argument about precise figures, nuclear electricity is now undoubtedly, in a wide range of circumstances, much cheaper than coal-based electricity, so those who do not opt for it will have to pay an economic penalty. Third, they stress that the realisation of nuclear energy programmes require Member States to make, and of course to sustain, a clear political choice. Fourth, they stress the need to speed up measures in the Community to install capacity for interim storage and reprocessing of irradiated fuel.

So much for the energy background. Turning to the Community's involvement in the transportation of radioactive materials, I would like to consider three points. Firstly, through our nuclear safeguards operations, we keep track of all movements of civil nuclear materials in the Community. In 1980 we recorded a total of 526 journeys involving 1049 tonnes of irradiated fuel from power reactors, these figures will rise substantially in the coming years as more power reactors come on stream. In 1980, 75 such reactors were in operation. By the end of the decade, about 80 new and bigger reactors are likely to be added, bringing the total to 155. As already said, nuclear energy production could thus more than treble between 1980 and 1990. We can expect the movements of irradiated fuels more or less to follow this trend. So, by 1990, the number of transports could rise to about 1600 a year in the Community, with a total volume of over 3000 tonnes – excluding any spent fuel coming into the Community from outside for reprocessing purposes. Although irradiated fuels are the most difficult nuclear materials to transport, many movements of radioactive materials occur, for example for use in research, medicine and industry. In terms of number of movements the latter represents over 90 per cent of the total.

Thus radioactive materials are already being transported within the Community on a substantial scale and the prospects are for continued growth. It is therefore appropriate to review the safety precautions and the efforts being made to strengthen them. Radioactive materials are not, of course, the only materials needed by modern society whose transport gives rise to problems for public health and the environment. Indeed, we have lived with many such problems for a long time. Take oil, for example. We currently consume in the Community about 500 million tonnes of oil per year. This has to be brought by sea and over land from the place of production to the final consumer. This huge quantity is normally transported safely, but accidents sometimes happen. We all remember the severe damage to coastal areas done by spills from wrecked oil tankers.

Gas is likewise liable to occasional accidents. I recall the tragic occasion in Spain a few years back, when hundreds of lives were lost in the gas flames from an overturned tank lorry. One could add many other examples of hazardous materials used in industry, notably chemicals. Indeed, the transport of such materials is an integral part of our industrialised society. We know that it must continue. Moreover, given that the Community contains many densely populated areas, such transportation can hardly avoid contact with urban zones. What we have to ensure is that it is subject to strict controls, which are properly applied. In the case of irradiated fuels, there has not been one serious accident during the time – more than a quarter of a century – that we have been transporting them. I think there are two main reasons for this.

One is that nuclear energy is, by its nature, presented in very concentrated form; so it involves much less transportation for the same amount of energy delivered. The annual production of a typical modern nuclear power plant of 1 GWe (1000 Megawatts) is equivalent to the energy from 1.4 million tonnes of oil. However, the nuclear plant only uses about 30 tonnes of fuel a year. In terms of quantities transported, the requirements of nuclear energy are therefore many orders of magnitude less than for the other main fuels. The other factor is that the safety requirements observed are very stringent indeed.

My second point is to consider who is responsible for safety in relation to transport of radioactive materials. The nuclear industry itself has a major interest, of course; it knows that its future depends on having and demonstrating a strong sense of responsibility. But it is the task of public authorities to lay down the basic regulatory requirements with which their industries must comply. In the Community, this responsibility lies with the governments of Member States. Movements of radioactive materials often cross national frontiers. So from the beginning there has been international cooperation in working out common guidelines on which national regulations could be based.

Due to the work of the International Atomic Energy Agency (IAEA), there is a wide international consensus about these guidelines (see Chapters 5 and 15). The principle behind the IAEA guidelines is that packages and flasks containing radioactive materials must be completely safe and able to withstand severe accidents. Flasks for transporting irradiated nuclear fuels must be able to withstand the impact of a fall of 9 metres and a temperature of 800°C for half an hour and still remain intact. These requirements largely explain why no accidents during the transport of radioactive materials have had severe radiation consequences to the public or the workers involved.

My third point concerns the role of the Community. Although Member States are responsible for the physical protection of nuclear materials at all stages of the nuclear fuel cycle and whilst the Community has no task to duplicate or interfere with the regulatory activities of the IAEA, there are nevertheless some things that the Commission can usefully do.

Given its duty to ensure both a regular supply of nuclear fuels and that nuclear materials must be able to circulate between Member States without undue obstacles, the Commission has to ensure that transport regulations through the Community are harmonised. Given the Community's role in relation to nuclear safeguards, commission inspectors impose checks at both ends of the journey to ensure that nuclear materials sent away duly arrive at the correct destination.

In addition, the Community has a responsibility for providing standards for the health and safety of workers and the public in relation to radiation hazards. In this connection, the Community finances a radiological protection R & D programme; the Community's annual expenditure of £9 million finances around 80 per cent of total R & D in this field throughout the Community. This work provides the basis for the regular updating of radiological-protection standards and especially their methods of implementation at each stage of the nuclear fuel cycle.

Finally, in the period 1979–82, the Commission has been cooperating with Member States in developing guidelines for transport safety. Over this period, the Commission has spent about £0.7 million to finance tests within UK, France and Germany on the ability of nuclear flasks to withstand severe accidents.

In 1982 the European Parliament reviewed community involvement in the transport of radioactive materials and required the Commission to set up a permanent working party with experts from Member States with the object of exchanging information and making proposals for new Community action. The working party has been established and proposes further work:

(1) to continue to ensure that the national legal and administration practices for Member States are eventually consistent;
(2) to further stimulate, coordinate and support the work of Member States in revising the IAEA guidelines; and
(3) to accelerate the process of bringing national legislation into line with the international guidelines.

By carrying out this work, I believe that the Commission can contribute to the consolidation and further development of high standards of safety throughout the Community for the transport of radioactive materials. The aim must, of course, be to ensure that the growing use of nuclear power to satisfy our energy requirements takes place in conditions which protect public health and the environment.

7 The Transport of CEGB Irradiated Nuclear Fuel

CENTRAL ELECTRICITY GENERATING BOARD

INTRODUCTION

The Central Electricity Generating Board (CEGB) has nine nuclear power stations – eight based on the Magnox reactor and one on the advanced gas cooled reactor (AGR). Between them they supply nearly 11 per cent of the electricity used in England and Wales. The amount of nuclear fuel used to make that electricity is equivalent to nearly 10 million tonnes of coal a year. A further three AGR stations are shortly due to become operational.

Each fuel element for a Magnox reactor consists of a bar of uranium metal encased in a tube of a magnesium alloy called Magnox. AGR fuel elements are made up of pellets of uranium oxide in stainless steel tubes, thirty-six of which are held together in a graphite sleeve.

Both types of element are manufactured by British Nuclear Fuels (BNFL) at their Springfields factory near Preston. The new elements, which are harmless enough to be held in the hands, are transported by road to the power stations in steel crates. After use in the reactor the spent elements, which are then highly radioactive, are removed and stored in a shielded area at the power station before being taken by rail in specially constructed flasks to BNFL's reprocessing works at Sellafield (Cumbria).

No accident involving the release of radioactivity has occurred in more than 7000 rail journeys since the first consignment in 1962.

FUEL ELEMENTS

The fuel bar for the Magnox nuclear power station at Sizewell (Suffolk), for example, is 1 metre long and 2.8 centimetres in diameter and weighs about 12 kilogrammes. Each of Sizewell's two reactors contains about 26,000 elements. Depending on their location within the core the elements remain in the reactor for periods of up to 11 years until the maximum amount of energy has been extracted.

This energy is obtained by means of the fission process, which means that the uranium atoms are split and heat is released. At the same time, fission produces new atoms. These are primarily the fission products, which are simply the separated parts of the uranium atoms, together with small quantities of other materials such as plutonium, known as transuranic elements. These new atoms continue to convert to other products even after the reactor is shut down, at the same time giving off radiation and small quantities of heat. Both radiation levels and heat output decrease with time.

This radioactive decay process is described by what is termed the half-life. Iodine 131 for instance has a half-life of 8 days. This means that the radioactivity decreases by half after 8 days. After a further 8 days the radioactivity has decreased by half again – that is, to a quarter of the original value; and so on. The more important radioactive products from a public hazard point of view are shown in Table 7.1.

TABLE 7.1 *Radioactive products*

Radioisotope	*Half-life (approx.)*
Krypton 85	10 years
Xenon 133	5 days
Iodine 131	8 days
Caesium 137	30 years
Strontium 90	28 years
Ruthenium 106	1 year
Plutonium	24,000 years

Iodine may be absorbed and concentrated in the thyroid gland, while caesium, strontium and ruthenium may be absorbed into various parts of the body. If an accident were to occur the public might, in extreme circumstances, have to be protected from inhaling or ingesting these substances. Krypton and xenon are gases and are not retained in the human body but, if they are released during an accident, they may cause direct radiation to the public. In the reactor accident at Three Mile Island in the United States it was the xenon release which gave the very small radiation exposures to the local population.

Irradiated AGR and Magnox fuel elements may be transported only after 90 days' storage. By this time the xenon and iodine have largely decayed so that they would be of little significance in any transport accident. But the presence of the remaining radioactive materials has to be taken into account in planning the arrangements for transporting irradiated fuel.

FUEL TRANSPORT

The statutory provisions for spent fuel transport stem from regulations and guidance laid down by the International Atomic Energy Agency (IAEA) and the responsibility for ensuring compliance in this country lies with the Department of Transport. For despatch to Sellafield, the elements are first loaded into an open-topped steel container called a skip. Each skip contains typically 200 Magnox fuel elements or 20 AGR fuel elements. In any one skip there will be fuel elements of varying radioactivity levels, with an overall limit being placed upon the heat produced by the continuing radioactive decay process.

The skip is loaded into a heavy steel flask under water, or within shielded handling facilities, and the flask lid is then replaced. On removal from the loading bay the flask is thoroughly washed down to remove any radioactive contamination. In preparation for transport special procedures ensure that the flask is correctly filled with water, the lid bolts are fully tightened, the seals are properly tested, and the radiological surveys of the flask show the regulatory limits are not exceeded.

THE NUCLEAR FUEL FLASKS

The flasks are designed and tested in accordance with the standards and requirements laid down by the International Atomic Energy Agency (IAEA). The standards cover normal use and accident conditions. The flasks have to be approved by the Radioactive Materials Transport Division of the Department of Transport and must comply with the British Rail standard conditions of acceptance for such goods. They are essentially massive steel boxes with thick walls to reduce radiation from the fuel elements, and typical data for a Magnox flask are as follows:

Dimensions	approx. 250 cm (8 ft) cube
Thickness of walls	37 cm (14.5 in)
Total loaded weight	50 tonnes
Lid weight	7 tonnes
Weight of fuel	approx. 2 tonnes
Number of fuel elements	200
Radiation levels	1 mRem/h at 1 metre

The lid is held in position by sixteen high-tensile steel bolts each 5 cm (2 in) in diameter. Between the body and the lid there are twin

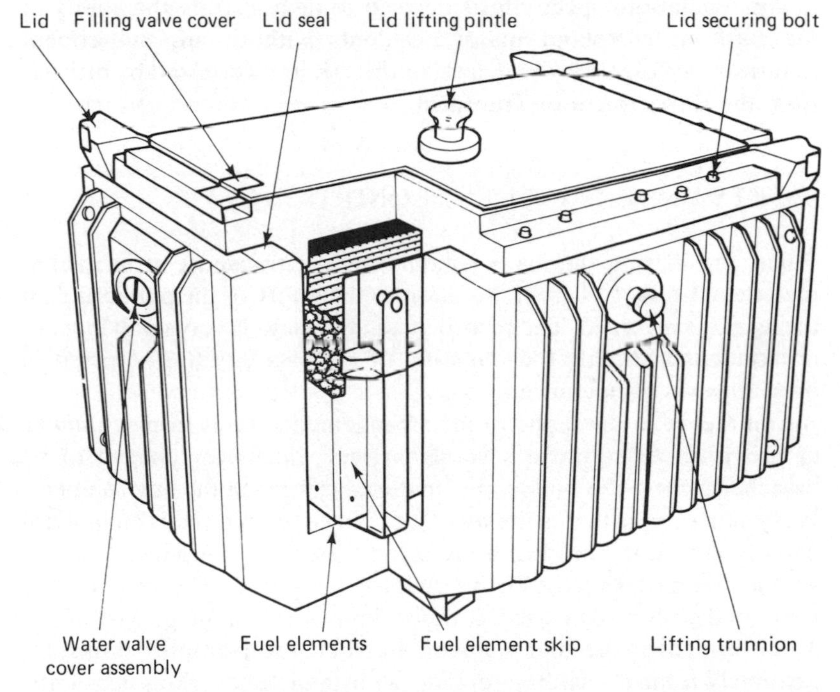

NOTE The flask's overall dimensions are: length 256 cm (8 ft 4.75 in), width 218 cm (7 ft 1.75 in), height 221 cm (7 ft 3 in).

FIGURE 7.1 *Transport flask for irradiation fuel elements from Magnox nuclear power stations.*

compression seals recessed in grooves. Figure 7.1 is a diagram of such a flask.

The flask designed for AGR fuel is similar in external dimensions but has a thinner steel wall with an additional liner of lead which is particularly efficient in absorbing radiation.

PROTECTION AGAINST ACCIDENTS

The regulations concerning the resistance of flasks to accidents have been prepared by panels of specialists convened by the IAEA. These specialists are engineers and scientists with wide experience in the transport of materials and nuclear equipment requirements from UK, USA, Canada, France, West Germany, Japan and elsewhere. The provisions are regularly reviewed by these expert bodies.

The regulations specify tests devised to demonstrate the ability of the flask to withstand major accidents without any subsequent significant leakages. Proving tests in the UK are witnessed by officials from the Department of Transport.

SAFETY UNDER ACCIDENT CONDITIONS

When considering the arrangements for the transport of irradiated fuel, the CEGB pays great attention to the safety of the public and all the workers involved. The record of achievement is very good indeed; no accident involving the releasing of radioactivity has occurred in over 20 years of operation.

Confidence in the safety of the arrangements stems from a number of provisions. There are several barriers which would have to be breached before the radioactive material can reach the environment. Particularly important is the fact that the active material is an integral part of the fuel. It is produced within, and largely remains incorporated within, the metal or the oxide pellets themselves. The fuel itself is contained within sealed metal tubes, known as cladding, part of the actual design of the fuel element. Further, the transport flasks are extremely robust, enabling them to withstand very serious accidents. They are very massive and have thick steel walls. The IAEA regulations call for demonstrations that the flasks can withstand both the impact in a drop from 9 metres onto an unyielding target, and the effect of a severe hydrocarbon fire. These demonstrations have been made in full scale fire tests and in drop tests on scale model flasks. Demonstration of the adequacy of the regulatory drop test has been carried out by full scale tests on flasks in the US. In practice the probability of severe impact or fire hazard is low as is shown by the statistics of railway operation. The British Rail system is very safe. The practical experience of 20 years is that flasks have never suffered an impact in transport, nor have they been involved in fires.

Although the above provisions make a release of radioactivity unlikely in the extreme, prudence requires emergency arrangements to cover any eventuality. To judge the requirements for response in an emergency, the consequences of damage to a flask have been considered.

Taking a hypothetical case in which a flask is damaged either in an accident or by deliberate act, we have considered what would be the result if a welded joint is cracked, the lid distorted, or the integrity breached in some other way. Some of the water covering the fuel elements may then be lost. This water could contaminate the ground in the vicinity, but would not pose any major threat to health. If the fuel itself remained undamaged, then the associated radioactivity would remain within the fuel cladding producing no additional problem. Even if the fuel were damaged, many hours would elapse before any release of radioactivity followed which was not entirely confined to the immediate vicinity of the flask. It is the objective of the emergency response to ensure that expert personnel and emergency services reach the scene during that period and take appropriate action.

EMERGENCY RESPONSE

A national plan, the Irradiated Fuel Transport Flask Emergency Plan, specifies the action to be taken following any incident involving a flask or flask wagon. The emergency procedures would be set in motion by a call from British Rail to the CEGB Alert Centre. If the incident were minor and it was immediately obvious that no damage could have been caused to the flask, the police and fire brigade would be informed but not called out. The local authority would not be informed of this type of incident since no hazard would be involved. Nevertheless, a Flask Emergency Team of specialists from the appropriate nuclear establishment would attend as a routine precaution.

The following statement outlines the principal duties and responsibilities of the parties if the incident was not obviously a minor one.

British Rail

(1) The train crew would take immediate action to isolate the train from other rail traffic and to keep people away from the flask.
(2) The crew would also initiate the established BR flask contingency arrangements by telephoning the Railway Control Office. The Control Office would inform and request assistance from the CEGB Alert Centre, giving the location and nature of the incident.
(3) BR Control would also inform BR police, the civil police, the fire brigade and, if necessary, the ambulance service.
(4) A BR coordinating officer (known as a Mishap Officer) would go to the scene to take charge of those activities confined to British Rail property.
(5) Lifting and righting operations would be carried out by BR staff in consultation with the Flask Emergency Team.

Central Electricity Generating Board

(1) The CEGB Alert Centre, available on a 24-hour basis, would call out a Flask Emergency Team from a CEGB nuclear power station, supplemented if necessary by additional specialist advisers.
(2) A Coordinating Officer at nuclear power stations is also available on a 24-hour basis. His duties in the event of a flask accident would include ensuring the despatch of the Flask Emergency Team. He would advise British Rail, police and fire brigade officers on action to be taken at the scene of the accident until the Flask Emergency Team arrived. He would also be responsible for ensuring that other organisations, including Government departments and the Health and Safety Executive, were informed of the incident.
(3) Arrangements also exist for the CEGB to make use of helicopters if necessary, and to call upon radiation protection experts in various parts of the country if their attendance would speed up the response.
(4) On arrival at the flask incident the Flask Emergency Team would examine the flask, carry out radiation monitoring in the immediate and surrounding areas, and call for additional health physics assistance if required. If any flask repair were necessary, the team would supervise the work and also arrange with British Rail staff for further transportation of the flask.

The Police (Civil and British Rail)

(1) The civil police would be alerted to the incident by British Rail, who would confirm that the appropriate nuclear establishment had been contacted and that a Flask Emergency Team had been despatched to the scene.
(2) The police would normally take the lead in coordinating any activities which might be required outside British Rail property, and would alert the local authority if appropriate, in accordance with normal procedures for emergencies. If the incident involved a fire, the fire brigade would take charge of the incident.
(3) The police would supervise the clearance of an area of 50 yards radius from the flask until the Flask Emergency Team had completed radiation surveys.
(4) The police would establish an incident control centre and assist in the provision of communication facilities in the case of any incidents which were likely to be prolonged.
(5) In the event of an airborne radioactive release the police would arrange for the further evacuation of members of the public if the Flask Emergency Team considered this to be desirable. Such an eventuality is of course unlikely in the extreme as, even if the flask were breached, the Flask Emergency Team would have more than sufficient time to take remedial steps.

The Fire Brigade

(1) The fire brigade would be alerted to the incident by the Railway Control Office.
(2) In the event of fire, the fire brigade would be in charge. A principal activity would then be to keep the flask cool by spraying with water or foam. Advice could be obtained from the Coordinating Officer at the nuclear station in the early stages, and from the Flask Emergency Team as soon as it arrived.

The Local Authority

(1) The local authority would be told by the civil police of any incidents affecting their interests and notified separately by the CEGB Coordinating Officer if the integrity of the flask was impaired.
(2) In the event of a flask incident severe enough to warrant the

evacuation of the local population, the local authority would need to provide assistance for them, as in any other occurrence necessitating evacuation.

The Water Authority

If the flask sealing had been impaired the CEGB Coordinating Officer would inform the water authority.

The Ambulance Service

The Railway Control Office would alert the ambulance service if assistance was required.

CONCLUSIONS

The public is protected from harmful effects due to spent fuel transport by a number of separate factors:

(1) The process is carried out under provisions laid down by the IAEA and regulated in the UK by the Department of Transport.
(2) There are three physical barriers which have to be breached before any release is possible.
(3) The BR transport system is very safe, as the statistics of accidents show, such that even slight damage to a flask will be a very rare event.
(4) The flasks are tested according to IAEA regulations to show their resistance to impact and fire accidents.
(5) Even in the event that the flask is breached, many hours will elapse before any serious spread of radioactivity is possible. This period is more than sufficient for the emergency personnel to reach the scene and take the necessary measures.

Note finally that the record is excellent. In 20 years' operation, involving 7000 journeys, there has been no occasion on which release of radioactive material from a flask has been even remotely possible.

8 The Regulatory Framework in the UK

RAYMUND O'SULLIVAN*

BASIC REGULATORY REQUIREMENTS

Regulating the safe transport of radioactive materials necessitates certain basic provisions which may be summarised as follows:

(1) control of the radiation emitted from the material;
(2) containment of the material;
(3) dissipation of any heat generated; and
(4) for fissile material, prevention of criticality.

The regulations for transporting radioactive material in this country are all based on the International Atomic Energy Agency's Regulations for the Safe Transport of Radioactive Materials.[1] The IAEA Regulations provide for each of the basic provisions by prescribing radiation level and release limits for both normal and accident conditions of transport. Rather than seeking to do this by controls such as special vehicles or routes, the Regulations are essentially directed to ensuring that safeguards appropriate to the nature and amount of the radioactive material are 'built in' to the design of the package in which the material is to be transported *on the premise that there could be a severe accident in transport*. The Regulations specify design performance standards which are independent of the mode of transport by which the package may be carried, e.g. road or rail. Where the radioactivity of the intended content exceeds a specified level, they include tests for demonstrating ability to withstand accident conditions and invoke the concept of independent assessment and certification of compliance.

RESPONSIBILITY

The Regulations assign responsibility for safety to three persons: *the consignor* of the material, *the carrier* of it and the *competent authority*. The consignor's responsibility, in essence, is to ensure correct preparation for transport. That is to say he must see to it that the package complies with the appropriate design, that it is correctly made up with the specified packaging and prescribed contents, that it complies with the conditions imposed in any competent authority approval certificate, that any necessary pre-transport measures have been taken, that it is correctly described in the transport documents, that it is properly labelled, and that any necessary instructions have been given to the carrier. As can be seen from the length of this list, the consignor's responsibility is an important and primary one. The Regulations require formal signed certification by the consignor before transport begins that the consignment does fully comply with the regulatory requirements. The Regulations minimise reliance on the carrier as far as possible. His responsibility is to handle, stow, move and segregate the consignment in accordance with the general requirements of the Regulations and any specific requirements in the transport documents. So much for the consignor and the carrier. There remains the third person with responsibility – the competent authority.

Competent authority

The IAEA Regulations define the term 'competent authority' as follows:

> Competent authority shall mean any national or international authority designated or otherwise recognised as such for any purpose in connection with these Regulations.[2]

There are three points to be made about this definition.

(1) The identity of the competent authority can vary from country to country depending on the particular legislative and administrative arrangements in force.

(2) It can also vary *within* countries depending on such factors as, for example, the mode of transport in question.
(3) The term is *not* exclusive to the transport of radioactive materials. Such materials form only one class of the hazardous goods that are transported and although the detailed safety arrangements vary depending on the particular hazardous characteristic of each class, the role of the competent authority in seeking to establish and apply effective safety standards is essentially the same.

In this country the Secretary of State for Transport is formally the competent authority for the purposes of all dangerous goods transport. In the particular case of radioactive materials the executive functions are carried out on his behalf by the Radioactive Materials Transport Division.

The functions of the national competent authority for the purposes of the IAEA Regulations can be summarised under the following three headings:

Regulation	– ensuring there is appropriate provision in the regulations to give effect to the IAEA requirements
Implementation	– independent assessment, certification and checking of compliance with the regulatory requirements
Information	– providing advice to consignors and carriers and giving information to central and local government and the public.

Before looking at each of these three functions in more detail, I must emphasise that the role of the competent authority under the IAEA Regulations is solely concerned with safety in normal and accident conditions of transport. It is *not* concerned with any other aspect of the transport operation; nor is it concerned with what happens to the radioactive material at either end of the journey. Moreover the competent authority's role is in principle the same whenever any radioactive material is transported: irradiated fuel is only one of the many radioactive materials that are transported for use in medicine, industry and research, or for disposal.

Provision of regulations

The first function of the Department of Transport as national competent authority is to ensure adequate regulatory provision to give effect to the IAEA Regulations. Two levels are involved, namely national and international. In both cases however the task is essentially to provide for effective implementation of the IAEA regulatory requirements, having regard to the particular legislative arrangements in this country and the characteristics of each mode of transport. The importance to safety of harmonising the regulatory requirements of different countries and modes cannot be overstated. Only in this way can satisfactory uniform standards be established and maintained and safe transport facilitated. The object of achieving such harmony was a primary reason for the United Nations entrusting the IAEA with the task of drawing up Regulations. The level of harmony now existing in the application of the IAEA Regulations represents a considerable success: the fact that over fifty Member States and the international regulatory authorities for each mode of transport have adopted them testifies to the degree of confidence in which the Regulations are held.

International regulations At the international regulatory level, the Department in essence provides the UK's voice at meetings of the international regulatory authority for each mode of transport, in addition to those of the IAEA itself. The main international regulations affecting shipments of radioactive material to and from the UK are those of the International Maritime Organisation (IMO) for the sea mode,[3] the International Civil Aviation Organisation (ICAO) for the air mode,[4] the European Agreement concerning the International Carriage of Dangerous Goods by Road (ADR),[5] prepared by the Inland Transport Committee of the Economic Commission for Europe, and Annex 1 of the International Convention concerning the Carriage of Goods by Rail (RID).[6] All of these are in harmony with the IAEA Regulations. I should also mention the Recommendations[7] of the United Nations Committee of Experts which provide the basis for the international regulation of all dangerous goods transport, which cite the IAEA Regulations in respect of radioactive materials. Indeed the extensive involvement of the United Nations in this field through various agencies, of which IAEA is one, is particularly noteworthy.

The Department's contribution to the work of these bodies consists of participation on the UK's behalf in the technical working groups and plenary sessions. These consider the proposals for the establishment and modification of regulatory provisions that Member States and the particular regulatory authority itself have made. The Radioactive Materials Transport Division takes a prominent role in these various committees and working groups and currently chairs the Working Group of IMO on Radioactive Materials Transport and the IAEA Technical Committee on Implementation of the Regulations. The Department is also a founder member of the informal group of competent authorities and their technical advisers known as the Radioactive Transport Study Group, which has met at approximately annual intervals since 1969 to discuss matters of mutual interest in the national and international administration of the IAEA Regulations.

National regulations At the national regulatory level, the Secretary of State for Transport has made Regulations[8] covering the carriage of radioactive material by road which are consistent with the current edition of the IAEA Regulations. These are supplemented by a Code of Practice[9] which supplies detailed technical information to help both consignors and carriers meet the requirements. In addition he has made separate Regulations,[10] based on the recommendations of the International Commission on Radiological Protection (ICRP),[11] and again consistent with the IAEA Regulations, to control the exposure to radiation of road-transport workers. The work of administering and enforcing all these regulations governing the transport of radioactive material by road in this country is carried out by the Department's Traffic Area Offices. On the railways, the IAEA Regulations are given effect by the Conditions of Acceptance of Dangerous Goods[12] that British Rail are empowered to make. Safety provisions on the railways are overseen by the Department's Railways Inspectorate.

In British ships and in national territorial waters, the Merchant Shipping Dangerous Goods Regulations[13] give effect to the IAEA Regulations by reference to the IMO requirements. Similar provision to give effect to the IAEA Regulations is contained in the Air Navigation Order.[14]

There is one future development that I should like to mention in this context. As some of you may know, the Health and Safety Commission are currently preparing regulations to implement the Euratom Directive on Ionizing Radiations. Since both the IAEA Regulations and the Directive are based on the recommendations of ICRP, the requirements in the proposed new Ionizing Radiation Regulations,[15] in so far as they concern transport operations, will also be consistent with those in the IAEA Regulations. I understand that a member of the Health and Safety Executive is attending the Conference and may have something to say on this subject during the Open Forum on the last day.

Regulatory reviews Before turning from this outline of the regulatory structure – which has necessarily been only a brief one in the time available to me – there is one important facet of the Department's regulatory function that needs to be mentioned. The IAEA Regulations have been and continue to be subjected to periodical review to keep them abreast of the changes in this rapidly developing technical field. The Department is responsible for participating in these reviews; for ensuring that changes made to the IAEA Regulations are duly reflected in our national regulations; and for the collaboration necessary to reflect them in the international transport regulations.

Implementation of regulations

The second function of the Department as competent authority for the purposes of the IAEA Regulations concerns implementation. As I have already indicated, the Regulations require the package designs for certain consignments to be subject to independent assessment and certification for compliance. It is the competent authority's responsibility to carry out this highly technical work. In this country it is done by a team of professional engineers and scientists in the Department's Radioactive Materials Transport Division. The consignments that require such assessment include of course irradiated fuel and I think it would be helpful if I spent a little time now explaining what this involves in some detail.

Design and application Let us assume that, for example, a potential consignor wishes to design a new transport flask to carry irradiated fuel. It may be several years before any 'hardware' actually exists and the designer will need to know to what safety standards the flask must eventually be made. The Department therefore affords and encourages full technical consultation with designers from the earliest stages of preparation. This is not to prejudice the assessment of the application for approval to the design which will ultimately be made but to avoid unnecessary work on the designer's part and to ensure that no actual or potential regulatory requirements are overlooked. It also serves to ensure both that the Department is aware of future designs for which its approval will be necessary and that, when application is made, it is received in a form containing all the information needed to carry out the assessment. In due course, and normally when the design has been finalised, a comprehensive safety case is prepared and application made to the Department for appropriate approval.

Safety case A typical safety case will consist of a complete specification of the design of the flask and its intended contents; evidence of compliance with the relevant IAEA requirements which will include, for example, stress, heat-transfer and criticality calculations, provisions for radiological protection; the operating, handling and maintenance instructions; details of emergency arrangements; and a full set of engineering drawings.

Testing Where testing is necessary to establish compliance of a design with the regulatory criteria, it is the responsibility of the applicant to carry it out. Tests are witnessed and if the Department is dissatisfied with the evidence of a test result or considers that further tests are necessary to establish compliance it can and does say so.

Assessment It is the applicant's responsibility to make the case that the design of flask complies with the Regulations. The Department's responsibility as competent authority is strictly limited to the assessment of that case: the Department does not make the case itself. The assessment work involves independent consideration of all aspects of the safety case. The Regulations[16] permit the applicant to demonstrate compliance by a variety of methods or a combination of methods and the Department must be satisfied with the methodology on which the evidence is based as well as the evidence itself.

Approval certificate When all necessary evidence of compliance with the IAEA Regulations has been presented and assessed to the Department's satisfaction, a certificate approving the design of the flask is prepared and issued. This certificate, together with the safety case on which it is based, is the property of the applicant.

The Regulations[17] prescribe the information that must be incorporated in certificates. It includes:

(1) the competent authority's identification mark which is unique to the design;
(2) brief description and illustration of the flask;
(3) specification of the contents;

(4) instructions on the preparation, use and maintenance of the flask;
(5) details of any supplementary operational controls;
(6) any necessary restrictions, e.g. to the mode of transport or type of vehicle to be used; and
(7) emergency arrangements.

Compliance assurance An integral part of the assessment and approval process concerns the measures that the applicant should propose to give assurance that, when it is actually constructed and used, the flask will comply and continue to comply with the Regulations. The applicant is required to provide details of these measures. A representative of the Department from time to time inspects flask components in course of manufacture and construction. The Department also makes similar visits to check that the quality-assurance procedures which ensure correct preparation of flasks for transport and their maintenance during use are complied with. Thus although, as I have indicated, it is the *consignor's* responsibility to fully comply with the terms and conditions of the Department's approval certificate, his compliance is subject to independent check from the time the flask is manufactured right up until the end of its useful life.

Administration The Department's function of implementing the IAEA Regulations involves a significant administrative workload. This mainly concerns the preparation and issue of certificates of approval in respect of package designs and shipments which originate in this country. In addition however certification of certain designs and consignments from abroad is required. It is worth emphasising, in passing, that where national certification is necessary for designs originating abroad, the Department requires the submission of a safety case accompanied by full engineering drawings just as for our own national designs. Such cases are subjected to full assessment by the Department's technical staff.

Information

I would like very briefly now to mention the third and final function of the Department as national competent authority. This function is to give advice and information on the regulatory safety standards. It involves two aspects. Firstly, providing technical and administrative advice to consignors and carriers on request and providing lecturers at specialist training courses. Secondly, responding to questions from central and local government, interested organisations and members of the general public. It is in this context that the Department is participating as an expert witness in the current Sizewell inquiry. A written Proof of Evidence[18] has been submitted and Departmental witnesses will be available when the inquiry considers the topic of transport safety.

CONCLUSION

In conclusion I should like to make two general points concerning the transport of irradiated fuel and the Department's role in regard to it. Firstly, safety in any field of activity can never be absolute. In highly technical and specialised fields such as the transport of radioactive materials it is necessarily an evolving process as experience grows, new techniques are developed and the results of the existing and planned research, be it at the international, national or local level, come to be added to the existing body of knowledge. The Department is always ready to consider on its technical merit the written evidence that any organisation may care to put to it in this respect.

Secondly, I should like to emphasise that the proper discharge of its regulatory functions requires the Department to be – and be seen to be – independent of all vested interests. That is why the Department does not attempt to publicise its work in this field or to influence public opinion on the safety issue. It is also why this paper describing the way in which the Department discharges these functions does not express any view of the merits of the case for transporting irradiated fuel or any other radioactive materials. The Department's role is to ensure that if such materials are to be transported they can only be moved under conditions which will properly safeguard the public.

NOTES

1. International Atomic Energy Agency's Safety Series No. 6, Regulations for the Safe Transport of Radioactive Materials (1973 Revised Edition (As Amended)).
2. Ibid., p. 3.
3. International Maritime Organisation (IMO), International Maritime Dangerous Goods Code – Class 7, Radioactive substances.
4. International Civil Aviation Organisation (ICAO), Technical Instructions for the Safe Transport of Dangerous Goods by Air (1983 edition).

5. European Agreement concerning the International Carriage of Dangerous Goods by Road (ADR), Annex A, 'Provisions concerning dangerous substances and articles' and Annex B, 'Provisions concerning transport equipment and transport operations' – Class 7 (1978).
6. International Convention concerning the Carriage of Goods by Rail (CIM), Annex 1, International Regulations Concerning the Carriage of Dangerous Goods by Rail (RID) – Class 7 (1978).
7. United Nations Economic and Social Council, Recommendations prepared by the United Nations Committee of Experts, Transport of Dangerous Goods (second revised edition 1981).
8. The Radioactive Substances (Carriage by Road) (Great Britain) Regulations 1974 (SI 1974 No. 1735).
9. Code of Practice for the Carriage of Radioactive Materials by Road, HMSO (1982 impression).
10. The Radioactive Substances (Road Transport Workers) (Great Britain) Regulations 1970, SI No. 1827 and the Radioactive Substances (Road Transport Workers) (Great Britain) (Amendment) Regulations 1975 (SI 1975 No. 1522).
11. Recommendations of the International Commission on Radiological Protection (adopted 17 September 1965), *ICRP Publication 9*, Pergamon Press (Oxford, 1966).
12. Dangerous Goods by Freight Train and by Passenger Train or Similar Service, List of Dangerous Goods and Conditions of Acceptance – Class 7, Radioactive Substances: British Rail publication BR 22426 (1977 edition).
13. The Merchant Shipping (Dangerous Goods) Regulations 1981 (SI 1981 No. 1747).
14. The Air Navigation Order 1980 (SI 1980 No. 1965).
15. The Ionizing Radiations Regulations (198–), *Consultative Document*, Health and Safety Commission (1982).
16. IAEA Regulations, p. 89.
17. IAEA Regulations, p. 105.
18. Sizewell B Inquiry DTp/P/1, Safety Arrangements for the Transport of Radioactive Material, *Proof of Evidence*, R. A. O'Sullivan (October 1982).

9 Emergency Procedures in London

DOUGLAS CREE

The Central Electricity Generating Board (CEGB) have said that over the twenty years that nuclear fuel flask traffic has been moved by rail no significant mishap has ever occurred (see Chapter 7). It therefore follows that the emergency services, by which I mean the fire brigade, ambulance and police, have not had to deal with serious incidents involving the movement of this fuel. Consequently our contingency plans for dealing with such mishaps are based on information provided by the authorities and from lessons learned in practice exercises. However, the emergency services in London have great experience in dealing with a wide range of catastrophes and it is around this experience that the new procedures are developed and tested. I shall concentrate on procedures used by the Metropolitan Police.

The Metropolitan Police district covers an area of some 780 square miles, extending for a radius of 16 miles from Charing Cross. In addition to Greater London it includes parts of the administrative counties of Essex, Hertfordshire and Surrey but excludes the City of London. It is a population of approximately $7\frac{1}{4}$ million according to 1981 estimates. It is divided into four administrative areas, two north and two south of the river Thames. Each area comprises six districts which are in turn sub-divided into divisions. In general, the boundaries of police districts are coterminous with those of local authority areas. Whilst each division has its own command structure and its own communication system, there is superimposed on these a central headquarters responsibility which ensures uniformity of policy and provides an efficient communications network. This is particularly important when action is required in connection with major emergencies which might affect large areas and require greater resources than are available locally.

Regarding action at the scene of a major incident, it is always difficult to explain how a large organisation with a complex structure, such as the Metropolitan Police, is galvanised into action, although experience proves that it is not difficult in practice. To assist I will explain the normal build-up in police action from the time an incident is reported.

Using his personal radio, the first officer to arrive at the scene would contact the local police station from which he operates. The communication channel for that division would then be opened, giving the officer simultaneous communication with all officers on that division. If necessary the officer would also be put in touch with the Information Room at New Scotland Yard with access to the force as a whole and call on any of its resources, including the mounted Branch and the River Thames Police. The River Thames provides 56 miles of highway through London which is generally fairly easy to move along and horses too, are particularly useful when road traffic is at a standstill. The officer could also summon a police helicopter to assist in a number of ways, such as providing illumination – they carry 92 million candle-power searchlights capable of illuminating Wembley Stadium sufficiently for a game to be played – or using its powerful loudspeakers to pass warnings and advice to the public – these are audible to a distance of five hundred feet depending on conditions. They also have navigational computers which can assess the wind speed and direction; they can also be fitted with television cameras relaying aerial pictures of the scene to the ground control point and providing, if necessary, a video record of the incident which could be useful to experts arriving later. The police are of course aware of the dangers in using helicopters in certain situations and they would only be sent up after appropriate experts have been consulted and they had given clearance.

On receipt of the first message from the officer at the scene, checks would be made to ensure that the other emergency services had been alerted and sent to the scene. They would arrive within minutes. In many instances they would have arrived at the scene before the police. The fire brigade will bring much needed technical expertise and would ensure the attendance of a scientific officer from the Greater London Council. The brigade also have protective clothing and devices for measuring radiation.

Meanwhile, at the local police level contingency plans will have been brought into operation with the senior officer on duty going to the scene and, as with all large organisations, calling out more senior members of staff as necessary. He would take with him a major incident box containing the paraphernalia necessary to put the plans into operation; he would establish a forward control post in the most suitable location from which the emergency services could be coordinated. The senior police officer would have a number of priorities to consider but most importantly he must be satisfied that all measures have been taken to minimise danger to the public. The basic measures to achieve this would include establishing cordons to prevent entry to the site of the incident and ensuring that there were clear pathways for the arrival and departure of emergency vehicles.

It is a fact of life that accidents of any sort attract crowds and that the crowds grow in proportion to the number of emergency vehicles seen or heard to arrive at the scene. This causes severe traffic congestion which in turn hinders the rescue operation. Increasing crowds also mean that more people are putting themselves at risk. There is also the problem of control of the news media whose speed of response almost matches that of the emergency services! Their natural and understandable desire to get a news story is not always compatible with the action being taken by the emergency services.

The next question to be considered is whether it will be necessary to evacuate the area. In deciding this, great reliance would be placed on the advice of the experts – the fire brigade, the ambulance and the scientists. If it were decided that it was advisable to move people, consideration would then be given to the extent of the area at risk. Again police would rely heavily on advice of the experts. The local knowledge of police will assist greatly in arriving at a decision. Evacuation is not always an easy option, particularly where housing estates with highrise flats or streets with multi-occupied properties are concerned. There are dangers in moving people; they could be put at greater risk. On occasions it might be preferable to leave them within the safety of their own homes. When the decision has been made a speedy but controlled evacuation of the area would begin. The local authority concerned would have been alerted and places of shelter provided. In the past London Transport had provided coaches as necessary. There may be occasions when the underground could be used to move large numbers. In addition, police maintain lists of private coach hire firms willing to assist in major emergencies. One fact which isn't always appreciated is that police have no statutory power to remove people from their homes. However, it is very rare indeed for anyone to refuse. To illustrate the points made it might be useful to give fairly recent examples of actual incidents where evacuation of moderately large numbers was undertâken at short notice in situations perhaps not dissimilar to those which might occur with irradiated fuel.

The first of these relates to the White Friars Glass Company incident in Wealdstone, north London. On 20 November 1980 at 7.30 p.m. the night-watchman on a factory site in north London discovered that liquid propane gas was leaking from a tank containing a large quantity of this highly inflammable material. The fire brigade attended at 7.31 p.m. followed by police. Clouds of propane gas were blowing towards an educational centre situated behind the factory. The senior fire brigade officer present asked for the centre to be evacuated. This was done and a hundred people were moved out of the area. The leak could not be stopped and at 10.15 p.m. the decision was made for more families to be evacuated. Police toured the area using loud hailers to warn residents that this was imminent. The local authority was alerted and accommodation made ready. The local London Transport bus garage was asked for assistance and provided six single deck buses with drivers. Present at the scene was a scientific officer from the Greater London Council. He consulted with the senior fire brigade officer present and a decision was made to evacuate an area of approximately ten acres. Local police went from house to house directing residents to a marshalling point nearby from where they were taken by bus to the shelter provided.

Some 1500 to 2000 people from approximately 425 households in that small area were moved that evening. Among those were many elderly people. The evacuation started at 10.35 p.m. and was completed by midnight. When the leak was eventually sealed the evacuees were returned to their homes in less than a hour. During this period police had patrolled the area to prevent crime, which is always a problem where houses are left in a hurry. They also set up an information centre at the local police station to assist anxious relatives. The official report of the incident made by the Health and Safety Committee stated that the evacuation proceeded efficiently.

The second incident was at a chemical works in Barking which is in the East End of London. This was a more serious incident which occurred on 21 January 1980 about 8.20 p.m. where a fire broke out in a chemical warehouse. A large contingent of fire appliances arrived at 8.28 p.m. followed by a chemical incident unit and then the police. At 8.37 p.m. there was a triple explosion in the warehouse, as a result of which a number of firemen were injured. At 9.16 p.m. the scientific adviser from the GLC advised that there was a large quantity of cyanide at the seat of the fire. Evacuation was necessary for a distance of one mile down-wind. This presented major problems. The area concerned included two large housing estates with some high-rise dwellings. Each estate was situated in a different local authority area. There was a total of about 7000 residents in the area concerned. As it turned out none of these problems caused any great delay. Fortunately, because of the wind speed, it was not necessary to evacuate the area to the north where there is a greater density of housing and through which runs a major trunk road. Once again London Transport readily offered their assistance and provided four double deck buses with drivers. Because of the urgency of this evacuation, ambulances, police vehicles and private cars were used to remove the families to nine rest centres set up by the local authorities concerned. A total of 1300 people were accommodated at the rest centres; the remaining families went to friends and relatives in the area. The evacuation of the two estates was completed by 11.00 p.m. and took approximately one and three-quarter hours.

It would, of course, be dishonest of me to claim that our contingency plans when put into operation always work as easily as these two examples. They do not. Those who were caught up in the chaos on Monday morning following the discovery of a bomb will bear witness to this. [Editor's note: This was an unexploded World War 2 bomb found in central London adjacent to the River Thames on the day before the Conference.] Any major incident in so central an area is bound to cause problems – so many people are at risk. The 'small bomb' referred to by the Deputy Leader of the GLC on television last evening was considered by the expert at the scene to be capable of projecting shrapnel for a distance of one thousand metres in an area with a large daytime population. With such advice from the expert and knowledge of the area concerned decisions had to be made. Was it necessary to evacuate the whole area at risk?

The decision was taken after consultation with the bomb disposal officer to have the streets adjacent to the river cleared, to have warnings sent to persons in charge of buildings in the area which were at risk to move personnel to the rear of the premises, that is away from the river, and in some cases to evacuate the buildings. It would, of course, have been a gargantuan task to remove such a large population which was growing by the minute as commuters arrived; but it would have been possible. According to the experts who presented papers yesterday there would be time after an incident involving irradiated fuel which would allow for a speedy but controlled evacuation of an area.

Our experience in dealing with terrorists' bombs in London shows that people respond well and do cooperate in clearing areas when so directed by police. The reassurance which Monday's incident should give is that police are not prepared to put lives at risk. The size or scale of the incident is not the test. It is the likelihood of injury being caused and in doing this we admit that we err on the side of caution. It is easy for some to criticise police action; but had we failed to do this and had injuries resulted, even one injury, I am sure that we would have been castigated for our lack of judgement.

I would now like to deal with the specific instructions issued to police or advice issued to police relating to the movement of irradiated fuel. These are based on advice from the CEGB and British Rail. In summary these are:

(1) to control access to the area around the flask observing the recommended evacuation area of 45 metres away from the flask in all directions and establish an instant control post upwind of the flask;
(2) to contact the British Rail mishap controller or his representative at the scene and liaise with him;
(3) to contact for further advice the Central Electricity Generating Board's nuclear power station or British Nuclear Fuels Limited establishment coordinating response to the mishap;
(4) to assist, if needed, health physicists and flask emergency teams from responding nuclear establishments with escort both to and from the location of the mishap;
(5) to arrange a landing site for the Ministry of Defence helicopter as necessary and transport of the team and equipment to the scene of the mishap;
(6) if the flask has been involved in a fire, to instruct the fire brigade to cool all accessible surfaces of the flask by spraying water; seek advice on the risk and possible need for further evacuation when the health physicist arrives;
(7) to ensure that personnel who approach the flask or move into the area downwind of the flask are noted for monitoring by the health physicist and inform him if casualties have been moved to hospital; anyone who might have been exposed to radiation would be taken to one of a number of designated hospitals in the London area which have specialist staff and facilities to deal with this type of problem.

These instructions are updated and the procedures revised if it is considered necessary. Since there are no experts within the police service who can assess the extent of problems related to the movement of irradiated fuel, we have to depend on the advice given by experts in the field. It is essential therefore that they can be called upon at short notice. In addition to experts from the power stations there are others from the CEGB headquarters and various scientific establishments in London who would be available in the event of severe weather conditions or if a long delay in the arrival of the expert from the relevant power station was anticipated. It is also important that we have liaison with local authorities and the other agencies involved. Police are represented on the London Emergencies Liaison Panel, which comprises representatives of the Greater London Council, the fire brigade, the London ambulance service and the City of London police. Police are also invited to meetings of the London Boroughs Emergency Planning Officers Committee. Additionally police liaise with Home Counties Emergency Planning Officers since the Metropolitan Police district covers areas beyond Greater London. For ease of communication in an emergency involving more than one police force there are direct radio and telephone links with the police headquarters of all the neighbouring counties concerned.

The liaison I have outlined is crucial to success of any planning to deal with major incidents. It is important to have an agreed policy. The cooperation extends even further, nationwide, with coordination being provided by the Home Office. They circulate advice to both Chief Officers of Police and the fire services. In addition the CEGB has printed two booklets containing more general advice to the public outlining the roles of the emergency services and incidents involving irradiated fuel.

To test the plans which have been drawn up exercises have been held in various parts of the country. These are run by the CEGB and British Rail and involve all the emergency services. No warning is given that the exercises are to take place, therefore the arrival times of the services are realistic. Since 1978 there has been a total of nineteen exercises held in various parts of England and Wales. Any lessons learned from these exercises have been incorporated in subsequent Home Office circulars. As I said at the beginning of this presentation there has frankly been no occasion when police in London have had to deal with a mishap involving the transportation of irradiated fuel. However, I am confident that the contingency arrangements which have been made by the emergency services in London to deal with such incidents are adequate, based as they are on considerable experience in dealing with a wide range of major emergencies and on comprehensive and clear advice from the CEGB.

10 Nuclear Fuel Transport in France

MARCEL J. BERTHIER

The transportation of irradiated fuel in France is carried out within the terms of the international regulations, particularly with regard to the safety of packaging. The French situation may, however, differ from that of other countries, in particular in that of administrative structure, the present and future volume of these shipments, and the perception of nuclear risk by the public. The most significant point is probably the rapid development of a nuclear industry involving the reprocessing of irradiated fuels, which necessitated the recent reinforcement of physical protection measures against acts of sabotage, and intervention measures in the event of accidents of a radiological nature.

For more than 20 years, the transportation of radioactive material has been regulated by special measures, which form part of the regulations that apply to hazardous materials in general. The first documents of this kind were set up in 1948 for Great Britain, then in 1955 for the United States. Since 1961 the International Atomic Energy Agency (IAEA) has been publishing detailed recommendations on the subject, and these have been adhered to. The French Government ruling adopted the recommendations of the IAEA, i.e. safety depends on the packaging, and the extent of security required is related to the potential risk.

From 1980, French legislation also took into consideration the need for physical protection against acts of sabotage. The resultant ruling is essentially applicable to fissile and toxic materials, in particular the transportation of irradiated fuel. Practically speaking, it concerns the protection of vehicles and the monitoring of their movements by radio.

As it stands, we have today an organisation which is dependent on the Ministry of Transport for safe packaging; and on the Ministry of

Industry for the physical protection of materials. Within the framework of its normal responsibilities of protection of people and belongings, the Ministry of the Interior is called on to intervene in these areas by means of two of its departments responsible for Civil Security and National Police. Finally, one characteristic of the French system is the important role of the department, in which the Police Commissioner of the Republic has overall responsibility for the supervision of transportation and intervention should an accident occur.

On French territory, the number and tonnage of irradiated fuel transportations are rapidly increasing. This situation results from a large nuclear power programme which might be considered ambitious, combined with the technical choice of reprocessing spent fuel – at least for this decade. The French nuclear industry is the second biggest in the world. Nuclear plants provide 40 per cent of the nation's electricity needs; this share could exceed 60 per cent by 1990. French nuclear power stations are almost all pressurised water and uranium enriched, with reactors of 900 MWe now, 1300 MWe in the future. The first 1300 MWe unit will come on stream in December of 1983.

The irradiated fuel in plants of this type is characterised by its high activity. It is reprocessed in the plant at La Hague, in Normandy. This plant is currently being extended to increase its processing capacity from the present 200 tonnes per year to 1600 tonnes per year towards 1991. From 1983–89, 6600 tonnes of irradiated fuel will be dealt with at La Hague. In 1982 the deliveries necessitated 100 road shipments and 130 rail shipments, the rail shipments often comprising several containers.

Rail is expected to be used more and more. Rail shipments comprise containers of 100 tonnes loaded with 6 tonnes of fuel. Road transport remains the only possible means for certain journeys and the total weight of the container is between 60 and 80 tonnes. Because of their weight, these vehicles are not authorised to use French motorways; this means delays and increases the risks, but no accident involving radiological consequences has been recorded to date. Road movements are not escorted by police unless the vehicle is larger than the authorised code of the road (this regulation applies to all very large vehicles, not just those which may be used for moving irradiated nuclear fuel).

Because of the increase in such transportations and more generally in the overall movement of radioactive material, an emergency plan has been drawn up to take into consideration the specific nature of radiological risk. The effectiveness of intervention in an incident or accident depends on the rapid alert of emergency services. Before each journey of irradiated fuel, the Ministry of the Interior provides each region through which it travels with detailed information. The information includes the weight of the materials and that of the packaging, the activity, the transport index and the particular procedure in case of fire. In each Department, the people informed include the Prefect, the Director of Civil Security, the Fire Inspector, the police and any other regional authorities who need to know, e.g. special traffic police. Any accident would entail – independently of the local services alert – the immediate alert of the Operational Centre of the Civil Security in Paris. Recently it was decided to have all such journeys monitored by radio by the transporters who can apply to a specialised branch of the Commissariat of the Atomic Energy. This measure will shortly come into effect for road transport.

As far as rail transports are concerned, any incident or accident is reported at the nearest station, which alerts the emergency systems. This chain system of the French railways seems extremely effective in its present form, although not entirely eliminating the possibility of acts of sabotage.

Once the alarm is given, what emergency measures would come into effect? The Civil Security would make use of its own area radioactive detection teams and mobile groups of national services, such as the Atomic Energy Commission and the Central Service of Radiological Protection. Nearer the event, and within a very short time, the first intervention would be that of firemen, equipped with military-type radiation meters, capable of carrying out simple measurements of dose rates and instigating urgent conservation measures. About 100 of its teams currently cover the main routes and have received suitable training.

At a second level of intervention more specialised teams who are also answerable to the Ministry of the Interior would be called on in case of need. These are also firemen, but using more precise detection equipment, better adapted to the risks encountered in peace time. There are, at present, in France twenty-two mobile units of this type, which cover the whole of the country and are more numerous in areas where there is a high concentration of installations or traffic flow. These teams, which can be transported by helicopter from eighteen Civil Security bases and are manned with officers who have received a practical training in radio protection at Atomic Energy Centre Headquarters, constitute the main operational resource.

Mention should be made of the existence of the important specialists from national services – those of the Central Service of Radiological Protection based in Paris, skilled in the evaluation of health consequences, and those of the ten centres of the Atomic Energy Agency.

How efficient are the measures for which the Interior Ministry is responsible? Without real accident experience it is difficult to be certain about the efficiency of firemen units and their equipment. Their officers are regularly trained and brought up to date at the Atomic Energy Headquarters but their daily activities are almost totally concerned with conventional risks. This emphasises the importance of land exercises and the upkeep of technique familiarisation. The units are capable of quick intervention and in recent years they have dealt with minor incidents very effectively.

Finally, a word about the French public opinion of the risk involved in the transportation of irradiated fuel. It is perhaps surprising that these activities have never seriously perturbed the public; they have confidence in the specialists, that this transport is no more worrying than any other; fortunately no accident has shaken this confidence. Opponents of civil nuclear power have at various times attacked both proposed centres and those already under construction, and their power lines. They have not taken shipments as their target, except in the very particular case of foreign fuel arriving at the Normandy reprocessing plant. Even the demonstrations against these shipments have virtually ceased, except for the maritime deliveries from Sweden and Japan.

However, this essential confidence would soon be shaken if an accident which entailed radiological consequences were to occur; even a minor accident would have disastrous effects on the politics of nuclear energy. This then justifies the existence of the stringent measures which have been outlined – measures unique in their kind for hazardous materials. Indeed, this might constitute a model for other hazardous materials, the transport of which gives more cause for concern than irradiated fuel.

11 The US Regulatory Framework

CHARLES E. MACDONALD*

The transportation of radioactive materials is regulated by both the US Department of Transportation (DOT) and the US Nuclear Regulatory Commission (NRC). To avoid duplication of effort, these two agencies partition their regulatory responsibilities by means of a Memorandum of Understanding. Although both agencies regulate packaging and shipment of all quantities of all radioactive materials, as a practical matter under the Memorandum of Understanding, the DOT mainly concerns itself with safety conditions of carriage of all quantities, performance standards of Type A packaging for small quantities, and packaging for low specific activity material. The NRC is primarily concerned with reviewing and certifying designs of packagings for all quantities of fissile materials and for significant quantities of other radioactive materials.

For the safety of shipments containing significant quantities of radioactive material, such as spent fuel, the NRC relies principally upon the engineered and designed features of the transport casks. The NRC reviews the designs of spent fuel casks for certification of compliance with the requirements (Title 10, Code of Federal Regulations, Part 71, Packaging of Radioactive Material for Transport and Transportation of Radioactive Material Under Certain Conditions). The review addresses the capability of the package design under both normal and hypothetical accident conditions to (1) retain its radioactive contents, (2) shield the environment from the radiation of its

* Although Charles MacDonald was unable to attend the GLC Conference, his paper is included because it outlines the official regulatory framework governing the transport of irradiated fuel in the US. This is a 'United States Government Work' under the US Copyright Act.

contents, (3) dissipate its internal heat to the environment, and (4) prevent the accumulation of a critical mass of fissile material.

Of the thousands of shipments that have been made during the past 30 years, none has resulted in an identifiable injury to the public through release of radioactive material. The success of the packaging strategy is illustrated by a situation involving a violent traffic accident that occurred on 8 December 1970 on a major highway near Oak Ridge, Tennessee. In this accident, the driver of a truck carrying a spent fuel cask swerved to avoid colliding with an oncoming vehicle, lost control, and overturned off the roadway. The driver of the truck was fatally injured. The spent fuel cask was thrown into a ditch, skidded, and tumbled more than 100 feet before coming to rest. No release of contents or release of radiation occurred. The spent fuel cask suffered minor damage to the outer surface and was placed on another trailer and taken to its destination. The spent fuel cask was later returned to service following repair of the minor damage and inspection.

Procedures applicable to the shipment of packages of radioactive material require that a package be labelled with a unique radioactive material label. In transportation, the carrier is required to exercise control over radioactive material packages, including loading and storage in areas separated from persons, and to limit the aggregation of packages to limit the exposure of persons. The procedures the carrier must follow in case of an accident include notification of the shipper and the DOT, isolating any spilled radioactive material from personnel contact, pending disposal instructions from qualified persons, and holding vehicles, buildings, areas or equipment from service or routine occupancy until they are cleaned to specified values. When an accident occurs, state and local governments are responsible for taking any actions deemed necessary to protect the public health and safety. To assist state and local governments, the federal Government has established an Interagency Radiological Assistance Plan under the coordination of the Department of Energy, which charges eight regional coordinating offices with the responsibility and authority for convening radiological assistance teams. Upon request, immediate action will be taken to respond to the emergency, including providing assistance at the scene.

An analysis in WASH-1238 ('Environmental survey of transportation of radioactive materials to and from nuclear power plants') provides additional information on this topic. WASH-1238 contains estimates of potential exposure to transport workers and the general public under normal conditions of transport. It also includes an analysis of the probabilities of occurrence and potential consequences of accidents in transportation. WASH-1238 is the primary data base which promulgated regulation 10 CFR Section 51.20(g), effective 5 February 1975. The environmental risk of radiological effects stemming from transportation accidents was found to be small.

The NRC also approves routes for the shipment of spent fuel based upon concern for deliberate acts to seize or damage the shipment. Under NRC rules, route plans for shipment of spent fuel must be reviewed and approved by the NRC staff to assure that the shipments will be conducted in accordance with Section 73.37 of Title 10, Code of Federal Regulations, Part 73, Physical Protection of Plants and Materials. The current interim protective measures are being reevaluated, based on analysis of the results of research that indicate that the respirable particle release for a successful act of sabotage is much less than earlier assumed.

Certain strategic quantities and types of special nuclear material require physical protection against theft during transit because of their potential for use in a nuclear explosive device (see 10 CFR 73.25, 73.26, 73.27, and 73.67). In January 1981, the DOT issued a final rule concerning highway routing and driver training requirements for radioactive material shipments. Under this rule, shipments made by truck would generally follow the most direct interstate route and would be required to avoid large cities where an interstate bypass or beltway is available. States are permitted to designate alternative routes when those routes are demonstrably as safe as the routes specified in the rule. The DOT routing rule was subsequently challenged in federal district court by the City of New York, the State of New York and others. The district court issued its final opinion and order on 5 May 1982. DOT was enjoined from implementing or enforcing the rule in so far as it required New York City to permit truck transport of spent fuel and other large quantity shipments of radioactive materials through densely populated areas. The district court refused to enjoin application of this rule to other jurisdictions. The case has been appealed to the United States Court of Appeals by the DOT. The court heard oral arguments on 14 February 1983. It is anticipated that the court will render its decision by early May. As a related matter, the NRC regulations require timely notification to the governor of any state or the governor's designee prior to transport of potentially hazardous nuclear waste, including spent fuel, to, through or across the boundary of that state. (This rule is in 71.5a of 10 CFR Part 71 and 73.72 of 10 CFR Part 73.)

As a matter of practice, the NRC has encouraged states to adopt regulations that are consistent with federal regulations and to join with the federal government in strengthening inspection and enforcement activities. Under the Hazardous Materials Transportation Act (HMTA) the DOT is charged with the responsibility for setting uniform material standards to assure safety in the transportation of hazardous materials. Under Section 112 of the HMTA (49 USC 1811), subject to the exception noted below, any requirement of a state (or political subdivision thereof) which is inconsistent with any requirement set forth in the Act, with regulations issued under the Act or with NRC safeguards regulations in 10 CFR Part 73, is preempted.

Any requirement of a state (or political subdivision thereof) which is not consistent with any requirement set forth in the Act, or in a regulation issued under the Act, is *not* preempted if, upon the application of an appropriate state agency, the Secretary of Transportation determines that such requirement (1) affords an equal or greater level of protection to the public than is afforded by the requirements of the Act or of regulations issued under the Act and (2) does not unreasonably burden commerce. Such requirement shall not be preempted to the extent specified in such determination by the Secretary of Transportation for so long as such state (or political subdivision thereof) continues to administer and enforce effectively such requirement.

To assure continued adequacy of measures required for the public health and safety, the NRC completed a re-evaluation of its regulations concerning transportation of radioactive materials. In the course of the re-evaluation, the NRC, in 1977, published a final environmental statement designated NUREG-0170 which included an examination of the transportation of radioactive material by all modes of transport. Considering the information developed, the public comments received, and the safety record associated with the transportation of radioactive materials, the NRC determined that its present regulations provide a reasonable degree of safety and that no immediate changes were needed to improve safety. The NRC published a statement to this effect in the *Federal Register* on 13 April 1981 as part of a public notice closing its rulemaking proceeding announced on 2 June 1975. Nevertheless, the NRC continues to study safety aspects of transportation of radioactive materials to determine where future cost-effective improvements for safety should be made.

12 Safety Problems and Government Regulations in the US

FRED MILLAR

INTRODUCTION

In promoting a new generation of US-design pressurised water (PWR) nuclear reactors, the UK atomic energy establishment seemingly wants to assert both that the UK operation of these will be safer than the ('cowboy'?) American PWR system and that the massive US commitment to PWR technology has meant that most of the problems in the system have been corrected. Evidence from the US experience with the regulation of nuclear waste transportation bears out neither of these assertions.

US transport of PWR spent fuel is already somewhat safer than UK spent fuel transport. Unlike in the UK, in the US concerned citizens have successfully insisted on Government publication of route information for nuclear waste shipments, on re-routing to avoid cities where possible and on armed security escorts where necessary. Since 1976 at least 182 US cities, counties and states have banned nuclear shipments; others have passed various restrictive laws or regulations. The legal and regulatory struggle in the US therefore involved the federal Government's attempt to impose more federal control over local governments.

There is increasing public and media awareness of several serious safety problems in US nuclear waste transport. Several of the seventeen nuclear spent-fuel shipping casks (flasks) have been found defective after having transported nuclear waste. The major railroad cask manufactured by General Electric, for example, has been withdrawn from service twice due to defective valves and design. No US casks in current use have been physically tested, and the Sandia 'crash test' films widely promoted by the Government have been credibly challenged as misleading propaganda.

The US railroads have embarrassed the federal Government since 1974 by testifying that nuclear casks have not been designed to survive actual crash conditions by rail. The railroads have insisted that nuclear waste shipments travel under 35 m.p.h. and with other special restrictions. The US Government has done preliminary studies of possible re-routing of all extremely hazardous shipments by rail so as to avoid highly populated cities.

The data base for predicting future accident rates is extremely limited. Nuclear spent-fuel shipments in the US have averaged only about 300 per year over the past 10 years. Until now most commercial nuclear spent fuel has stayed in water storage pools at reactor sites; a 'dramatic increase' in nuclear transport volumes is predicted by federal studies, to 9000 per year in 20 years. New on-site storage techniques such as dry cask storage make unnecessary almost all near-term spent-fuel shipments, and encourage local permit ordinances that require shippers to show the safety and the need for the proposed shipments.

A draft 1981 study from the National Academy of Sciences characterises as 'primitive' the current US regulatory programme for spent-fuel transportation, and suggests that serious upgrading of safety measures and more federal cooperation with state and local officials will be necessary if impasses are to be avoided. Official Government route maps document serious regional inequities in nuclear transport risks, given current federal waste disposal programs targeted on western states.

The DOE transport of military spent-fuel and weapons-production-related wastes has begun to generate critical interest, despite the much greater secrecy and federal dominance in this area. State and local officials in a few states have begun to challenge the DOE on safety and environmental grounds.

While enough similarities exist to justify an extended comparison between the respective spent-fuel transportation systems in the commercial sectors of the US and the UK, there are several major political and technical differences that make comparison difficult. But even these differences are illuminating, and their implications might give pause to those who look carefully into the prospect of adding a new generation of US-design PWR nuclear reactors for the UK.

The first major difference is that there is no reprocessing of commercial spent fuel in the US and consequently much less spent-fuel transport on a routine basis. The reprocessing of UK spent fuel, if I understand the situation, means that approximately three shipments of spent fuel travel by rail through London every two weeks. Presidents Ford and Carter in 1976 cut off reprocessing as an option for US nuclear utilities, in large part to avoid the diversion dangers inherent in a transport scheme involving plutonium shipments moving frequently around the countryside. Several state and local authorities in the US, moreover, have acted on their own to prevent spent fuel shipments from travelling through New York, Boston and other highly populated cities.

The second major difference is that nearly all US spent-fuel shipments go by truck, not by rail. The main reason is that the private-sector US railroads have been notably reluctant to carry spent fuel and risk a major accident that could tie up their primary capital asset, their right-of-way, for months or years at $100,000 per hour. The much larger crash forces generated in a potential railroad accident are not taken into account by the US NRC and IAEA regulatory standard of a 30-foot drop test, which translates into a 30 m.p.h. crash.

Despite all the serious problems which exist in the US transportation of spent PWR fuel, the UK seems now to be transporting its own spent fuel more recklessly than the US in several respects:

(1) In the UK the routes are kept secret, whereas in the US citizens have insisted upon the routes being made public so they can evaluate the risks they are being exposed to. Some questions they raise concern the adequacy of the casks and of emergency response capabilities, whether the safest route and mode of transport have been chosen, etc. The federal Freedom of Information Act in the US (the virtual opposite of the UK's Official Secrets Act) has been a powerful tool for concerned citizens to pry vital information out of Government agencies.
(2) In the UK there currently seems to be no requirement as in the US for armed escorts to accompany shipments through highly populated areas.
(3) In the UK there currently seems to be no ability for local authorities to regulate nuclear shipments (or any other extremely hazardous shipments) as has been done in the US. The extreme

centralization in the UK of nuclear matters in the nationalized CEGB and British Rail seems a formidable obstacle to a greater democratic participation in spent-fuel transportation issues.

(4) In the UK there seems to be no adequate system of prenotification of elected officials and emergency response personnel as in the US for spent fuel shipments.

As will be obvious from my discussion of the ongoing struggles between federal authorities and local and state authorities as to who should regulate nuclear transportation, the US federal authorities seem to have been no less arrogant than UK authorities; but the UK authorities seem to have been more successful in being arrogant.

The overall politics of nuclear waste transportation in the US is following what seems to be the typical scenario of American nuclear issue politics, as exemplified in the waste disposal issue:

(1) The first stage is that the nuclear industry/federal Government try a quick-and-dirty 'solution' to the problem. The US Atomic Energy Commission (now DOE), for example, tried to put high-level nuclear waste in the ground in a big hurry in an allegedly 'perfect' salt formation near Lyons, Kansas, in 1971.

(2) The second stage is that state geologists and other local officials demonstrate serious problems with the alleged solution. The proposed Kansas site was shown to be clearly unsuitable, and federal officials were promptly thrown out of the state, their credibility in shreds. Subsequent confrontations in other States resulted in similar federal humiliations.

(3) The third stage is that two beneficial results emerge from the confrontations. One is more technological conservatism on the part of the Government: the US will have a geological disposal site for spent fuel in operation no sooner than 1995, and in the meantime careful studies are required of the waste form, the canisters, the rock media and the geological regions of the country. The second benefit is that federal officials decide they must work more carefully with state and local officials. For example, the new federal nuclear Waste Policy Act of 1982 gives state officials a veto over waste-site selection; this veto can be overridden only by a vote of both House and Senate in the US Congress. By passing laws and filing lawsuits challenging federal policy, some states have successfully forced the federal Government to negotiate concessions in nuclear waste policies.

In the nuclear transportation issue, the quick-and-dirty solution consists of two assertions by the nuclear establishment: (1) the casks are virtually invulnerable to accident; (2) the federal Government by right has preemptive domination over state and local laws. Environmental critics and state/local officials have seriously challenged both arguments, so we are at stage two, in my estimation, in the political scenario. Confrontation prevails on all sides.

US CONFLICTS OVER NUCLEAR ROUTING

An as yet unpublished draft study by the prestigious National Academy of Sciences (NAS) on radioactive waste issues details very serious transportation problems which will impact almost all the states and criticizes the current plans for nuclear spent fuel transportation and disposal. The 1981 draft study by the NAS Panel on Social and Economic Aspects of Radioactive Waste Management characterizes as 'primitive' the federal regulatory framework for transporting high-level radioactive spent fuel, and the study predicts serious 'impasses' between state and federal officials if state officials are not given greater voice in regulating the safety of the projected 75,000 nuclear truck shipments through their states. Without 'drastic revision' of current federal regulation, the NAS study says, 'the probability of serious accidents will increase.' There has been a substantial neglect of the transportation problems for nuclear spent fuel:

> The transportation system would involve most of the states, whether they were the site of nuclear power plants or not. The system would be required to have a high degree of reliability and to operate under the close scrutiny of a concerned public and local officials.

The volume of truck shipments will be significant:

> If the number of power plants in operation rises from the current 60 to the 150 authorized, the number of shipments to off-site locations will increase. These shipments would begin in the mid-1980s and would increase to nearly 9,000 shipments per year by about 2005, if they were all by truck.
>
> If all the spent fuel were shipped by truck the rate of growth would be much more rapid, increasing at an annual rate of about 500

truckloads per year. By 2004 there would be of the order of 9000 individual shipments per year. The last statistic is important. If all of these shipments were sent to a single point – either an AFR, a reprocessing plant, or a geologic repository – they would arrive at an average rate of one per hour all year round.

The impact of rail shipments could be clearly unacceptable:

The rail transport costs used here assume that rail cars originating at various reactors would probably be assembled at depots or marshalling yards. Thus, at any given time there would be a number (perhaps a large number) of rail cars carrying spent fuel waiting in marshalling yards across the country. Some of these yards – Chicago, for example – are likely to be in densely populated areas. Moreover, a single car may wait a week or more. This means that for all practical purposes railroad marshalling yards would become de facto short-term AFRs. In a fully operating system hundreds of rail casks might at any given time be 'stored' in these yards.

A centralized nuclear waste disposal system is seen as both likely and undesirable:

Despite the fact that the Department of Energy professes to still have regional repositories under consideration (US Dept. of Energy 1981), all indications point to the emergence of a relatively centralized waste disposal system for spent fuel, with one or several waste repositories (or a repository with multiple shafts) sites concentrated in the west or southwest. The Panel concludes that this choice will result in a waste system which is costly in its transportation requirements, enlarges inequities among regions, is potentially vulnerable to operational bottlenecks, and places significant institutional burdens upon states. The western siting of a single repository will exacerbate all these problems, particularly through the creation of a waste transport system which is national in scope and funnel-shaped in configuration.

The NAS study states bluntly that:

Federal/state conflicts remain unresolved. While the federal Government possesses needed formal authority to implement waste disposal system, the states possess ample means to challenge federal authority over (particularly) non-radiation issues and could create substantial delay and continued conflict. Although 'concurrence' presupposes that agreement will be reached, in fact impasses are likely and there currently exists no effective means for overcoming such impasses.

The ORNL maps have been created by federal Government experts using a computer model based on routing data supplied by a major US nuclear waste trucking firm, and also taking into account the relevant federal and local nuclear waste routing regulations. The maps show the nuclear route patterns for various likely nuclear waste sites (see Figure 12.1).

The Environmental Policy Institute extrapolated the ORNL routes

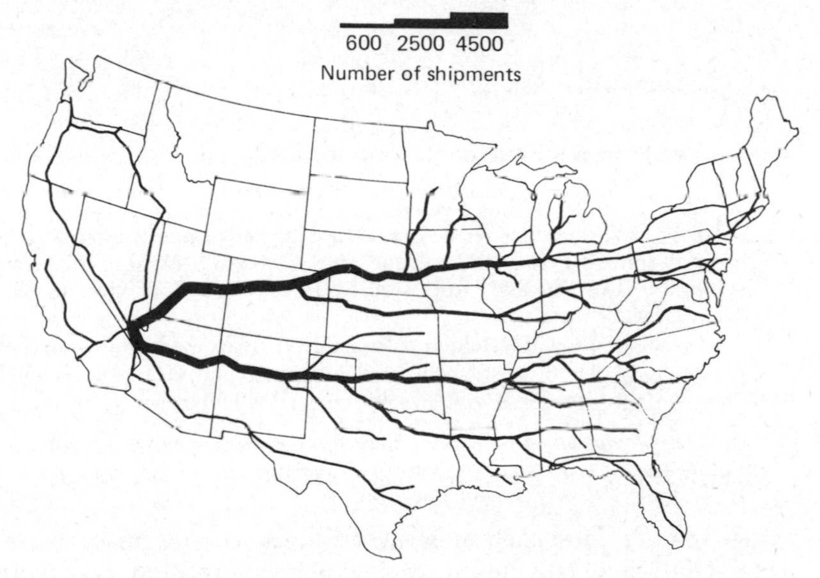

NOTE This ORNL map shows a southern Nevada waste storage or disposal site creating a nuclear waste funnel with massive waste corridors from the eastern and midwestern reactors, along Route I-40 from Tennessee to Nevada and along I-80, I-80/90, I-80, I-76, I-70 from Pennsylvania to Nevada.

SOURCE National Academy of Sciences, 'Social and economic aspects of radioactive waste disposal,' 1981 draft report.

FIGURE 12.1 *Projected annual spent fuel shipments to a western storage site in 2004 (basis: truck shipments from all reactors (for demonstration purposes only))*

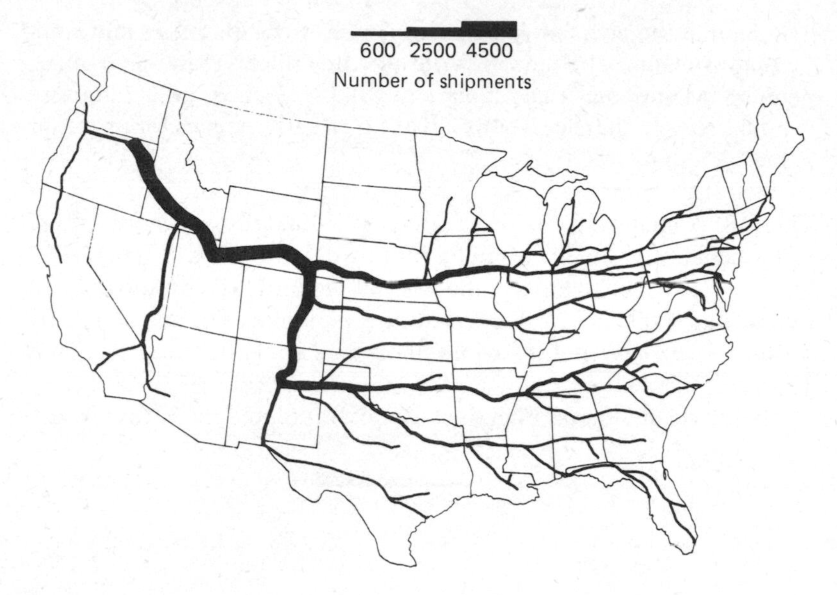

NOTE This EPI map is based on the Oak Ridge National Laboratories map shown in the 1981 draft report of the National Academy of Sciences. This map shows nuclear waste truck routes converging on the proposed high-level waste site at Hanford, Washington. Waste funnels are formed primarily along Route I-40 from Tennessee to Albuquerque, New Mexico; along I-70 from Pennsylvania to Denver; and along I-90/I-80 from New York State to Salt Lake City, Utah.

The main assumption EPI added to the ORNL map is that the southern reactor shipments would turn north at Albuquerque on I-25 and join the northern reactor I-80 waste stream at Cheyenne, Wyoming.

FIGURE 12.2 *Projected annual spent fuel shipments to a western storage site in 2004 (basis: truck shipments from all reactors (for demonstration purposes only))*

maps to show the most likely highway 'waste funnel' patterns for two of the top DOE candidate sites for geological high-level (and spent fuel) waste disposal, one at Hanford, Washington, and the other at Moab, Utah (see Figures 12.2 and 12.3). One pattern that stands out is that the eastern and midwestern states with many nuclear reactors are likely to be shipping their wastes across and into what the US Defense Department (in planning atmospheric nuclear explosions in southern Nevada) called the 'virtually uninhabited' areas of western states. On the way, however, shipments will traverse the largest western cities (St Louis, Omaha, Des Moines, Denver) that the Interstate highways were obviously built to connect. And the 'virtual uninhabitants' of

Utah and Nevada have recently emerged from their ghostly non-existence to halt such additional nuclear burdens on their areas as the racetrack MX-missile.

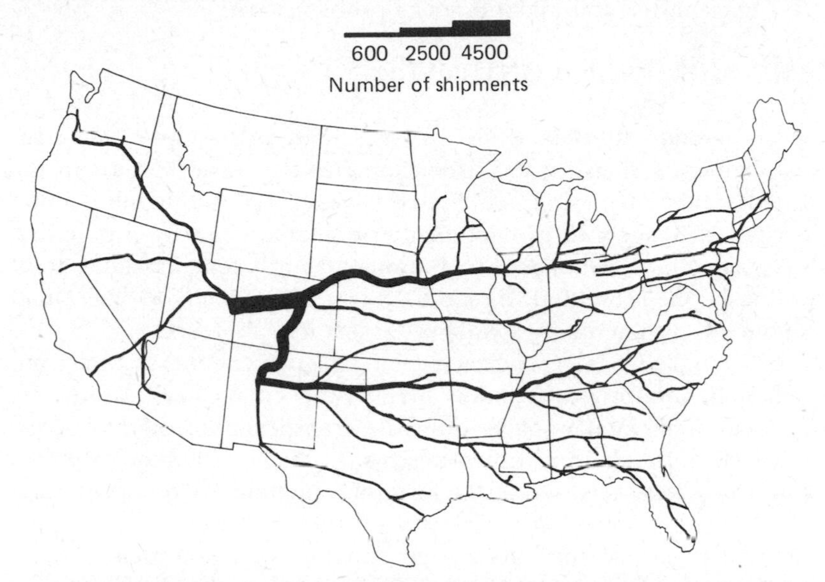

NOTE This EPI-generated map shows nuclear truck cargo routes converging on the proposed nuclear waste site near Moab, Utah. The map is a minor adaptation of the Oak Ridge National Laboratory map showing the potential waste site in southern Nevada, which would create a nuclear waste funnel with massive waste corridors from the eastern and midwestern reactors, along Route I-40 from Tennessee to Nevada and along I-80, I-80/90, I-80, I-76, I-70 from Pennsylvania to Nevada.

The only major change is to guess that southern nuclear waste shipments would turn north from I-40 at Albuquerque, then take I-25 north to Denver and there join the I-76 and I-70 waste streams. (The shorter highway route from Albuquerque to Moab would require use of non-Interstate routes 666 and 163.)

FIGURE 12.3 *Projected annual spent fuel shipments to a western storage site in 2004 (basis: truck shipments from all reactors (for demonstration purposes only))*

The route maps have generated enormous media interest across the US. At least two conclusions emerge from consideration of the NAS-type analysis:

(1) In the US citizens are probably not going to tolerate the most

hazardous nuclear waste shipments without some real improvements in safety and regulation.
(2) When democracy is genuinely allowed to intrude in any area of nuclear decision-making, sticky problems arise.

THE SABOTAGE CONTROVERSY

It is paradoxical that while the UK has much more sabotage-consciousness than the US (for example, the warning signs in the London tube cars about unattended packages), it is only the US that has acted strongly to protect highly populated cities from potential sabotage of a nuclear spent-fuel shipment. Early federal studies took seriously the potential threat to cities and identified spent-fuel shipments as potentially attractive targets for saboteurs.

In 1976 the US NRC contracted with an independent research firm, Sandia Laboratories, to conduct an assessment of the health and safety and environmental impacts from the transportation of radioactive materials through urban environments. In May of 1978 the results of that study were released in the form of a preliminary report entitled 'Generic environmental assessment on transportation of radioactive materials near, or through a large densely populated area,' SAND 77–1927 (Sandia Report). The Sandia Report found that shipments of irradiated spent-fuel rods from nuclear reactors represent 'the largest single source of radioactivity routinely shipped,' and therefore constitute 'an attractive target for sabotage or theft.' It was further found that 'Access to shipments of spent fuel would be possible for an adversary intent upon sabotage or theft,' and that 'it can be expected that any such attack will take place in densely populated areas.'

The Sandia Report concluded that if such an attack were successful, the radioactivity released to the immediate area would result in human deaths within one year numbering in the '10s'. Within weeks of the incident, there would be resultant illnesses numbering in the '100s–1000s'. In addition, '100s' of persons would eventually die because of cancer caused by the incident. The non-radiological effects of such an accident, e.g. the effects of the explosives used, were estimated to rise as high as 140–50 human deaths. The purely economic costs of this type of incident were estimated at $2.1 billion.

The Sandia Report and previous NRC staff documents, NUREG-0194, 'Calculations of radiological consequences from sabotage of shipping casks for spent fuel and high-level waste,' February 1977, and NUREG-0170, 'FES on the transportation of

TABLE 12.1 *Summary of consequence estimates*

Source	Population density	Early fatalities	Early morbidities	Latent cancer fatalities
3-element truck cask	2,000 p/mi²	0.4/6.3	Not calculated	220/270
3-element truck cask	42,000 to 115,000 p/mi²	26/44	1,000/1,500	450/550
10-element rail cask	42,000 to 115,000 p/mi²	130/1,200	660/7,600	1,600/7,500

NOTES The first numbers represent average values. The second numbers are maximum values.

Assumed release: 100 per cent noble gases, 1.6 per cent caesium, 1 per cent solids at respirable material.

SOURCE *US NRC SECY-79-298* (18 April, 1979).

radioactive material by air and other modes,' December 1977, postulated essentially the same release fractions – 100 per cent noble gases, 1 per cent solids as respirable material and about 1 per cent caesium – resulting from a high-explosive breach of the cask. Table 12.1 summarizes the estimated sabotage consequences contained in these documents.

While directing confirmatory research on the possible results of breaching by explosive charge a spent-fuel cask, in 1979 the US NRC hastily published new interim regulations 'to protect spent-fuel statements against diversion and sabotage'. The 1979 rules required nuclear shippers to prenotify the NRC, to obtain NRC approval for proposed routes, to develop vehicle-immobilization devices, liaison with police forces along the routes, and other procedures for coping with threats and emergencies of a 'safeguards' (security) nature. They also required the shipments to avoid travelling within 3 miles of any US city over 100,000 in population, unless transit through a populated area was 'unavoidable.' A list of 146 US cities was provided in NUREG-0561, the NRC's 'Interim Guidance' document that explained the new regulations ('Physical protection of shipments of irradiated reactor fuel'). In 1980 the NRC modified its new rules (but left them still on an interim basis) and published Revision 1 of NUREG-0561. The major revision is that the 1980 modified rules now allow shipments to go through cities, provided Interstate highways are used and armed escorts accompany the shipments. A new list of 184 urban areas is included. The relevant NRC office is the Office of Nuclear Material Safety and Safeguards.

So on the federal level there was some relaxation of protection of cities in 1980, in part so that NRC regulations might conform to the merging US Department of Transportation (DOT) regulation, HM-164. Subsequently, two multimillion-dollar confirmatory studies were conducted in 1981 by DOE and NRC, both designed to evaluate the effects of a sabotage attack on a spent-fuel shipment. A critique by Lindsay Audin of the DOE's Sandia test is given in Annex I.

The overall result of the two 1981 studies is that the release fraction of solid respirable material is now predicted to be only one-tenth or one-hundredth of the 1 per cent release earlier assumed. (The two studies differ by an order by magnitude.) The new test results may lead to further relaxations of the federal regulations on shipment through cities.

The US NRC has done a study designed to identify groups that are potential sabotage threats to spent fuel shipments in the US and could find none. Nevertheless, NRC paid Sandia $30,000 to develop two 'war gaming' types of board game – one called 'Skirmish', the other 'Ambush' – to help train whatever law officials might have to respond to an attack on a spent-fuel shipment.

The sabotage issue has often been cited in public hearings on proposed local or state laws to restrict nuclear spent-fuel transportation. Since the federal regulations are seen as erratic and unsatisfactory, several local and state laws have required additional escorts or re-routing to avoid high populations. The bottom line for many of the US citizen groups and elected officials' discussions has been that regardless of the technical data on how only a minimal amount of 'respirable' radioactive spent-fuel can be blown into the air by an explosive device, the mere threat to do so could be a credible one and could cause severe public anxieties or even panic.

In Germany the heavy cast-iron CASTOR cask has been developed to survive a severe missile test in which a one-ton projectile is fired at 650 m.p.h. at the side of the cask. There seems no current move in the US towards developing a more sabotage-resistant cask. Our experience in the US is that the sabotage possibility has consistently been used by the federal agencies to justify excessive secrecy. In 1980 the US NRC, which is the equivalent I think of the UK Nuclear Inspectorate, went to the Congress of the United States and tried to get a federal law that would keep secret the routes of nuclear waste shipments, using the rationale, and this is for spent fuel specifically, that the PWR spent fuel is potentially a target for saboteurs and that route information would be useful for saboteurs. Well, we went to the Congress and absolutely ridiculed that decision.

In the first place we said, 'Look, this is supposed to be a democratic country and people have a right to know what risks they are being exposed to; therefore, they should know what the routes are.' Secondly, we said, 'The notion that keeping the routes secret from saboteurs will be effective is absolutely ludicrous', and we quoted an NRC official who in an earlier hearing in Charlotte, North Carolina, had asked a series of questions. He said,

> Isn't it true that these spent fuel shipments for PWRs move in giant 30 ton casks, steel and lead, on overweight trucks, or in hundred ton casks, by rail? And isn't it true that they move on public highways and public rail lines? Isn't it true that they are marked clearly 'radio-active' on the outside? And isn't it true that they move between clearly identifiable nuclear power plant sites and clearly identifiable waste sites, storage tanks? I mean isn't it going to be a little bit like elephants tiptoeing through the tulips if we try to keep the routes secret?

Congress accepted our argument. We got an amendment passed in 1980, that forces routes to be made public, and also forces the Nuclear Regulatory Commission to prenotify the governors of the states through which these spent fuel shipments would move. The governors are then free to pass that information on to whatever local officials, safety emergency response officials, they choose. That is an example of the kind of struggle that Americans have engaged in with their federal officials.

THE SAFETY CONTROVERSY

I understand that in the UK, as in the US, the official line is that the nuclear spent-fuel flasks (we say casks) are virtually invulnerable even in very serious accidents. Apparently CEGB maintains that although the flasks have not actually been tested, they have been designed to survive a crash at about 30 m.p.h. into an unyielding surface. Furthermore, CEGB maintains that there are no unyielding surfaces in the UK rail transport environment. While this assertion is no doubt technically true, the image that we are left with is huge steel-and-lead flasks tumbling from derailed railcars and demolishing everything in sight: bridges, buildings, other trains – a sort of instant urban renewal programme.

How likely is a serious nuclear spent-fuel transport accident? There certainly seems to be a relatively low probability of a large release of spent fuel and/or radioactive 'crud' from a cask accident. But is the probability so low as to make unnecessary any other precautions to mitigate the possible consequences, such as re-routing shipments to avoid cities? That seems a political question, with no clear technical framework that all can agree on.

In the US we have been very lucky in that there has been so far no serious release of nuclear spent fuel, as far as we know, due to a transportation accident. No 'Three Mile Island on the highways', so to speak. We not only lack historical data in understanding what might happen, for example, if a nuclear spent-fuel cask is caught in an extended rail accident fire, but since really useful laboratory experiments seem out of the question, the US federal experts cannot even agree on what would constitute the release mode and pathways for a worst-case cask release. Too much crucial data is lacking, they concluded.

There is no controversy over the enormous hazard that nuclear spent fuel poses should even 1 per cent of the contents of a spent-fuel cask be released due to accident or sabotage in an urban area. It is the probability of serious radiation release that is disputed. Although everyone agrees that nuclear truck and train accidents will certainly occur with greater frequency, the nuclear industry relies on the federal government studies that predict a serious radiation release from a nuclear waste transport accident *once every 25 billion years*. This kind of estimate, however, is based on an historical data base that even government researchers admit is scanty and unreliable for confident predictions. Nuclear spent-fuel shipments in the US have averaged only about 300 per year over the past 10 years. Until now most commercial nuclear spent fuel has stayed in water storage pools on plant sites; a 'dramatic increase' in nuclear transport volumes is predicted by federal studies, to 9000 per year in 20 years.

The impact of a major (1 per cent) release would be severe, according to government studies in a low-probability, high-consequence accident. Sandia Laboratories of New Mexico estimated in 1980 that a major release of spent fuel in an urban area could result in $700 million damages. A major release of plutonium could result in billions of dollars of damage in an urban environment, not to mention the poisoning of people and loss of life. Yet the Price-Anderson Act limits the liability of the nuclear industry to just $560 million, and its application to some spent-fuel shipments is not clear.

Impacts on human life are not so easily quantified. Estimates of short-term deaths from a major nuclear transportation release range from a few hundred to tens of thousands. Latent cancers can develop in those people exposed to far less than immediately fatal doses. This could result in tens to hundreds of thousands of latent cancer fatalities. So the massive steel-and-lead casks must be enormously strong if disaster is to be averted in the inevitable accidents.

Over several years citizen researchers have gradually amassed evidence, mostly by careful study of Government documents, that the spent-fuel shipping casks are not as invulnerable as the public has been led to believe. Progressively it was learned that:

(1) Seven of the fifteen spent-nuclear casks in use in 1980 were taken out of service due to structural defects, *after* having been used to carry spent fuel.
(2) Nuclear Regulatory Commission regulations require the casks to be able to withstand fires of 800°C for 30 minutes; however, according to the Department of Transportation, the average temperature of a transportation accident fire is 1000°C, and some of these fires have burned for days.
(3) Government reports show that a 30-minute fire at 1000°C could cause the seal and pressure valve of a cask to fail. The NRC does not test new casks, but only requires an engineering analysis of the design. Such analysis has missed several serious defects, such as the pressure relief valves on the four General Electric IF-300 rail casks which were found able to open to release radioactive steam but unable to close again as they were designed to do.
(4) The cask welds have never been X-rayed for flaws, and casks in use cannot be checked due to permeation by radioactivity.
(5) Government reports show that a side-on collision of only 12.5 m.p.h. could cause the cask cavity to open.

Since the primary reliance for transportation safety assurances has always been placed in cask survival even in accident conditions, these revelations were damaging.

Additional credible challenges were raised concerning the adequacy of emergency response capabilities. This history of serious radioactive transport accidents reveals that local police and fire personnel, who have primary responsibility for accident response, are often uninformed about the radioactive contents of the cargo and usually unprepared for any serious problems.

31 March 1977: A train carrying radioactive uranium hexafluoride derailed near Rockingham, North Carolina, triggering fires and scattering the cylinders of uranium in the flaming wreckage. Confusion marked the emergency response, as seventeen agencies arrived at the scene. Workers were called on and off their emergency efforts as agencies disagreed as to whether a radiological hazard existed or not. Report from the Oak Ridge Radiological Assistance Team, 13 April 1977.

27 September 1977: A tractor-trailer carrying 20 tons of radioactive uranium 'yellowcake' crashed, spilling 5 tons alongside the road in southeast Colorado. It was *12 hours* before professional health specialists arrived, and 3 days before clean-up operations were begun, as the involved parties bickered over who had responsibility (Report by Critical Mass Energy Project, 31 October 1977).

Eighty per cent of US firefighters are volunteers. Even in conventional hazardous-materials transport accidents, such as recent emergencies in Waverly, Tennessee; Somerville, Massachusetts; and Mississauga, Ontario, the emergency response is still 'trial and error': firefighters who responded to a Somerville, Massachusetts, rail tank car phosphorus trichloride spill (in 1980) say they are no better prepared for that sort of emergency today than they were then, according to a published report. Somerville Fire Chief Charles Donovan pointed out to a *Boston Globe* reporter that the DOT emergency-response guidebook still says to use water on the chemical, adding, 'Boy, we learned the hard way that was the wrong thing to do.' The resulting acid cloud forced the evacuation of 30,000 people, sent 70 to hospitals, and corroded paint and metal on emergency vehicles. No one was killed but several responders apparently still suffer respiratory ailments (*Hazardous Materials Transportation*, May 1981).

The nuclear establishment's safety assurances were even further damaged when citizens learned to challenge the visually powerful but misleading public relations film (known as the 'Sandia crash test' film) that nuclear utility and DOE spokespersons showed in hundreds of public meetings and on TV programmes whenever nuclear transportation safety was questioned. The dramatic and well-promoted Government (DOE) film allegedly 'demonstrates' the safety of nuclear casks, based on the only full-scale transportation accident testing of nuclear waste casks, carried out at Sandia Laboratories in Albuquerque, New Mexico. The actual written report of these tests, 'Full scale simulations of accidents on spent-nuclear-fuel shipping systems', contains the following statements that contradict the way the film was used:

> Little information exists on the response of spent-nuclear-fuel casks and their transport systems when subjected to severe transport accidents.
>
> These tests are *not* intended to validate present regulatory standards.
>
> Accident scenarios which could not be conducted properly . . . *were not considered*.
>
> Due to the high costs of modern casks, it was necessary . . . to use *older and obsolete equipment*.

Each test was conducted only *once*. Fresh nuclear fuel was used in the tests; spent nuclear fuel – which the casks would contain when in use – is *one million times* as radioactive as fresh fuel. The pressure inside the casks with fresh fuel is 10 p.s.i., while with spent fuel the pressure is *300* p.s.i.

Perhaps the most telling comment on these and similar cask tests came from a nuclear industry source, Union Carbide researchers, who said, 'Admittedly, the tests (on a similar cask) are far from conclusive. The value lies in their public appeal.'

As in Europe, the US nuclear establishment relies not on actual cask tests, but on four test standards to which the casks are supposed to be designed and manufactured. These standards have not only been challenged as inadequate by American railroads, but the Government has been embarrassed by recent revelations of serious failures in its design-screening process. The major US rail cask for spent fuel, the General Electric IF-300, is a $5 million cask with pressure relief valves said to cost $25,000 each. These valves were approved in their design, but it was discovered in 1982 that they do not operate correctly. Soon thereafter it was discovered that the cask had not been built according to its design. So much for NRC regulation of designs.

The main point to underline here is that there have emerged enough credible safety problems with the seventeen US spent-fuel casks that the nuclear industry's and federal Government's safety assurances are more and more frequently disbelieved. State and local authorities are frequently inclined, therefore, to take the regulatory responsibility into their own hands, in line with their traditional 'police' authority for public health and safety.

A new book by the Council on Economic Priorities provides a thorough examination of the safety problems that have been uncovered concerning the manufacture and operation of spent fuel casks in the US (*The Next Nuclear Gamble*, Council on Economic Priorities, 84 Fifth Avenue, New York, New York 10011). Part of their analysis is an estimate of the consequences of a very severe release from a spent-fuel cask in an urban area:

> The NRC has further assumed that economic damages, ranging up to $4 billion, are only caused by the release of contaminated water coolant. The agency has not analysed other scenarios, such as a much larger release of radioactivity caused by a fire involving a tank car of high-temperature combustibles. Because of shortcomings in NRC calculations, the extent of contamination in urban and rural accidents has been calculated as part of this study. The results of any accident scenario depend on the assumed weather conditions at the time and the amount of radioactivity release. Under moderately stable meteorological conditions, with little mixing of air layers, almost all the radioactivity will be deposited in a cigar-shaped plume extending up to six miles away and will exceed EPA limits by a factor of 1,000. If not cleaned off, even after 100 years the residual radioactivity in this area will exceed EPA limits.
>
> Under more turbulent weather conditions at the time of the accident, about 60 per cent of the radioactivity will be deposited in a plume that will extend as far as 37 miles. Even at the 37-mile point, the radiation level will exceed the EPA limits by a factor of ten. If people are allowed to reside in the area contaminated by a radioactive plume, the number of delayed cancers and health effects will range into the hundreds of thousands. On the other hand, evacuation will be very costly. While the NRC has estimated up to $4 billion in damages, more realistic use of New York City property values places the property loss in the tens of billions of 1982 dollars. This does not account for the costs of delayed cancers, litigation or local government costs of handling the emergency.
>
> The NRC study claims that high accident consequences due to population density are unique to New York City. Yet, other US cities like Boston, Chicago, Pittsburgh, St. Louis and San Francisco,

have comparably high population densities in the central business districts during daytime business hours.

Neither the NRC nor DOT has written an environmental impact statement or calculated the consequences of accidents in rural areas. The contamination plume from an irradiated fuel accident in a rural area is similar to that for a city. Under slightly unstable weather conditions, the plume could easily encompass 139 square miles.

THE CONTROVERSY OVER WHO SHOULD REGULATE NUCLEAR TRANSPORTATION

The New York City Board of Health in 1976 enacted regulations that effectively banned the hundreds of truck shipments of nuclear spent fuel that were planned to go through Manhattan from reactors on Long Island. The federal DOT decided immediately to promulgate a federal rule that would make such local regulation illegal. But it took several years, and in the meantime scores of other localities and states enacted similar restrictions on nuclear shipment, as well as various regulations requiring prenotification for shipments or restricting times of travel, and in some cases imposing outright bans.

The existing federal law governing this issue is the Hazardous Materials Transportation Act of 1974. While the Act explicitly allows some federal uniformity for safety reasons (e.g. same colour placards in all states for a specific hazardous cargo), it also explicitly allows room for state and local regulations on local transport hazards.

The most formidable obstacle to the state and local regulation of nuclear transportation in the US is the nuclear industry's lengthy and largely successful work in persuading federal officials (DOT) to try to preempt (render invalid) the local regulations. In the overall politics of hazardous-materials transportation, the politically potent shippers and carriers of chemicals and other dangerous cargoes have laboured to set in place a *clearly* token federal programme so that the fifty states can be preempted and discouraged from strong regulation of hazardous shipments. (One statistic here: US DOT is responsible for regulating 412,800 tank trucks that regularly transport various hazardous materials; the agency has nine full-time inspectors in the whole US.)

On 19 January 1981, with less than 24 hours remaining in the life of the outgoing Carter administration, the US Department of Transportation handed down one of its most controversial regulations governing the shipment of hazardous materials on the nation's highways. The federal rule dictated new regulations for the truck shipment of radioactive materials and set minimum training requirements for drivers of the vehicles which carry them.

Although the regulations, officially known as 'HM-164' (HM stands for Hazardous Materials regulation) did not go into effect until 1 February 1982, they soon added fuel to an ongoing controversy about who should be responsible for supervising the shipments of such materials. Safe-energy activists and environmentalists had anticipated the DOT action for more than a year, and the publication of the final rules caused little surprise. But the ruling delivered a new opportunity for safe-energy activists to strengthen ties with state and local officials.

The DOT publicly argued that the new rules were necessary to free up a backlog of radioactive shipments awaiting storage or disposal. But department spokesman Lee Santman told the press that the regulations would provide a 'building block' in the planned federal programme of dealing with nuclear waste. That programme was expected to result in the shipment of thousands of packages of highly radioactive spent-fuel rods from the nation's commercial nuclear reactors to temporary storage at centralized federal facilities. The shipments would traverse most of the nation's major Interstate Highways and would pass through dozens of major population areas.

Representatives of the nuclear and chemical industries said the new DOT rules were welcomed by the companies who must ship hazardous materials. Primarily, the industries were pleased with the portions that prohibit permits and fees now levied by state and local governments through whose jurisdictions shipments pass. The revenue generated by these fees have been used mostly to pay for emergency-response training in preparation for a possible accident.

At the centre of the controversy over the new DOT regulations is the authority of state and local governments to determine when and where these shipments will be routed. The new rules would preempt more than 200 state and local laws passed in the last few years to regulate the transport of radioactive materials. DOT officials admitted in the new rules that one of the intended effects of HM-164 was to render void any state or local law which would 'prohibit or have potential to delay' nuclear shipments 'between any two points served by highway'. An example of the effect this would have is that without some legal action by a New York State agency, highly radioactive spent-fuel shipments would once again be permitted to move through downtown Manhattan. A 1976 ban on such shipments passed by New York City would be wiped out by the DOT assertion of federal dominance over such shipments.

Strong objections to the DOT plans came from citizens and state and local government officials. Opposition culminated in formal statements of opposition from several prominent environmental groups as well as from the National Association of Counties, the National League of Cities, the National Association of Attorneys General, and the National Conference of State Legislatures.

The provisions of the lengthy DOT rules were published in the 19 January 1980 Federal Register and included:

(1) Opening up the entire 42,500 rule Interstate Highway System as 'preferred' routes for large-quantity radioactive shipments. Beltway routes around cities, where available, are designated as preferred to Interstate Highways through cities.
(2) Preemption of state and local laws which would hinder or delay shipments. 'The public risks in transporting these materials by highway arc too low to justify thc unilateral imposition by local governments of bans and other scvere restrictions,' the DOT regulations explained
(3) Providing for a process where states or localities which object to certain federally designated Interstate routes are required to work through an appropriate state agency to provide alternate 'preferred' routes. Such a challenge puts the burden on local officials to initiate an elaborate process which includes a thorough 'technical review' of suggested alternatives, extensive consultation and negotiation with affected localities and adjacent states, and possible defence in court.
(4) Adhcrcncc to an 'implementation package' furnished by DOT to state and local governments for use in designating alternate 'preferred' routes. Calculations to demonstrate suitability of each alternate route must include the 'normal' and 'accident' risks of radiation exposure. Determination of the risks on alternate routes requires consideration of such factors as population density, travel times, accident rates and 'special facilities' (schools and hospitals). Ironically, the DOT's designation of Interstate Highways as the primary 'preferred routes' did not consider these factors.
(5) Submission of route plans by nuclear transporters to the DOT within 90 days after the shipments have occurred. State and local officials would not be notified until after that.

(6) Training for drivers who will operate vehicles carrying nuclear shipments only in the form of 'written instruction' every 2 years.

This heavy-handed federal attempt to preempt local and state regulation was patently more a political response to the perceived proliferation of local legislation than a real safety regulation as mandated by the federal Hazardous Materials Transportation Act of 1974. It was also viewed as a likely precedent for further federal preemption on other hazardous materials besides nuclear. HM-164 was promptly challenged in court suits filed by the Ohio Attorneys General and the legal officers of New York City and New York State. In May 1982 a New York federal district court judge issued a ruling severely criticizing the new DOT rule and allowing the New York City law on nuclear shipments to stand. Other district and appeals courts ruling in Boston and New York, regarding both nuclear and other extremely hazardous shipments, have been decided in favour of regulations enacted recently by those cities to protect densely populated areas from extremely hazardous truck shipments.

The overall US legal situation on hazardous materials transport remains unclear. Opinion varies widely on what kind of compromise might be worked out in the long run. Currently, however, states and cities are continuing to enact new regulations, such as those in Michigan, Vermont and New York State that have effectively blocked spent-fuel shipments from a Canadian research reactor into the US.

An absolute ban on nuclear shipments probably would not stand up in court. Therefore, most of the recently drafted nuclear shipment ordinances, like the 1976 New York City law, are sophisticated permit ordinances. In order to get a permit from local authorities, nuclear shippers must give rigid safety assurances, such as the adequate testing of casks or inspection of rail tracks, and must show a 'need' for the shipment.

MOST NUCLEAR IRRADIATED FUEL SHIPMENTS ARE UNNECESSARY

The 'need to ship' angle is a key one. Most nuclear utilities currently planning shipments in 1982–84 cannot show that their shipments are necessary. Either their water storage pools have years to go before filling up (e.g. 1993 for Cooper Nuclear Station in Nebraska) or they can use on-site dry cask storage, which can be licensed and built within two years. Virtually all nuclear spent-fuel shipments are therefore unnecessary because every US plant has at least 2 years' storage in its pools.

Many kinds of issues besides safety risks and 'need to ship' can be raised in local hearings on proposed nuclear transport ordinances: testing of the shipment casks, emergency-response training and preparedness, nuclear insurance coverage, federal safety regulation, etc. And new grassroots support for transport ordinances can be found among people who live dozens of miles away from the closest nuclear plant but whose communities are on the 'approved' nuclear waste transport routes that cross thirty-one states.

Nuclear critics in the US have no intention of trying to block permanently the transportation of spent fuel. Much of it is in very unsatisfactory places, such as the Three Mile Island waste on an island in the flood-prone Susquahanna River, or in temporary storage pools at reactor sites or in (leaking) storage tanks in the US weapons programme.

The transportation issue in the US is one in which citizens have a limited but significant ability to achieve some democratic control in nuclear waste decisions, which historically have been tightly controlled by the few. These decisions must be made with public safety uppermost, not as merely 'business decisions' made by private corporations.

It has often been said that individual cities or counties in the US who pass bans or restrictive regulations on spent-fuel transport are simply parochial and beggar-thy-neighbour in shifting risks from their own community to some adjacent one. But given a 'primitive' federal regulatory scheme, states and localities think they have acted in their own self-interest and that it is up to other jurisdictions to get educated on the issues and act accordingly. And in fact a ban imposed on one route or part of entry has often led to corresponding bans on the alternate routes or ports.

Only one port on the East Coast, for example, the heavily militarized Norfolk port in Virginia, seems willing to accept weekly spent-fuel shipments returning from research reactors in Europe, Japan and South Africa for reprocessing by the DOE in South Carolina. Several other ports have banned these shipments, which are of weapons-grade highly enriched uranium. It seems that only from scattered local and state actions that impose burdens on the nuclear establishment can citizens get the attention of that establishment and persuade them that there must be some serious changes in regional and national policy.

Across the US, citizen groups critical of nuclear waste transportation found a similar strategy that took maximum advantage of the somewhat decentralized authority for hazardous materials regulation. An exciting series of local ordinance campaigns to restrict nuclear waste transportation has shown that grass-roots groups can enact new legislation despite powerful federal and industry opposition.

Public meetings on proposed nuclear transport have been packed in Creston and Burlington, Iowa, Lincoln, Nebraska, Biloxi, Missouri, Sacramento, California and Birmingham, Alabama. New state laws regulating nuclear transport have been passed in New York, Michigan, Mississippi and South Carolina. Nuclear transport ordinance campaigns have even succeeded in usually conservative jurisdictions such as Charleston, South Carolina, Savannah, Georgia, Miami, and sixteen Cleveland suburbs. Stiff local laws failed by only the narrowest of margins in the Birmingham and Burlington city councils; otherwise they would have local bans No. 183 and 184 in the US.

The nuclear industry has scrambled to contain this movement, sending its flying hit squad of PR men to debate environmental experts in hearings well covered by local and regional media. Environmental activists have, however, learned effective ways to attack industry and federal safety assurances and to rebut charges that local regulation is unconstitutional.

With no permanent disposal sites for high-level nuclear wastes, US nuclear power plant storage pools are rapidly filling up with highly radioactive spent fuel. The nuclear industry's inclination is to meet this waste crisis with stop-gap measures that involves transporting thousands of spent-fuel shipments by truck or rail from near-full temporary storage pools to other temporary storage pools not yet filled up.

'Permanent' geological disposal for nuclear waste is *at least* a decade away, so the potential is real for a national shell game of nuclear waste transportation involving 75,000 truck-loads of spent fuel over the next 20 years, according to a recent draft report by the National Academy of Sciences. Some communities may even be exposed to the same risky nuclear shipments twice: once as they go to a temporary pool and then, some years later, when they go to a permanent site.

The new opportunity for environmental groups is to cut off the transport option for nuclear utilities and force them to use on-site dry cask storage, which most nuclear critics and boosters alike see as clearly safer and cheaper than the transportation-and-water-pool-storage combination. A 1981 DOE study, the Johnson Report, said that dry cask storage is safer than water pool storage and could be available for use by 1985. A new on-site water storage pool typically costs $1100 million, and takes 8 years to construct; transportation between pools may cost from $5 to $15 million. But an on-site dry cask storage system costs about $30 million.

While DOE has been pushing the fifty-seven nuclear utilities to adopt dry cask storage, the utilities are seeking a federal bail-out in the form of a federally operated away-from-reactor temporary storage pool. Such a bail-out is included in the new Nuclear Waste Policy Act of 1982, but in a measure that limits the federally provided AFR program to 1900 metric tons.

The new Nuclear Waste Policy Act of 1982 clearly recognizes widespread citizen opposition to nuclear waste shipment. The Act directs DOE to 'minimize' the transportation of spent fuel and to maximize temporary on-site storage by various methods (Section 135). Existing state and local laws and federal regulations on nuclear transport are explicitly left untouched by the Act (Sections 9 and 137). Transportation to a federally operated AFR is clearly to be 'last resort'. In selecting waste repositories, DOE must consider waste transportation impacts on a regional and local basis *both* in the EA/EIS stages (with public hearings) (Section 112) *and* in the 'consulation and cooperation' process, where DOE negotiates with state and local officials about 'offsite concerns', including specific transportation issues (Section 117). The passage of new state and local laws, especially tough permit ordinances, regulating nuclear shipment will clearly be the most effective way of ensuring that DOE takes seriously the concerns of citizens and elected officials.

CONCLUSION

It would seem to be very difficult for CEGB and British Rail to avoid scores of spent-fuel shipments annually through London. It seems clear, however, that shipment through London is a test case, political-ly, for the nuclear *status quo*, just as New York City might seem a crucial test case in the US. Federal officials in the US seem determined to engage in protracted court battles in their attempts to override New York City's ban, despite some concessions to state and local officials in HM-164. The federal official seems to be carrying the water for the nuclear industry, in a desperate effort to crush the efforts of citizens' groups and one-federal officials to impose some public rationality on the waste transport system.

It is simply unacceptable to increasing numbers of Americans that there are no federal agencies which have accepted the responsibility to choose the safest route and the safest mode of transport for a particular set of shipments; that the casks have not been tested; and that adequate emergency-response training and equipment are not available. In the absence of serious safety and regulatory changes on the federal level, local and state officials will no doubt continue to exhibit what I characterize approvingly as an obnoxious concern and an irritating insistence on being part of the game in nuclear transport decisions.

ANNEX 1 ANALYSIS OF THE SANDIA SABOTAGE TEST LINDSAY AUDIN

In order to evaluate the effects of a sabotage attack on a spent-fuel shipment, Sandia Laboratories subjected a 25-ton lead and steel cask to the blast of a high explosive device in early 1981. The cask, which contained no water, held a zirconium clad fresh fuel assembly. Sandia issued a study of the test entitled 'A safety assessment of spent fuel transportation through urban regions'.[1] Despite its preliminary nature, this report has already been cited as 'proof' that sabotage does not present a serious danger to the public.[2]

The testing facility encased the cask in a 10-foot diameter steel chamber sealed at both ends and fitted with sampling mechanisms to measure the concentration of uranium dispersed in the air. The explosive device was detonated outside the chamber and the blast narrowed through a passage (which was closed at the end of the explosion to contain any dispersed materials).

The explosion created a blast and flame that penetrated the steel and lead layers of the cask, dispersed about 12 pounds of uranium and about 5 pounds of cladding, and distributed about 3 grams of uranium in particles small enough to be inhaled (called 'aerosols'). Approximately one of the 12 pounds was scattered in particles small enough to cling to the surfaces of the chamber.

From these results, Sandia inputted the quantities of aerosolized particles into the CRAC computer program to determine the effects of a similar amount of inhalable *irradiated* fuel on an area as densely populated as an average part of Manhattan. According to Sandia, only one latent cancer fatality and no early fatalities or illnesses would result.

These results were compared to the assumed release fraction of a sabotage incident described in NUREG/CR-0743, the NRC-sponsored study of 'Transportation of radionuclides in urban environs'. That analysis assumed 0.07 per cent of the spent fuel would be dispersed as aerosols in an explosion, while this test found only 0.0006 per cent distributed in inhalable form.

There are several factors, however, that could severely narrow the applicability of this test and its results to other forms of sabotage attack. Questions also exist regarding the simplifications used in the test and their effects on the amount of material released. Following are several of the factors that make this test inapplicable to a broad range of potential sabotage scenarios.

First, the explosion was an exterior blast of a high explosive, not a contact explosion as could occur with an artillery shell or rocket. Such a mass puncturing the cask at a high velocity, accompanied by a contact explosion, could have also penetrated the far side of the cask (left intact in the Sandia test) carrying or forcing much more of the simulated spent fuel outside of the cask.

Second, since the primary mechanism of uranium aerosol dispersal was due to the high temperature of the blast flame, more could have been scattered in breathable form if an incendiary explosive, such as a phosphorus warhead, were used. A completely independent test, performed by another federally sponsored laboratory, found that an aerosol release could be ten times greater than indicated by the Sandia test.[3]

Third, the cask was dry, yet it is often the case with fuel that has remained in the fuel pool for several years that it contains water. Water would have absorbed some of the heat, but also distributed the shock to the entire interior of the cask which contains a quantity of 'crud' (small radioactive particles from the surface of the fuel and contaminants from the spent fuel pool). Loosening of the crud and vaporization of the water could lead to dispersal of more respirable materials than indicated in Sandia's conclusions. The escaping steam could also remove more of the uranium aerosols and larger particles. Finally, the presence of water would convey the shock wave to the vent valve and end cap of the cask, opening either or both of them and creating another passage for particle dispersal.

Fourth, had the fuel been hot (as is typical with spent fuel less than two years out of the reactor), more of the blast's thermal energy would have been available to aerosolize the uranium, thereby potentially making more of it inhalable and increasing the consequences. While such hot fuel is often contained in a water-filled cask, many of the casks in use today move hot fuel in a gas-filled medium (temperatures in the centre of the fuel may be as high as 750°). A truly worst-case analysis would have involved heating the fuel to such a temperature prior to the test.

Fifth, the consequences cited by Sandia (one latent cancer fatality) only accounted for the inhalable fraction of *inhalable spent fuel*. No accounting was made of the release of the 'crud' (the *only* material assumed to be released in the hypothetical accident in NUREG/CR-0743 that would require a $2 billion clean-up) or that portion of the fuel which becomes gaseous when heated (called 'volatile fission products'). Neither substance was present during the test. In the past, only a very small percentage of caesium (which vaporizes at 1239°F) was assumed to be released during a sabotage or accident scenario. The complete loss of containment, simultaneously accompanied by temperatures hot enough to vaporize steel (over 1600°F), would certainly lead to vaporization of much of the caesium in the dispersed fuel and possibly a portion of the fuel remaining in the cask (which would be heated by the blast). An incendiary shell might therefore increase the total consequences of the incident by vaporizing a significant amount of caesium.

Sixth, three questions also arise concerning the mechanics of the blast and its thermal effects:

(1) the cask used had heat radiating fins on its exterior; most of today's truck casks have no such fins. How much of the thermal and mechanical energy of the blast was dissipated as a result of the presence of the fins?
(2) the blast was focused on one spot on the cask. Would an explosion pressing on the rest of the outer shell cause the cask to buckle, opening additional pathways to release?
(3) did the presence of fins on the far side of the cask, or any supports beneath the cask, reduce the possibility of puncture or buckling of that side of the cask? (Most spent fuel casks are supported at the ends, not in the middle.)

CONCLUSIONS

All of the above items combine to lead to four general conclusions:

(1) A lead and steel cask can, indeed, be breached without the presence of a high velocity shell or an incendiary warhead.
(2) Destruction of cold, dry, unirradiated fuel does not yield as large a percentage of aerosolized material as assumed in NUREG/CR-0743.
(3) The health effects of a sabotage incident would involve dispersed 'crud' and volatile fission products, yet neither have been evaluated by any full-scale test, including this one.
(4) This single test did not simulate all of the effects of a worst case sabotage incident and cannot be used as basis for concluding that such an incident would have minor health consequences.

Ignored in the presentation was the cost to decontaminate an area near a dispersal. The assumed release fraction of solids leading to a $2 billion clean up bill was 1 per cent. The test *did* verify that such a release fraction was possible since *over 1 per cent* of the fuel was dispersed, despite the mitigating factors previously mentioned (e.g. cold-fuel, no incendiary, etc.)

Since terrorists and governments overseas have demonstrated their agility with rockets and bombs against nuclear facilities to which they are opposed, any analysis of a sabotage incident on a spent-fuel cask must be more sophisticated than the Sandia test, assuming it is to be taken seriously.

NOTES TO ANNEX

1. 'A safety assessment of spent fuel transportation through urban regions', by R. P. Sandoval of Sandia Laboratories and George J. Newton of the Lovelace Foundation for Medical Evaluation and Research (August 1981).
2. WNET television program, 2/16/82, comment by Mark Mills, executive director of Scientists and Engineers for Secure Energy.
3. 'Shipping cask sabotage source term investigation', Battelle Columbus Laboratories (December 1981), NUREG/CR-2472 (BMI-2089).

13 New York City – A Case Study in Defence of Radiological Public Health

LEONARD SOLON

GENERAL BACKGROUND

On 15 January 1976, the New York City Board of Health, an agency charged by legislative mandate with protection of the public health of residents of the City, acting upon the recommendation of the scientific staff of its Department of Health's Bureau for Radiation Control, passed an amendment to the Radiation Control (Article 175) section of the New York City Health Code. The thrust of this new regulation, enacted after the required document publication and circulation, and a subsequent public hearing, was to severely restrict shipments into and through the City of certain categories and quantities of radioactive materials including plutonium, uranium highly enriched in the fissionable isotope uranium-235 and irradiated (i.e. spent) nuclear fuel elements. A copy of the regulations is given at Annex I.

More specifically, the ordinance applies to the following:

(1) Plutonium isotopes exceeding 2 grams or 20 curies,
(2) Uranium enriched to a level where the fissionable isotope uranium-235 exceeds 25 per cent of the total uranium content.
(3) Actinides or nuclides whose atomic number is 89 or greater, in aggregate activities greater than 20 curies.
(4) Radioactive substances characterized as 'large' institutions require a Certificate of Emergency Transport. When such a source

is brought into the City, the issuance of an authorizing certificate means that a suitable police escort along with health physics surveillance is assured until the radiation apparatus has arrived at its ultimate destination.

On the other hand the authority to issue a Certificate of Emergency Transport for a necessary medical shipment means the capacity to withhold such authorization for shipments dangerous to public health and without compensating social value.

PLUTONIUM SHIPMENTS BY AIR[1]

Among the materials covered is reactor-produced plutonium. Prior to the enactment of this regulation, shipments of plutonium containing of the order of 100 pounds of mixed isotopes of plutonium – 238, 239, 240, 241 and 242, with aggregate activities of the order of 500,000 curies – were shipped by air into and out of New York City's John F. Kennedy Airport.

One will recall the enormous and justifiable furore by the world scientific community which led to the discontinuance of atmospheric weapons testing by the only strategic nuclear powers at the time. The total plutonium activity released to the atmosphere by all nations having acknowledged achieving fission explosions to this hour, i.e. the United States, the Soviet Union, the United Kingdom, China, France and India, is about 400,000 curies. This is about the same, or somewhat less in fact, than the contents of a single shipment into Kennedy Airport which occurred six times in the year preceding the enactment of our regulation. Dispersion of even a small fraction of the contents of one of these shipments as the result of an air crash, concomitant fire, and high winds within the City of New York – air crashes which regrettably occur with the small but lamentable inevitability dictated by Poisson statistics – could have brought cataclysmic results bringing death or serious injury to thousands of New York City residents. The inhalation or ingestion of milligram quantities of plutonium would result in almost certain death of an individual within weeks or months. Inhalation of a quantity one-hundredfold smaller – tens of micrograms – would result in that person having a special susceptibility to fatal pulmonary cancer in one or two decades. To recapitulate: 1 milligram, death; 10 micrograms, lung cancer; one air shipment, 40 million (4×10^7) milligrams or 40 billion (4×10^{10}) micrograms.

BROOKHAVEN NATIONAL LABORATORY SHIPMENTS

However, the principal reason for the genesis of the New York City radioactive materials regulation was consideration of the problems presented by the transportation of spent nuclear reactor fuel. Prior to the enactment of the regulation, Brookhaven National Laboratory, an important United States Department of Energy Research Institution, dispatched through the City of New York a limited number of highly radioactive shipments of irradiated fuel elements comprised principally of 93 per cent enriched uranium-235. Shipments were principally from Brookhaven's High Flux Beam Reactor – a reactor used in a variety of scientific research programmes. Each shipment, involving of the order of 300,000 curies of mixed fission products, was carried by truck in a specially engineered shielded cask.

The routes employed by Brookhaven National Laboratory for shipment of nuclear reactor spent fuel for more than a decade through New York City passed consistently through residential neighbourhoods having the highest population densities in the City, and indeed, among the highest in the country. I have not found any record or evidence that the United States Department of Transportation, the United States Atomic Energy Commission, or its dual successor agencies, the Nuclear Regulatory Commission or Energy Research and Development Administration, any other United States agency, and certainly not Brookhaven itself, took an exception to, or suggested alternate routes through less congested areas. The prevailing position of the federal authorities was to leave the matter of routes up to New York City officials alone. Clearly, the matter of radioactive materials' transportation routes was regarded as a purely local responsibility for New York City until the Department of Health and Board of Health exercised their public health responsibilities. These intrinsically hazardous shipments, completely unrelated to national security or defence, have been interdicted by the Health Code regulation.

SHIPMENTS BY LONG ISLAND LIGHTING COMPANY

In addition, at that time, Long Island Lighting Company (LILCO), an electric power utility company operating in Long Island's Nassau and Suffolk counties, east of New York City, had three nuclear power generating stations either under construction or planned, which would have increased reactor spent-fuel shipments through the City to a frequency approaching one per working day. This figure applies to reactor spent fuel alone and did not include high-level radioactive waste. This comes about in the following way.

At Shoreham, Suffolk County, on the north shore of Long Island, 56 miles east of the New York City line, Long Island Lighting Company has nearly completed a boiling water nuclear power station with a generating capacity of 819 electric megawatts (MWe). This reactor will employ fuel only slightly enriched (between 3 and 5 per cent) in the fissionable isotope uranium-235. It should be mentioned parenthetically that fresh unirradiated reactor fuel of this low enrichment does not constitute either a radiological or nuclear criticality hazard and is not influenced by the New York City transportation regulation.

However, when the Shoreham reactor becomes fully operational, probably in 1984, and begins shipping its irradiated fuel to reprocessing centres, the City is supposed to become host to between an additional thirty-five and seventy truck shipments each year through New York City streets, each shipment carrying several million curies of mixed fission products and tens of thousands of curies of plutonium and other actinides.

In addition, Long Island Lighting had in an advanced state of planning two pressurized water reactors, each of 1150 electrical megawatt capacity, at Jamesport, Suffolk County, some 18 miles from Shoreham. These were originally scheduled to have begun operation in 1983 and 1985 respectively but have been cancelled or indefinitely deferred.

These reactors would have added an additional 150–160 shipments annually of spent fuel through City streets – again each shipment carrying of the order of 2.5 million curies of mixed fission products and between 30,000 and 40,000 curies of plutonium and other actinides. Thus, adding up the totality of shipments from Brookhaven, Shoreham and Jamesport, there were to have been between 200 and 250 shipments of these cargos of potential malignancy-inducing materials traversing New York City each year – approximately one per working day.

The New York City Health Code regulation turned out to be a precise regulatory mechanism for achieving compliance with US Government statutes and US Congressional intent in protecting the health and safety of New York City residents by controlling dangerous shipments of this character.

CONSEQUENCES OF CASK FAILURE[2–8, 12]

What could be the result of an accident or sabotage event which resulted in a breach of spent-fuel container integrity?

It is evident that the potential for thousands of prompt or latent cancer deaths and injuries exists in a single breach of container integrity where 1 per cent of the radioactivity was released either through sabotage, accident or human error. These catastrophic numbers are not surprising. A 1 per cent release from these spent-fuel casks means something of the order of 100,000 curies of mixed fission products and actinides released in an uncontrolled way into the local environment of New York City. Anyone who has worked seriously in radiological protection recognizes such an event as a public health disaster of the first magnitude.

The potential economic cost and social disruption has been addressed with varying conclusions in different studies. Whether one is talking about London, New York, Tokyo, or Paris, a major release of mixed fission products and actinides in a densely populated area would generate long-term unprecedented problems which, to this observer's knowledge, no municipal jurisdiction anywhere is prepared to confront adequately.

JUDICIAL AND LEGISLATIVE HISTORY[9–11]

The issues which were raised by the enactment of the New York City Health Code regulation on 15 January 1976 have not been confronted exclusively, or even mainly, in the scientific, public health and engineering arenas.

The most decisively important developments in the matter have taken place in litigation with the City of New York, and many local jurisdictions who have emulated its action in whole or part, on one side. In the adversary role has been the United States Department of Transportation, the United States Nuclear Regulatory Commission, some of the very great American institutions of higher learning, and large electric generating utilities employing nuclear reactors for power.

A capsule review of this history is given here. At present the matter is unresolved and further adjudication is necessary.

On the same day, 15 January 1976, that the Board of Health enacted its regulation, the United States Attorney for the Southern District of

New York filed a suit in the federal court requesting a preliminary injunction to prevent New York City from enforcing its regulation. The motivation for the requested judicial relief was the status of the truck shipments of spent-fuel elements from Brookhaven National Laboratory's High Flux Beam Reactor (HFBR) which, as mentioned previously, had been using New York City streets en route to their destination at a then United States Energy Research and Development Administration (ERDA) nuclear fuel reprocessing and waste disposal centre at Savannah River, South California, in the south-eastern United States. On 23 January 1976 the request for a preliminary injunction was denied by Judge Inzer B. Wyatt in the United States Court for the Southern District in New York City.

On 1 March 1977, Associated Universities Incorporated (AUI) filed an application for an 'inconsistency' ruling asking the United States Department of Transportation for an opinion or determination that the City Health Code regulation was inconsistent with federal statutes, especially the Hazardous Materials Transportation Act and regulations derived from it. Essentially, what was advanced was the argument of federal preemption of state or local jurisdiction statutes provided for in the American Constitution under Supremacy Clause and the Commerce Clause and under the Atomic Energy Act of 1954.

What is Associated Universities (AUI) which sought to overturn the New York City regulation by petition to the US Department of Transportation? For the record, Associated Universities or AUI is the group of nine institutions of higher learning which serve essentially as the board of directors of Brookhaven National Laboratory. These universities are Columbia, Cornell, Harvard, Johns Hopkins, Massachusetts Institute of Technology, Princeton, Pennsylvania, Rochester and Yale. I personally found it incompatible with the liberal, humane and scientific tradition of these very great institutions to associate themselves with this legalistic health-threatening artifice to reverse the City regulation.

THE AUI PETITION REJECTED

The Associated Universities petition, if it had succeeded, would have had the effect of reopening the streets and highways of New York City to extraordinarily hazardous shipments of radioactive materials, once again jeopardizing the health and lives of city residents and workers.

Brookhaven National Laboratory is a federal laboratory on federal property with a resident federal area manager. The United States

Department of Energy (operating under its former designation as Energy Research and Development Administration), the United States Nuclear Regulatory Commission (NRC) and the United States Department of Transportation made their clear opposition to the City Health Code regulations known in a vigorous way in formal written submissions to the New York City Board of Health prior to its adoption by the Board of Health.

On 4 April 1978, the Department of Transportation issued a decision in favour of the City of New York which stated: 'Section 175.11 of the New York City Health Code, as amended through 15 January 1976, is not inconsistent with requirements of the HMTA or with requirements in regulations issued to date thereunder.' The burden of the decision was that because a routing requirement had not yet been established under the Hazardous Materials Transportation Act, the latter did not preempt the City Health Code. However, the Department of Transportation (DOT) did not wait long to try to abrogate the New York City regulation, which it found unacceptable. On 17 August 1978, the DOT issued a notice of rulemaking inviting comment 'on the need and possible methods for establishing routing requirements pertaining to highway carriers of radioactive materials'. On 31 January 1980, the DOT issued its proposal rule confining it to highway routing and stating that it would postpone consideration of railroad transportation. Barge or ship water-borne shipments were not addressed at all.

The rule language acknowledged that its origin had as its basis the New York City Health Code's 'interdiction' of Brookhaven's irradiated nuclear fuel. The final Department of Transportation rule, which is frequently encountered in the literature under its DOT docket number designation HM-164, was published on 19 January 1981 and was to become effective on 1 February 1982. This new rule, which would have totally preempted and obliterated the Health Code regulation, has been strenuously opposed by the City in the federal courts. On 5 May 1982, United States District Judge Abraham Sofaer in a lengthy decision found for the City of New York. Major parts of the HM-164 were declared to be 'arbitrary, capricious, an abuse of discretion', and in violation of several federal laws dealing with protection of the environment and transportation safety.

In very brief summary, the court decision held, *inter alia*:

(1) The DOT did not examine the possibility of sabotage as a source of transportation risk.

(2) The feasibility of alternative modes, in particular barge or rail, was not explored in terms of reducing public health and safety risk.
(3) A catastrophic highway accident in a densely populated urban area was not addressed adequately in terms of significant environmental impact.
(4) Numerous errors were made in the quantitative probability analysis of accidents leading to the release of radioactive materials to the environment.

CURRENT STATUS OF LITIGATION AND THE NEW YORK EXPERIENCE

The aforementioned May 1982 decision in favour of the City of New York was vigorously challenged by the US Department of Transportation and eleven large American utility power companies in an appeal to the United States Court of Appeals – a judicial body immediately superior to the US District Court and just below the US Supreme Court. Finally, oral arguments before the court were made on 14 February 1983. As of this date, 15 March 1983, no decision has been rendered.

An effort is made here to focus the New York City experience in a way which will be hopefully constructive in the direction of augmented radiological health and safety in nuclear fuel transportation.

Risks related to spent-fuel transport

The working group reports of the Final International Nuclear Fuel Cycle Evaluation (INFCE) must be regarded as source documents of the very first rank in addressing problems of nuclear energy for peaceful purposes. It is fortunate that with the penetrating analyses done in so many other areas of spent-fuel management, the risks related to spent nuclear fuel transportation are essentially dismissed as very small (INFCE *Summary and Overview*, p. 210, para. 2.3, 'Transportation techniques').

It is clear from the New York City experience, and the technical studies which followed, that the potential health and safety hazards of irradiated nuclear fuel transportation deserve the most meticulous scientific scrutiny.

The action taken by the New York City Board of Health in restricting spent-fuel shipments was not based upon a history of prior accidents. Rather, the regulation had at its foundation an assessment of the anticipated cataclysmic consequences of a major radioactive release from accident or deliberate sabotage.

Evaluation points to a large number of prompt deaths and latent cancer fatalities as well as clean-up costs ranging into the billion-dollar region from major radioactive material dispersions.

Highway, rail, water and intermodal transportation alternatives

To the author's knowledge, a detailed scientific study has not been done vectored specifically to the risk-benefit, cost-benefit aspects of water or intermodal spent-fuel transportation. It would appear that spent-fuel transportation by barge or ship, for maritime countries like the United Kingdom or Japan, would certainly be strongly preferable to land transport by rail or highway in minimizing potential radiation exposure. As one American study has pointed out (NUREG-0170, *Final Environmental Statement on the Transportation of Radioactive Material by Air and Other Modes*, Nuclear Regulatory Commission (December 1977), pp. 4–25; para. 4.3.4.1, 'Transport by barge'), '. . . with relatively few people exposed during movement and a few exposed at each terminal, population exposure is expected to be negligible'.

The concept is expanded in the same report on pp. 6–9 ff., para. 6.2.7, 'All feasible irradiated fuel by barge' – there is a small decrease in radiological impact by employing barge transportation in conjunction with rail transportation (25,040 person-rem compared to 25,360 person-rem from the baseline case pp. 6–1). There is a slight increase with barge transportation (25,700 person-rem with the same baseline number).

The concluding paragraph (pp. 6–11) states: 'The fact that transportation costs are so much lower for barges than for other modes makes this alternative certainly worth additional investigation. Barge transportation of irradiated fuel may be a viable alternative, at least for some specific reactor sites if not as a nationwide scheme'.

A report which finds against water transport on the grounds of vulnerability to attack is NUREG-0324, *Safeguards Systems Concepts for Nuclear Materials Transportation*, System Development Corporation for USA Nuclear Regulatory Commission (September 1977) (Cf. 5.3.4, 'Water transport', pp. 5–7).

Informed citizen involvement

Probably the most single important parameter in addressing health and safety problems of advanced technology, at least in the United States, and perhaps in all democratic nations, is a restoration in citizen confidence that they are being given the whole truthful story by their government and its scientific advisers. The New York City regulation irrespective of what inferences some of our well-intentioned supporters or respected adversaries have drawn, is not anti-nuclear. It is, however, strongly pro-public health. Many of us in the scientific community see a compelling need for and support the development of nuclear energy sources along with the scientific pursuit of other energy options. However, the nuclear option must evolve with careful and objective respect, and full disclosure of the very serious public health and environmental questions involved. This includes, in particular, the nuclear fuel transportation issue. Transportation is almost certainly the most vulnerable part of the entire nuclear fuel cycle from the point of view of accident or sabotage.

Since this is a complex scientific issue with numerous public health implications, informed public education, and indeed active participation, is absolutely essential if the nuclear option is to have the required degree of citizen endorsement and support.

NOTES

1. 'Final environmental statement on the transportation of radioactive material by air and other modes', NUREG-0170, US Nuclear Regulatory Commission, Washington, DC (December 1977) (National Technical Information Service, Springfield, Virginia, 22161 USA).
2. 'Transportation of radionuclides in urban environs: draft environmental assessment', N. C. Finley *et al.*, Sandia National Laboratories, Albuquerque, New Mexico, 87185 NUREG/CR-0743 (July 1980) (US Nuclear Regulatory Commission, Washington, DC, 20555).
3. 'Identification and assessment of the social impacts of transportation of radioactive materials in urban environments', C. Cluett *et al.*, Battelle Human Affairs Research Centers, NUREG/CR-0744 (July 1980) (US Nuclear Regulatory Commission, Washington, DC, 20555) (National Technical Information Service, Springfield, Virginia 22161).
4. 'Review and integration of existing literature concerning the social impacts of transportation of radioactive materials in urban environs', C. Gordon *et al.*, Rice University, Texas, for Sandia National Laboratories, Sand 78–7017 (May 1978).

5. 'Severities of transportation accidents', R. K. Clarke *et al.*, Sandia National Laboratories SLA–74–0001 (July 1976).
6. 'Conceptual design of a shipping container for transporting high-level waste by railroad', P. L. Peterson and R. E. Rhoads, Pacific Northwest Laboratory, Richland, Washington, 99352, United States Department of Energy, PNL-2244 UC-71 (December 1978) (National Technical Information Service, Springfield, Virginia 22161).
7. 'An assessment of the risk of transporting uranium hexafluoride by truck and train', C. A. Oeffen *et al.*, Pacific Northwest Laboratory, Richland, Washington, 99352, US Department of Energy, PNL-2211 UC-71 (August 1978) (National Technical Information Service, Springfield, Virginia 22161).
8. 'Consequences of postulated losses of LWR spent fuel and plutonium shipping packages at sea', S. W. Heaberlin *et al.* Battelle Pacific Northwest Laboratories, PNL-2093 UC-71, United States Department of Energy (October 1977) (National Technical Information Service, Springfield, Virginia 22161).
9. *Judgment* 81 Civ. 1778 (ADS) United States District Court Southern District of New York, the City of New York, Plaintiff against The United States Department of Transportation *et al.* (5 May 1982), 539 F. Supp. 1237 (S.D.N.Y. 1982).
10. *Radwaste*, Fred C. Shapiro Random House, New York (1981) (Chapter 9, 'Transportation').
11. 'Federal actions are needed to improve safety and security of nuclear materials transportation', Comptroller General (E. B. Staats) Report to the Congress of the United States EMD–79–18 (7 May 1979) (United States General Accounting Office, Washington, DC, 20548).
12. *The Next Nuclear Gamble, Transportation and Storage of Nuclear Waste*, Marvin Resnikoff (March 1983) (Council on Economic Priorities; New York, N.Y. 10011).

ANNEX: HEALTH CODE TRANSPORTATION REGULATIONS

§175.111 Transportation

(a) The person responsible for the transportation of a radiation source outside of an installation shall take all necessary precautions so that no individual is exposed to ionizing radiation exceeding in amount the maximum permissible dose rate specified in section 175.05 of this Article. Sources shall be containerized and packaged, warning labels shall be affixed to the outer container holding the radiation source, and the airplane, ship, vehicle or rail carrier concerned placarded in conformity with the regulations of the U.S. Department of Transportation, U.S. Atomic Energy Commission or other related federal or state agencies regardless of whether the shipment is being made intracity, intrastate or interstate.

(b) When a radiation source which may involve a high degree of hazard in the event of an accident is to be transported into, within, through or out of the City, the shipper or carrier shall first notify the Department of the date of shipment, type and quantity of radiation source involved, method of transportation, route, starting point, destination and such other information as the Department may require. So much of such information as is then available shall be furnished at least two weeks prior to such shipment. The unsupplied information shall be furnished no later than two hours prior to arrival of the radiation source in the City or in the case of a shipment originating in the City no later than the time of its departure from the point of origin. The radiation source shall be transpoted through the City over such route or routes, or at such time or times of the day, consistent with the public health and safety and the convenience of the shipper or carrier, as the Department may direct.

(c) This section shall not apply to radiation sources shipped by or for the United States Government for military or national security purposes or which are related to national defense. Nothing herein shall be construed as requiring the disclosure of any defense information or restricted data as defined in the Atomic Energy Act of 1954 and the Energy Reorganization Act of 1974, as amended.

(d) The following radiation sources may involve a high degree of hazard in the event of accident:

(1) Any quantity of radioactive material specified as a 'Large Quantity' by the United States Atomic Energy Commission in 10 CFR Part 71, 'Packaging of Radioactive Material for Transport';

(2) Any quantity, arrangement and packaging combination of fissile material specified by the United States Atomic Energy Commission as a 'Fissile Class III' shipment in 10 CFR Part 71, 'Packaging of Radioactive Material for Transport'; or

(3) Any shipment of radioactive material that is required by the appropriate regulating agency to be accompanied by an escort for safety reasons.

(e) Motor vehicles transporting the radiation sources designated in subsection (b) shall follow the 'Truck Routes' listed in Article 16 of the Traffic Regulations adopted by the Commissioner of Traffic of the City of New York except that any such motor vehicle may be operated on a street not so listed for the purpose of delivery, loading or servicing, provided it leaves and returns to a listed street by the most direct route.

(f) In addition, in the case of shipments of radiation sources in the following categories the provisions of subsections (g) and (h) below shall be complied with:

(1) More than ten times the quantity specified as a 'Large Quantity' of:

(i) Mixed fission products, including that contained in, but not limited to, a shipment of spent nuclear fuel, or

(ii) Liquid or gaseous radioactive material;

(2) Fissile Class III;

(3) Required to be escorted for safety reasons; or

(4) Any intrastate (New York) or intracity (New York City) 'Large Quantity' shipment of radioactive material which does not meet the U.S. Department of Transportation packaging, labeling and placarding requirements in all respects.

(g) The shipper or carrier shall notify the Department and shall arrange with the Department for an escort through this City at a time to be designated by the

Department. He shall report the point of entry into the City or point of origin with the City and such other information as is required by this Code.

(h) If the destination is within this City to a pier, airport or other delivery point and facilities are not available during the night for unloading upon arrival, the motor vehicle shall be parked out-of-doors and not close to any building. The driver shall remain with the vehicle until the materials are unloaded.

(i) The following exemptions, subject to the specified conditions, will not result in undue danger to life and health:

(1) All shippers or carriers of 'Large Quantity' radiation sources may be exempt from the requirement of reporting their proposed route, provided they follow the 'Truck Routes' specified in subsection (e) above.

(2) All shippers or carriers of 'Large Quantity' radiation sources ten times or less than the designated quantity may be exempt from supplying the Department with the information prescribed by this section, provided they follow the 'Truck Routes' specified in subsection (e) above and leave the City without making a delivery and without any loading, servicing or layover within the City.

(j) Vehicles passing over any bridge under the jurisdiction of the Department of Highways shall meet the provisions of Local Law No. 49 of 1954, section 683a2-1.0 of the Administrative Code of the City of New York governing the maximum dimensions and weights of vehicles.

(k) Transport of radioactive material within the City by a physician for use in the practice of medicine is exempt from the requirement of this section, provided that such material is not transported via mass transit facilities.

(l) Notwithstanding the foregoing provisions of this section, a Certificate of Emergency Transport issued by the Commissioner or his designated representative shall be required for each shipment, to be transported through the City or brought into the City of any of the following materials:

(1) Plutonium isotopes in any quantity and form exceeding two grams or 20 curies, whichever is less;

(2) Uranium enriched in the isotope U-235 exceeding 25 atomic per cent of the total uranium content in quantities where the U-235 content exceeds one kilogram;

(3) Any of the actinides (i.e., elements with atomic number 89 or greater) the activity of which exceeds 20 curies;

(4) Spent reactor fuel elements or mixed fission products associated with such spent fuel elements the activity of which exceeds 20 curies; or

(5) Any quantity of radioactive material specified as a 'Large Quantity' by the Nuclear Regulatory Commission in 10 CFR Part 71, entitled 'Packaging of Radioactive Material for Transport.'

(m) Each shipper or carrier who transports radioactive materials or radiation source into, within, through or out of the City shall secure all such materials or source against theft, loss or any unauthorized removal from the vehicle, mode of transport or place in which they are stored.

NOTES: Subsection (b), (g) and (i) were amended by resolution adopted on December 15, 1977 to additionally apply their provisions to carriers of radioactive sources or materials operating within the City.

Subsection (c) was amended by resolution adopted on January 15, 1976 to conform its provisions with subsection (1) adopted by the same resolution.

Subsection (1) was added by resolution adopted on January 15, 1976 to require the approval of the Commissioner or his designated representative through the issuance of a Certificate of Emergency Transport for the transport or the bringing into this City of specified large quantities of plutonium, enriched uranium and other actinides and spent reactor fuel elements which would present a great hazard to public health in this densely and highly populated City. It is intended that such Certificate will be issued for the most compelling reasons involving urgent public policy or national security interests transcending public health and safety concerns and that economic considerations alone will not be acceptable as justification for the issuance of such Certificate. Such Certificates are also intended to be issued for hectocuric and kilocuric cobalt-60 and cesium-137 teletherapy sources employed in therapeutic radiology and biomedical research or educational purposes and for medical devices designed for individual human application (e.g., cardiac pacemakers) containing plutonium-238, promethium-147 or other radioactive material. This subsection is not intended to apply to small quantities of specified radioactive materials intended for therapeutic radiology and biomedical research or educational purposes.

Subsection (m) was added by resolution adopted on December 15, 1977 to require that security measures be taken against theft, loss or unauthorized removal of radioactive materials or radiation sources transported in the City.

14 Risk Assessment

TONY COX, PETER J. KAYES AND BETHAN MORGAN

THE ROLE OF TECHNICAL RISK ASSESSMENT

This paper is concerned solely with the so-called 'technical' or 'objective' risks associated with urban transportation of irradiated fuel, meaning the risks of such things as adverse physical health effects or costs associated with decontamination operations. It is not addressed to other 'subjective' effects which are associated with the way the risk is perceived, for example the psychological effects of anxiety. This limitation is imposed with no intention of belittling the importance of psychological impact, either in its direct impact or in its implications for risk or terrorist attack; rather, it is used to obtain an answer to a part of the problem that is amenable to analysis and also is an essential starting-point for the education processes that should occur to ensure that the psychological impacts are no more, nor less, than what the 'objective' situation warrants.

Technical risk assessment is a method of analysis which has been developed over the past 20 years to permit rational evaluation of the safety of activities having a large intrinsic hazard potential but high reliability. It has been applied to aircraft crashes, dam failures, chemical plant explosions, nuclear plant and gas pipelines, to mention only some of the more obvious examples.

In all of these cases, if the designer has not omitted any important factor and has not committed any errors concerning the factors he knows about, or if he has carried out prior tests that truly simulate the operational situation, the installation will stand a very good chance of operating throughout its life with no major disaster occurring. This happens, in fact, most of the time. The question that has to be answered is: just what level of risk does this activity entail, and is it sufficiently low, having regard to the scale of the possible consequences?

The method by which risk analysis approaches this question involves four main steps:

(1) identification of all possible failure modes;
(2) estimation of the probability of occurrence of each failure mode;
(3) quantification of the consequences of each failure or, where relevant, of the consequences and probability of each possible scenario following each failure;
(4) summarising of the results of steps (2) and (3) in a form that is understandable and useful.

Such an analysis necessarily involves examination of a wide range of failures. There is a general tendency for the larger disasters to have a lower expected frequency than smaller ones, but this cannot be taken for granted. Even if this is true, it is not a simple matter to determine at what point the frequency is low enough in relation to the magnitude of the event to enable one to say it is 'safe'. For these reasons, it is correct to retain the full spectrum of possible accidents in the analysis and not to exclude any of them on such subjective grounds as, for example: 'it is so unlikely as to be effectively impossible'.

A risk analysis of this 'classical' type will inevitably generate a very large mass of intermediate results, describing the likelihood and consequences of a large number of scenarios. These results are almost unusable in this form and must be summarised in a way that suits the question at issue. For example, in land-use planning problems, it is the geographical distribution of risk that is most relevant (e.g. the way that risk to an individual person would vary as he or she moves from place to place in the vicinity of an inherently hazardous plant). This is often represented in 'Risk Contour' form. For the 'yes/no' question of the acceptability of building a new installation on a specific site, the total integrated impact on the whole community will be one of the most important measures. This can be represented by a graph (the 'F–N' curve), showing the frequency F with which accidents causing N or more fatalities would occur.

These different forms of risk presentation serve quite different purposes and retain or reject different parts of the original information contained in the intermediate results. For example, the 'geographical risk distribution' throws away all information about the size of multiple-fatality accidents; it is only concerned with risks to an individual. By contrast, the 'F–N' curves contain no information about the location or identity of any one particular victim of the accidents which it represents.

In different circumstances, a risk may be 'high' by one of these measures and 'low' by another. In the case of transport operations, it is often found that risks to individual people living near the route are extremely low because the source of risk is spread out along the whole route; the chance of the accident occurring near to one particular fixed individual is naturally very low. On the other hand, if the accident occurs, it must occur somewhere. Therefore, the integrated societal impact may in principle be high, while the individual risk is low. For reasons such as these, great care has to be taken in evaluating the results of these calculations, and for the present purpose, the 'F–N' curve method has been selected as being the most appropriate.

A final point that must be mentioned is the problem of uncertainty in calculated numerical values for risks of very rare events. These numbers are, of course, very approximate both because of the difficulties of predicting accident consequences and because of the inadequacies of available statistics on which frequency estimates can be based. However, studies of the uncertainty itself have been made and generally show that, if no deliberate systematic conservatism or pessimism has been introduced, the overall error will be about a factor of ten each way to the ninety-five percentile value of the final calculated risk. Usually, the 'consequences' part of the calculation alone accounts for a factor of about two, and the 'probability' part accounts for the remainder. This degree of uncertainty might seem large, but is in fact usually adequate for decision-making purposes.

Thus, to summarise these introductory remarks, we can say that technical risk analysis is a tool which must inevitably be used when dealing with 'low probability/high consequence' events; it is imprecise and only provides a part of the total picture yet its contribution is an essential precursor to rational debate and effective decision-making. We now turn to the specific case of urban transport of irradiated fuel to see where the principles set out above lead us.

INTRINSIC HAZARD POTENTIAL OF IRRADIATED FUEL

For the purposes of this paper attention will be restricted to Magnox fuel and PWR fuel, but different cooling periods will be considered, to reflect such factors as differences in UK and USA practice, possibility of error in identifying fuel assemblies, and need to transport some short-cooled fuel for Post-Irradiation Examination (PIE).

Table 14.1 gives an approximate comparison of fission product inventories for the more important species. The data have been obtained from different sources, so there are some minor inconsistencies in the table, which, however, do not affect the argument. The key points to note are as follows:

TABLE 14.1 *Comparison of fission product inventories for different irradiated fuels (Units: Ci/Te(uranium))*

Fuel type	*Magnox*	*PWR(PIE)*	*PWR*
Burn up:	5 GWd/Te	33 GWd/Te	33 GWd/Te
Cooling period:	90 days	150 days	5 years
Nuclide			
Co 58	–	2,300	0
Co 60	–	110	0
Kr 85	1,400	11,000	6,900
Cs134	6,700	260,000	42,000
Cs137	16,000	110,000	89,000
Ru106	47,000	420,000	19,000
I 131	35	3	0
Xe133	1.5	0	0
Actinides (Total)	17,000	138,000	114,000
Totals (including others not shown)	0.38×10^6	3.9×10^6	0.61×10^6
Typical tonnage (U) per flask	2.5	0.462	3.23
Typical total Ci per flask	0.95×10^6	1.7×10^6	2.0×10^6

(1) Some of the species that tend to dominate the early deaths calculation in the radiological consequence model (e.g. notably I131 and Xe133) decay rapidly during the cooling period, so that for both Magnox and PWR fuel the early deaths are sharply reduced if cooled fuel only is transported.

(2) The residual danger in cooled fuel is associated with caesium and the actinides (e.g. Pu, Cm) which are respectively gamma and alpha emitters. These species could be inhaled if aerosolized and in the case of Cs could contribute to whole body irradiation if deposited on the ground. These can cause both early and late deaths in principle, although in the present case it is found that only the late deaths are significant.

(3) For a balanced analysis it is necessary to include all species, rather than concentrate on iodine as is often done, otherwise the benefits of cooling will be over-emphasised.

(4) In general, the total radioactive inventory is similar for all three forms of fuel that are likely to be transported in the UK, i.e. about 1 to 2 million curies per flask.

A fundamental problem arises in considering the 'hazard potential' of the radioactive inventory, in that the capability for causing harm depends not only on the amount of radiotoxic material present but also on the effectiveness of the available mechanisms for administering it to the population. One occasionally sees calculations of the total lethal potential of particular materials which are based on the assumption that these distributive mechanisms are perfectly efficient, i.e. that they administer the whole inventory to very large numbers of people at individual dose levels which are just on the minimum lethal quantity. From what is known about the processes that govern the transport of gases and particles and the absorption of radiation in the environment and in the body, it is certain that this situation is not even remotely approached in any case of interest. Therefore, such calculations are highly misleading. Realistic estimates of the efficiency of the distributive mechanisms are discussed in the following section.

MECHANISMS AND CONSEQUENCES OF RELEASES

The physical form of the fuel must now be considered. In all cases, the fuel is basically a solid material, metal in the case of Magnox and a sintered oxide (ceramic) for the PWR. This material is contained within a metallic can and gases including Kr85 and Xe133 accumulate in the interior spaces of this can (the 'clad-gap' inventory). Under operational conditions, small quantities of Cs and other radionuclides may also migrate from the fuel into the gap. If the element is punctured, these substances will be released – rapidly in the case of

PWR fuel for which the pins are pressurised, and more slowly for the unpressurised Magnox cans.

However, the clad-gap inventory constitutes only a tiny proportion of the total radioactivity in the consignment. On its own, it presents an extremely low risk. The other fission products are fixed within the solid fuel pellets themselves. Therefore, if we are to satisfy ourselves concerning the probability of accidents involving a significant fraction of the total hazard potential, we must postulate release mechanisms that volatilise or pulverise some or all of the original solid material. The chief mechanisms that have so far been considered in this context are external fire, impact and chemical explosion.

These mechanisms may act separately or in combination and the number of possible scenarios is large. For the purposes of the present discussion, we have considered two of the more serious of these in some detail, and two others in a relatively crude manner. These cases are serious impact followed by prolonged intense fire, deliberate attack using explosives, serious impact without fire, and intense fire without impact.

Unfortunately, very little published work exists on the behaviour of irradiated fuel elements under extreme conditions of these kinds; instead, the emphasis has been on the performance of the flask itself. The information that is required is an estimate of the fraction of the inventory of each species that would be dispersed in a form that can be carried by the wind (i.e. as a vapour or as a fine aerosol of, say, less than 10 microns maximum particle size). In the early days of this subject, release fractions were more or less guessed. Figures of the order of 1 per cent were often quoted, and even 10 per cent has been suggested. However, as often happens in this type of situation, the initial guesses tended to err on the side of caution and subsequent thought and analysis has tended to yield lower estimates. Unfortunately, we know of no published analysis of release fractions for transport of Magnox fuel and for this reason the remainder of the present analysis refers to PWR fuel only.

Serious Impact and Fire

The most thorough analysis of the release fraction for this case that we have identified in the public domain is that for PWR fuel by Elder (1981) who considered the available experimental data on five chemical and mechanical release mechanisms that could enable radionuclides to enter the transport flask cavity from the fuel pins:

TABLE 14.2 *Estimates of release fractions for severe impact and fire (PWR fuel)*

	Class of radionuclide					
	Noble gases (mainly Kr, Xe)	*Iodine*	*Caesium/ ruthenium*	*All other fission products*	*Actinides*	*Crud (Co60)*
Release fraction (after Elder, 1981)						
180 days cooling	0.3	0.1	4×10^{-4}	1.5×10^{-6}	1×10^{-6}	0.025
4 years cooling	0.3	0.1	3×10^{-4}	1×10^{-6}	1×10^{-6}	0.0025
Total release of activity (Ci)						22
150-day fuel (1 assembly)	1500	0	145	2.01	0.06	28
5-year fuel (7 assemblies)	6700	0	145	0.89	0.37	0

(1) clad-gap release (mostly Kr, Xe),
(2) vaporisation (mostly Cs),
(3) leaching (when flask coolant enters fuel pins),
(4) oxidation (when a large flask breach allows air to enter causing oxidation of UO_2 to U_3O_8, heating and particle formation),
(5) crud release; this is release of active corrosion products, etc., from the outside surface of the fuel pins.

After some detailed analysis of these effects in the context of eight different accident scenarios, Elder estimates release fractions for six classes of radionuclide. His results for the case of 'severe flask impact followed by two hours fire at 1010°C' are summarised in Table 14.2. The figures are not very sensitive to the cooling time of the fuel, and therefore we may assume that the figures for a 4-year cooled fuel apply equally well to the 5-year fuel proposed to be transported in the UK, and that the 180-day figure applies also to the 150-day fuel.

Applying these fractions to the inventories listed in Table 14.1, for the PWR fuel, we obtain the activity releases given in the last two lines of Table 14.2. In themselves, these figures are not direct indicators of biological impact because their physical behaviour and biological importance vary widely. For example, the noble gases, while accounting for most of the activity released, are relatively unimportant because they do not lodge in the lung when inhaled, nor do they accumulate preferentially in any part of the body, nor do they get deposited on the ground and thereby contribute to whole body irradiation. They therefore contribute insignificantly to late deaths. It also emerges, as shown by Clarke (1983), that they do not contribute significantly to early deaths either, in this particular case.

Such analyses show that caesium is the most significant nuclide for this type of event, although it should also be remembered that, since late fatal cancers are the dominant effect, the long-lived actinides which are mostly alpha emitters can have significant inputs when integrated over long time periods. For this particular case, Clarke's (1983) work indicates an expected number of late fatal cancers of 1.9 for this case when the event is assumed to have occurred at a particular location on the rail route for Sizewell B PWR fuel through Greater London.

Clarke also presents a range of outcomes depending on different weather conditions. The probabilities of exceeding different numbers of late fatal cancers, given that the release has occurred are shown in

Table 14.3. These values are later used (with results for other scenarios) to build up an approximate 'F–N' curve (see below).

TABLE 14.3 *Probability of late fatal cancers*

Probability	*Number of fatal cancers*
0.99	0.15
0.50	0.52
0.10	5.8
0.01	14.0
0.001	18.0

Deliberate Explosive Attack

As far as we are aware, the most thorough published analysis of the release fractions that could be attained in this case is that contained in the 'Draft environmental assessment of transportation of radionuclides in urban emissions' by Sandia National Laboratories (N. C. Finley *et al*., 1980). The details of their analysis were not reported for security reasons but it is stated that high explosive attack was analysed and that mechanisms of rupturing fuel pins and fragmenting the fuel pellets were evaluated. These researches estimated the release fractions for this scenario shown in Table 14.4. These fractions were

TABLE 14.4 *Release fractions*

	Noble gas	*Respirable aerosolised solids* (%)
Upper estimate	0.25	0.2
'Baseline'	0.1	0.07
Lower estimate	0.1	0.02

applied by Finley *et al.* to the case of a truck-mounted flask containing three PWR fuel assemblies cooled for 150 days and the resulting source terms were input to the consequence analysis model METRAN, which is used for much the same purposes as the NRPB's MARC programme described by Clarke (1983). The results for a number of releases assumed to occur in a 'typical hyperurban' location are set out in Table 14.5. Bearing in mind the inventory difference (three short-cooled assemblies compared with the seven long-cooled ones of the UK PWR flask), and the higher release fractions assumed for the more important species, these results seem to be wholly consistent with those of Clarke 1983 and confirm the general conclusion that the risk of early death is not very important compared with that of late fatal cancers.

TABLE 14.5 *Application of release fractions*

	Early deaths	*Late fatal cancers*
Upper estimate	2.8	104
'Baseline'	<1	40
Lower estimate	<<1	12

Lesser accidents

In order to complete the risk picture, we have considered two lesser accidental scenarios, severe impact alone and severe fire alone. Since the consequences of the combined event 'impact with fire' are quite small, we have simply scaled down the late fatal cancers results from that case, using approximate release fractions. This results in about 0.1 expected value for late deaths for the serious impact and 0.01 for the fire.

All of these results presuppose that the event in question has already occurred. We now turn to the estimation of the probability of these events.

ACCIDENT PROBABILITY

In order to obtain the probabilities of accidents it is necessary to define the various scenarios.

Serious impact

The following figures are based on the proposed Sizewell B PWR shipments only. There are three main possibilities to consider in this category:

Derailment of the flask vehicle, which then hits an unyielding obstacle. The rate of bogie vehicle derailments on the British Rail network is about 2.4×10^{-7} per bogie vehicle mile, giving about 1×10^{-3} per year for the projected Sizewell B flask traffic. The fraction of journey at risk due to the presence of relevant solid obstacles (bridge, tunnels), is about 10^{-2}, a figure which has been worked out by CEGB and presented in their Sizewell evidence, and which we accept. We have allowed for some further attenuating factors to cover such factors as impact speed, energy partitioning between flask and obstacle, and point of impact, amounting to about 0.075 in all. Therefore, our final risk estimate for serious impact after derailment is 7.5×10^{-7} per year. This accident may occur at any point along the 395-mile route and may be more likely to occur in rural areas than in urban ones due to higher speeds, although we have not taken this factor into account.

Derailment of the flask vehicle, with subsequent impact by another train. The probability of derailment is as stated above. Following a similar line of argument, we derive the coincidentally identical figures of 7.5×10^{-7}/year for this case.

Collision of the flask vehicle with another train. There were twenty-eight collisions with other trains in 1981 (Railway Inspectorate), of which twenty-three concerned freight vehicles. This gives a collision probability of 3.81×10^{-4}/year and 1.5×10^{-3}/year respectively for all trains and freight only. Multiplying these by 0.3 for the speed factor (Taig, 1980) and assuming a probability of 0.01 that serious damage will occur to the flask vehicle itself, this gives frequency of about 1 to 5×10^{-6}/year (3×10^{-6} used in subsequent analysis).

The total probabilities for all scenarios considered here for serious impact is therefore 4.5×10^{-6}/year.

Fire

The average number of fires on BR freight trains is about 44/year, giving a probability of 1.71×10^{-3}/year for the Sizewell B traffic. However, most of these fires would be very limited in severity and duration, and so some attenuating factors must be considered.

The probability that the flask wagon itself will be affected is 0.1 (one wagon out of ten say). The probability that the fire will be serious enough to damage the flask is very low since there would be little combustible material available. A figure of 0.01 is assumed, to allow for locomotive fuel or presence of unauthorised material. Therefore the probability is (rather conservatively) estimated to be 1.7×10^{-6}/year.

There are other possible fire scenarios such as that of a train colliding with another train or storage tank carrying flammable material. Both of these events have probability components associated with the scenarios considered under serious impact, with further reducing factors. It is therefore difficult to foresee probabilities higher than that quoted above for fire without prior impact.

Sequential impact accident and fire

Each of the serious impact cases has been considered with regard to the probability of a subsequent serious fire. The probabilities of fire vary from case to case depending on the type of other vehicle that may be involved. The final total for this case is estimated to be 8.3×10^{-8} per year. These estimates were made on a 'best estimate' basis, meaning that no conscious pessimism or optimism has been introduced.

The results are very low because the scenarios considered necessarily involve combination of a number of adverse circumstances before releases of radionuclides at the rates postulated above would occur.

Sabotage

The expected frequency of deliberate attack has not been estimated in this paper because it is too much subject to behavioural aspects that are not amenable to analysis by purely 'technical' means. Instead, we proceed by assembling an 'F–N' curve for the accident case alone, and then will indicate what maximum frequency of successful sabotage attack is allowable if the overall risks are to be kept within acceptable limits.

OVERALL ASSESSMENT OF RISK

For the reasons I gave at the outset, it is societal risk rather than individual risk that is most critical for transportation activities. For this reason, we have chosen to present our results in 'F–N' curve form. This is necessarily approximate because the analysis presented in this paper has involved several simplifications in order to classify the argument. The resulting figures must therefore be regarded as preliminary.

The results for accident cases only have been plotted in the 'F–N' plane in Figure 14.1. Also shown on this figure are the criterion lines

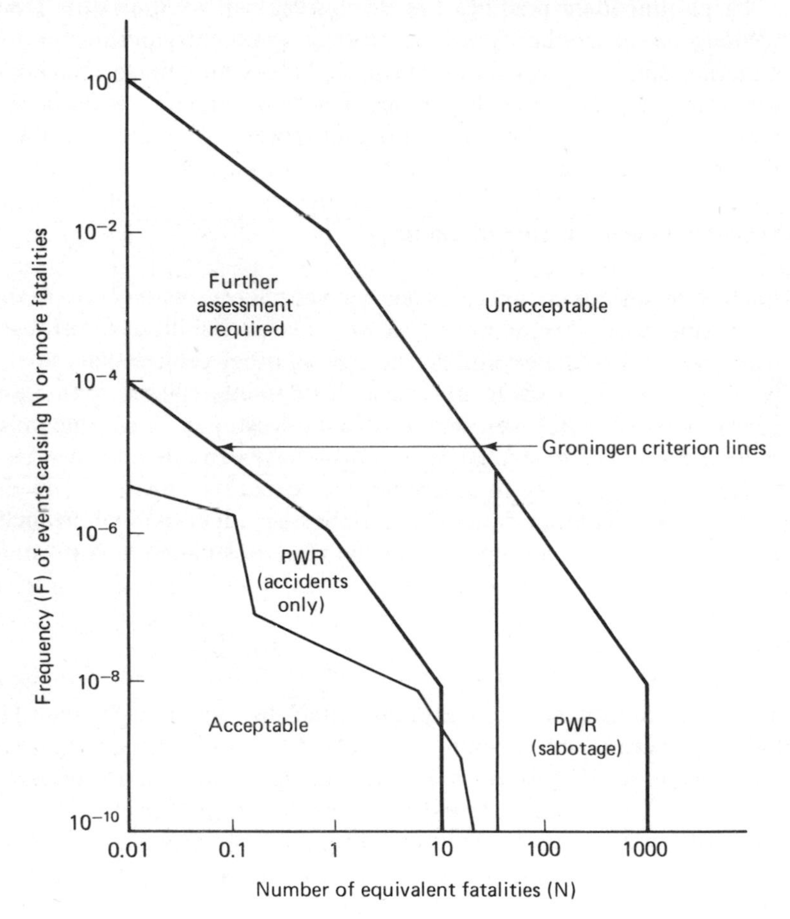

FIGURE 14.1 *Estimated accident risks to external population due to projected UK PWR irradiated fuel traffic, plotted in the F–N plane, and compared with Groningen criteria*

published by the Provinciale Waterstaat Groningen, Netherlands (1979). These criterion were derived from consideration of the natural background of multiple fatality accidents and the middle of the central band lies about two orders of magnitude below this background. The 'acceptable' line is regarded by most risk analysts as being extremely stringent; in practice, we have found that most 'major hazard' installations which we have analysed to date have risks which are estimated to be in the upper part of the central band 'further assessment required.'

The plotting of the results for Sizewell B fuel takes no account of the fact that the consequences have been estimated for urban sites for the release, while the probabilities relate to the entire route, much of which is rural. This therefore introduces further pessimism in the results, although because late deaths are dominant, this effect may not be very great.

By this yardstick therefore we have provisionally concluded that the risks to the external public from accidents to the Sizewell B irradiated fuel traffic are sufficiently low as to be of no concern. (It must be noted that risks to emergency response personnel have not been considered in this analysis.)

It remains only to mention the sabotage aspect, which is not included in this graph. The Sandia METRAN analysis discussed above was carried out for the case of sabotage attack on a truck-mounted flask carrying three assemblies of 150-day cooled fuel. By scaling on the dominant caesium content, we have concluded that the impact in an urban area would be very similar to that for the seven-assembly, 5-year cooled fuel appropriate to Sizewell, and rather more than that for one assembly of PIE fuel. Therefore, this event will generate fatalities in the approximate range ten to a hundred. If the risk of this event is to be within even the 'further assessment' band of the Groningen criterion, then the probability must be reduced to below about ten per year. It is a matter for future study to examine whether this performance can be demonstrated.

REFERENCES

Clarke, R. H. (1983) 'Assessing the consequences of accidental release of activity during irradiated fuel transport', Int. Conf. on the Urban Transportation of Irradiated Fuel, Connaught Rooms, London (12–14 April 1983).

Elder, H. K. *et al.* (1981) 'An analysis of the risk of transporting spent nuclear fuel by train', Battelle Pacific Northwest Laboratories, Rep. PNL-2682.

Finley, N. C. *et al.* (1980) 'Transportation of radionuclides in urban environs Draft Environmental Assessment', Sandia National Laboratories for US NRC, NUREG/CR-0743.

Provinciale Waterstaat Groningen (1979) 'Criteria for risks related to dangerous goods', in 'Pollution control and the use of norms in Groningen', PWG, Groningen (April 1979).

Taig, A. R. (1980) 'Radioactive materials packages and the British Railway accident environment', 6th Int. Symp. on Packaging and Transportation of Radioactive Materials, Berlin (10–14 November 1980).

15 Potential Consequences of Accidents

ROGER CLARKE

The National Radiological Protection Board (NRPB) is an independent statutory body which was created by the 1970 Radiological Protection Act. Its responsibilities are to advance knowledge about the protection of mankind from radiation hazards and to provide information and advice to individuals and organisations in the UK with responsibilities relating to protection against radiation hazards. The NRPB coordinates and represents UK interests in radiation protection internationally; in the UK, it advises on the adoption of international standards and defines emergency criteria. It is an advisory body as opposed to a regulatory one. It conducts research, provides services to industry and makes assessments of situations involving exposure of the public both through normal operations and under accident conditions. The present paper is concerned mainly with its assessments methods.

INTRODUCTION

Assessment of the radiological consequences of accidental releases of radionuclides provides an important input to development of design criteria, siting philosophy and emergency planning arrangements for nuclear installations. Since the probability of an accident occurring is very low, assessment of those potential accident sequences foreseen by designers has to be undertaken using theoretical modelling procedures. NRPB has been involved in accident assessment studies for many years and has published the results of a number of studies.[1] The procedure adopted for estimating the radiological consequences of hypothetical accidental releases consists of a series of models interlinked in a suite of computer programs known as MARC – Methodology for Assessing Radiological Consequences.[2] The methods enable predictions to be made of the transfer of radionuclides, which have been released to the atmosphere, through the environment to man. The MARC suite can be used to analyse the health and other consequences of hypothetical accidents anywhere within the United Kingdom. Support for the development of MARC has come from several Government Departments and particularly the Health and Safety Executive's Nuclear Installations Inspectorate for whom reports on specific topics have been produced.

NRPB has applied these programs, under contract, to releases specified by the Central Electricity Generating Board to assess the health-related consequences of degraded core accidents in the PWR proposed for Sizewell and the results have been published.[3] A contract was placed with the Board by Friends of the Earth Trust Limited to assess the dosimetric and health consequences of specific accident sequences in particular meteorological conditions for the proposed Sizewell PWR. Further, the Board has undertaken contract work for the Greater London Council (GLC) on the health consequences to the population within the GLC area of accidents at the proposed PWR, and on transport flask incidents which might occur within the GLC area.[4]

Irradiated fuel transport flasks are used for the movement of irradiated nuclear fuel by road and rail in the United Kingdom. These flasks are subject to international and national regulations and to procedures specified by the carrier or consignor. Despite good safety records, in order to assist with design criteria and planning emergency response arrangements, it is useful to evaluate the radiological consequences of postulated transport accidents.

The probability of a release, its magnitude and location, are crucial to the estimation of consequential radiological risks. For the purposes of this paper a hypothetical release is assumed to occur in the Greater London area and the radiological consequences have then been evaluated using MARC.

CONSEQUENCES OF RELEASES OF RADIOACTIVE MATERIAL

Following an accidental release of radionuclides to the atmosphere, the radionuclides would be dispersed from the release location in a spreading plume, or cloud, in a manner determined by the characteristics of the release and the prevailing weather conditions. The potential consequences for public health from an accidental release would also depend on the distribution of the population over which the plume dispersed and on the effectiveness of any countermeasures which may be taken. Direct exposure to a passing plume would cause external irradiation of the whole body and internal irradiation following inhalation of radionuclides from the plume. The dispersing radionuclides would be diluted in the atmosphere but some would become deposited on to land, roads and buildings, thus leading to various other routes of exposure. These would include external exposure from deposited radionuclides and internal exposure following consumption of foodstuffs contaminated directly or derived from contaminated areas.

In general the impact of an accidental release of radionuclides will be expressed in terms of the health detriment to the exposed population and the effects of any countermeasures invoked to control the further exposure of the population.

Before proceeding with a description of the procedure to assess the consequences of accidental releases, it is useful to review briefly the biological effects of irradiation and the basis for estimating risk coefficients for use in radiological consequence modelling.

Radiation effects

If the whole body is exposed to a very high dose of radiation death may occur in a matter of weeks. An instantaneous dose of about 4 Gy (400 rads) would be fatal for some 50 per cent of those irradiated. Higher doses will be tolerated by individual organs or tissues if irradiated in isolation, doses of several tens of Gy (several thousand rads) commonly being used in radiotherapy to kill malignant cells. If the same total doses are received over a longer time, there may be no signs of early injury and the damage which has occurred may be manifested later in the irradiated individual or in his descendants. These are known as late effects.

The most important late effect appears to be the induction of cancer in a small proportion of those irradiated. The fundamental processes by which cancer is induced by radiation are not fully understood, but a great incidence of various malignant diseases, i.e. cancers, has been observed in a number of human population groups who had been exposed a number of years previously to various high doses of radiation. These groups include, among others, the Japanese survivors of Hiroshima and Nagasaki; patients treated by X-ray irradiation for calcification of the spine; children irradiated to treat ringworm in the scalp; pelvic irradiation of women for a variety of reasons; workers who, in the course of their occupation, ingested quantities of the naturally occurring radionuclide, radium.

Few people so exposed contract cancer, but each person has a probability of contracting it that depends on the level of radiation dose received. The situation may be expressed as a mathematical probability. If the number of people in an irradiated group and the doses they have received are known, and if the number of cancers eventually observed in that group exceeds the number that would be expected in a similar non-irradiated group, the excess number of cancers may be attributed to radiation and the risk factor can be calculated. If, for example, 50,000 people were each to receive 2 Sv (200 rem) to a particular organ, and if 100 more cancers of that organ were to appear in that group than in a similar but unexposed group, the risk factor would be 100 divided by (50,000 $\times$ 2), i.e. 1 in 1000 or 10^{-3} Sv^{-1}.

The information on human risks from radiation is reviewed periodically by the United Nations Scientific Committee on the Effects of Atomic Radiation (UNSCEAR) which was established in 1955 and reports to the United Nations General Assembly. In the US the National Academy of Sciences has a Committee on the Biological Effects of Ionising Radiation (BEIR) which also reports on risk factors.

Not all cancers are fatal. The total risk of inducing cancer by uniform whole body irradiation is about three times the risk of inducing a fatal cancer. Because of its overwhelming significance, risk of fatal cancer is of prime concern in radiological protection. The risk factors for fatal cancer following uniform whole body irradiation are about 10^{-2} Sv^{-1} (10^{-4} rem^{-1}), the cancers appearing over several decades. This means that if 10^4 people were each to receive a whole body dose of 1 Sv (10 rem) there would only be 10 excess fatal cancers attributable to the irradiation while the natural incidence of all forms of cancer would suggest that 2000 people in that population group would die of cancer.

Another important late effect of radiation is hereditary damage, the probability of which depends on the dose received. The damage arises following irradiation of the gonads – involving sperm cells in males and egg cells in females. The exact processes by which mutations occur are not known, but the effects can range from minor skin blemishes to serious physical or mental handicaps. No conclusive evidence for hereditary effects attributable to natural or artificial radiation has been found amongst human offspring. Resort has to be made to animal studies, mainly mice exposed to ionizing radiation. The average risk of serious hereditary damage in humans, inferred from animal studies is about 10^{-2} Sv^{-1} (10^{-4} rem^{-1}) in the first two generations and twice this in all future generations. If allowance is made for the fact that only a fraction of the population is still to produce offspring at any given time, the above figure reduces to about 4×10^{-3} Sv^{-1} (4×10^{-5} rem^{-1}). The incidence is comparable with the figures for fatal cancer above.

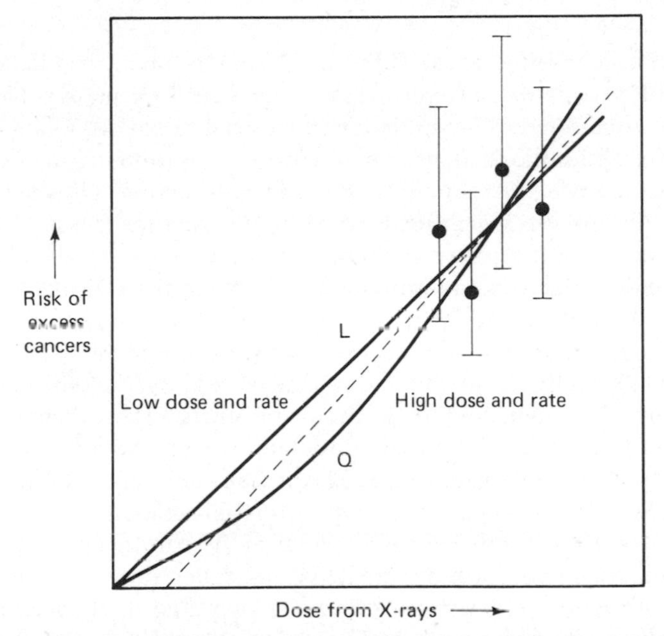

FIGURE 15.1 *Estimating risk at low doses from data at high doses*

The risk factors described above are determined, as far as cancers are concerned, from data derived at relatively high radiation doses delivered in short periods. Such situations are not normally encountered except in severe accidents. People are usually exposed at lower doses in normal operation and many hypothetical accident situations led to the majority of people being irradiated at levels of dose or dose rate several orders of magnitude lower than those from which the risk data were derived. This is illustrated in Figure 15.1, which shows that for radiological protection purposes, it is assumed that all doses no matter how small lead to some risk, i.e. there is no threshold, and the risk factor is the slope of the straight line. This means that the assumption is made that if 10^4 people receive 10^{-1} Sv (10 rem), with a resulting excess of 10 cancers, the same excess of 10 will be seen if 10^6 people receive 10^{-3} Sv (10^{-1} rem). This leads to the concept of collective risk or total health detriment for late effects in the whole population, regardless of the actual distribution of doses amongst individuals. It is a simplification that may overestimate the true consequences.

Radiation of natural origin

Radiation naturally pervades the whole environment irradiating all people to a greater or lesser extent. There are four main sources of natural radiation: cosmic radiation, terrestrial radiation, radon decay products, and radionuclides in diet. Cosmic radiation originates from outside the earth's atmosphere and it is the interaction of this radiation in the atmosphere which leads to external radiation doses. As they penetrate the atmosphere some of these radiations are absorbed so that the dose decreases as altitude decreases. In the UK, the average dose is about 0.31 mSv (31 mrem) per year from cosmic radiation.

Terrestrial radiation arises because all materials in the earth's crust contain naturally occurring radionuclides mainly on the elements Uranium, Thorium, and their decay products. These can lead to external y-ray exposure of the population from the activity in the ground and in building materials. The average dose in the UK is about 0.38 mSv (38 mrem) per year from terrestrial radiation.

One of the radioactive decay products of the natural uranium series is radium which produces a radioactive gas, called radon. When radon enters a building either from the floors or walls, the concentration builds up because it is unable to disperse as it would in the atmosphere. The immediate decay products of radon are solid radionuclide particles which attach themselves to dust particles in the air. When they are inhaled they irradiate the lung. The annual dose from radon products is equivalent to about 0.8 mSv (80 mrem) dose to the whole body, and variations around the UK will often be ten times higher than this.

Radionuclides in diet arise mainly because of potassium and uranium decay products in foodstuffs. The average dose from UK diet is about 0.37 mSv (37 mrem) per year.

There is thus an average total of 1.86 mSv (186 mrem) per year received by each member of the UK population from radiation arising from natural sources. To this can be added an average total body dose of 0.5 mSv (50 mrem) per year from medical procedures in the UK and about 0.01 mSv (1 mrem) from fall-out from atmospheric testing of nuclear weapons, both of these sources being artificially produced radionuclides or radiation.

METHODOLOGY

MARC has a modular structure consisting of a series of interlinked models which are used to predict the transfer of radionuclides, released in an accident, through the environment to man, the subsequent dose distributions and health consequences in the population, and the impact of restrictions which may be placed on the utilisation of contaminated land. Modelling this transfer requires an understanding of atmospheric dispersion, the processes of removal from the atmosphere leading to deposition on the ground and the subsequent behaviour in the terrestrial environment.

A variety of radiological endpoints can be calculated to provide a measure of the overall impact of an accidental release. Those which are considered to be most useful in the context of risk analysis and in analyses of siting, emergency planning and design criteria can be summarised as:

(1) the incidence of particular health effects in the exposed population and its descendants;
(2) the number of people who may be affected by the application of countermeasures;
(3) the area of land contaminated above particular levels, especially areas where restrictions may need to be placed on its use or on foodstuffs derived from it;
(4) the economic cost of the introduction of countermeasures, particularly restriction on the use of contaminated land and property.

Endpoints (2) and (3) are clearly necessary for the calculation of economic costs; however, in themselves they provide a useful measure of the overall impact. An example of the results using these endpoints is given below.

The major countermeasures which can be introduced to protect the population after an accidental release are: sheltering, evacuation, banning of foodstuffs, closing of areas and, where appropriate, administration of stable iodine tablets. In an irradiated fuel transport accident the radioactive iodine content and the distribution of stable iodine tablets would not be warranted. The decision to introduce a given countermeasure should in general be made on the basis of reducing levels of individual risk.

A schematic diagram of the structure of MARC is shown in Figure 15.2, which identifies the general calculational procedure and the various endpoints. Within each module a number of models may be incorporated, the choice of appropriate model in any study depending on the particular application. The essential input to MARC is a description of the release of radionuclides. It must include such parameters as the quantity of each nuclide released and its physico-chemical form, together with the duration and height of release and the heat input associated with that release. These data are commonly referred to as the 'source term' and are input to the atmospheric dispersion module.

If an accidental release actually occurs estimates can be made of atmospheric dispersion, and of associated doses to the particular population group affected. However, in assessing potential consequences of accidents that have yet to occur there is no way of defining a unique population at risk.

The location of the release, the wind direction and the prevailing meteorological conditions affect the consequences of a given accidental release. For any location there will be a probability distribution of both wind direction and meteorological conditions: the consequence of any postulated accidental release has therefore to be described as a probability distribution. Within MARC each of the effects is evaluated as a probability distribution and the following characteristics specified: mean or average, median, and various percentiles. These probabilities are conditional upon the accident occurring. The total risk to any given individual or the societal risk is given by the product of the frequency with which the incident is predicted to occur and the probability of harm conditional upon the release. In terms of acceptable risk to the individual, or society, it is essential that account is taken of the frequency with which the event is predicted to occur.

A typical example of a distribution of consequence is shown in Figure 15.3. This is an arbitrary set of results shown in different formats to highlight different facets of the results. The consequences could be any of those identified and are shown as probabilities assuming the accident has occurred. In an integral presentation, as

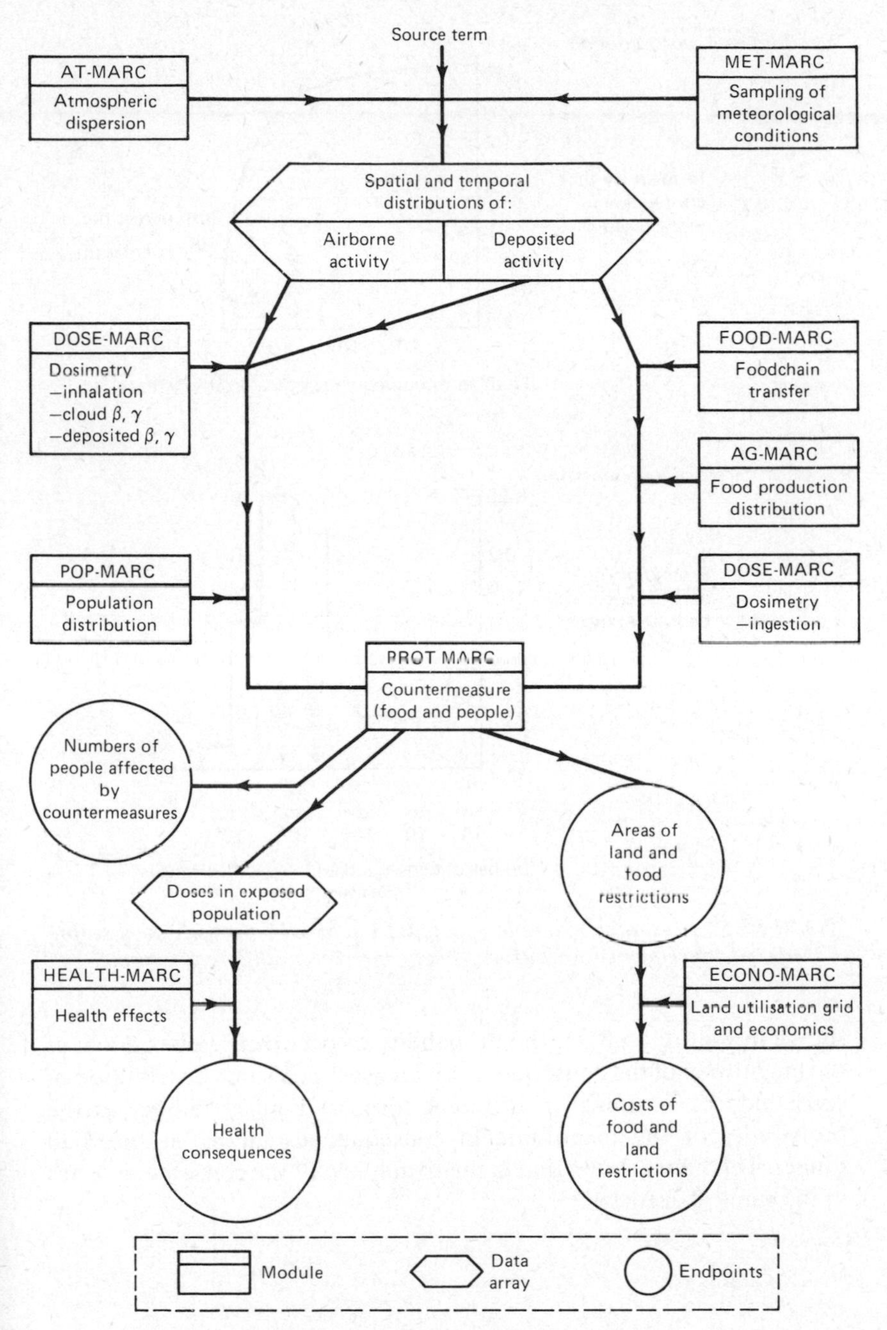

FIGURE 15.2 *The structure of MARC*

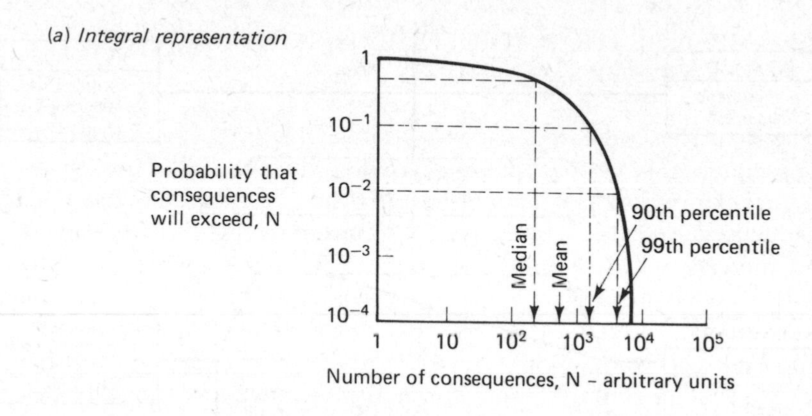

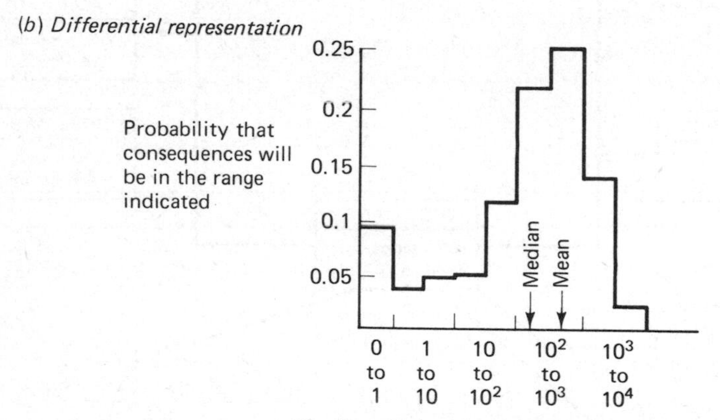

FIGURE 15.3 *Illustrative example of output from MARC showing the distribution of consequences (arbitrary units) expressed in different formats*

shown in Figure 15.3(*a*), the probability of occurrence goes down as the magnitude of the consequences increases. The same distribution of consequences is shown in different form in Figure 15.3(*b*) as the probability of the magnitude of consequences falling in specified ranges. The figure shows that in this example half the consequences are in the range 10^2 to 10^3.

EXAMPLE OF THE RADIOLOGICAL CONSEQUENCES OF A TRANSPORT ACCIDENT

The Board has evaluated, under contract to the GLC, the radiological consequences of various transport accident scenarios involving irradiated fuel transport flasks moved by rail in the London area.[5] One of the sequences involved a severe impact followed by a fire. The assumed location and release parameters for this sequence were specified by the GLC and the radiological impact was evaluated for the GL population assuming, at the request of the GLC, that no countermeasures were invoked. This work has now been extended by the Board to cover the effects to the whole UK population and to include those countermeasures which are appropriate. The source terms and release characteristics are given in Table 15.1. No consideration has been given to the frequency with which such an accident may occur and the results presented are conditional upon the occurrence of each release. In

TABLE 15.1 *Characteristics of release*

Conditions	*Source definition*	*Height of release*	*Duration of release*
Severe impact, 2-hr fire then release	Point	2 m	1.5 hrs

TABLE 15.2 *Release fractions for the postulated accidental releases*

Fractional release of airborne radioactivity				
Noble gases	*Caesium/ Ruthenium*	*All other fission products*	*Actinides*	*Cobalt*
0.3	$3.0\ 10^{-4}$	$1.0\ 10^{-6}$	$1.0\ 10^{-6}$	$2.5\ 10^{-3}$

NOTES The specified release fractions of the flask inventory are assumed to be released uniformly over the specified release duration (see Table 15.1).

All elements except the noble gases are assumed to be released as a 1 μm AMAD aerosol with each element in an oxide form.

SOURCE G. N. Kelly, C. R. Hemming, D. Charles, J. A. Jones, L. Ferguson, S. M. Haywood and R. H. Clarke, 'Degraded core accidents for the Sizewell PWR: a sensitivity analysis of the radiological consequences', NRPB-R142 (1982) (London: HMSO).

assessing the absolute significance of the results, account must be taken of the frequency with which such releases are predicted to occur.

The characteristics of this release which was specified by the GLC were taken largely from Elder.[6] The transport flask is assumed to suffer a severe impact followed by a 2-hour fire at about 1000°C. The release is assumed to occur at an effective height of 2 metres and for a duration of 1.5 hours. The predicted fractions of radionuclides released from the flask to the atmosphere in a respirable form are given in Table 15.2. It was assumed that the radionuclides were released uniformly over the specified release duration. The noble

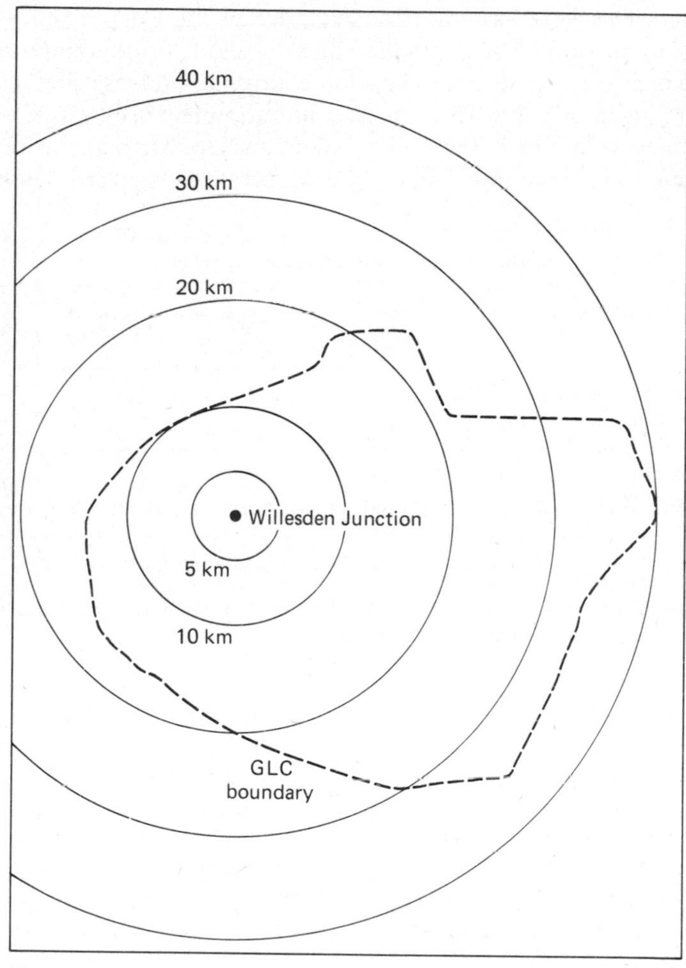

FIGURE 15.4 *Location of release with respect to GLC boundary*

gases and more volatile elements may be preferentially released during the early stages of the release; the simplifying assumption of uniform release will, however, have little significance for the predicted consequences.

The location of the postulated release was specified by the GLC for their contract to be a marshalling yard at Willesden Junction (grid reference TQ193842). The transport flask was assumed to be stationary in the yard on a siding. This marshalling yard is on the west side of London and its location relative to the GLC area is shown in Figure 15.4. The same location was used for this duty.

The frequency distribution of meteorological conditions attributed to the release location was obtained from some 35 years of hourly measurements recorded at a meteorological station at Heathrow Airport, which is about 15 kilometres from Willesden Junction.

The UK population with respect to the assumed release location at Willesden Junction was evaluated from a compilation of the UK population on 1 km square grid,[7] derived mainly from the 1971 Census. The population distribution was evaluated as a function of distance direction from the assumed release location, assuming the population density to be uniform within each grid square. A simplified approach was adopted to modify the derived population distribution over distances up to 0.5 kilometre from the release location. Examination of the Willesden Junction area showed that it was reasonable to assume no one lived within 100 metres of the assumed released location. Some additional areas were also assumed to be unpopulated based on an inspection of the general layout of the marshalling yard. Details of these assumptions are shown in Figure 15.5.

The methodology adopted is essentially that used in earlier studies of the radiological impact of these, and other, releases on the UK population;[8] to the extent practicable the models and parameter values used in this and the earlier studies were kept the same to enable the respective results to be compared directly. The methodology has been described fully elsewhere[9] and more detailed consideration is given only to those facets of the methodology where changes were made from that previously adopted.

Countermeasures were assumed to be introduced in line with the advice given by the Board in ERL2.[10] However, for the accident sequence analysed here, the magnitude of individual doses and timescales were such that no countermeasures were invoked. Since the population of the entire United Kingdom is now included in this work, transfer of radionuclides through foodchains has been included.

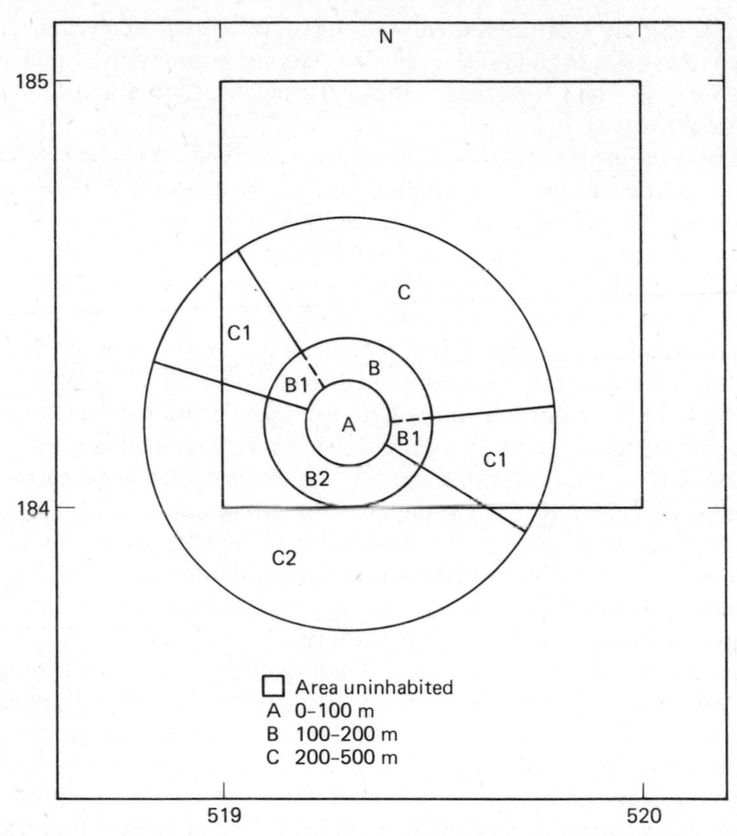

NOTE The population in area A was added to the populations of areas B, B1 and B2. The populations of areas B1 were summed to the population of B2 while the population of area B was distributed in area C. The population of areas C1 were distributed in areas C and C2.

FIGURE 15.5 *Redistribution of population in vicinity of the release location*

Characteristic quantities of the probability distributions of health effects in the UK population, conditional upon the release are summarised in Table 15.3. The probability distribution of fatal cancers is also illustrated in Figure 15.6.

The magnitude of the release is such that the predicted doses are far below the thresholds for the incidence of early injuries. The health impact of the release is therefore limited to late effects, of which cancer in the exposed population and hereditary effects in their descendants are the most important. The predicted incidence of these effects

TABLE 15.3 *Characteristic quantities of the distributions of consequences conditional upon the occurrence of the release*[a]

Health effect	*Number of health effects, N*						*% Probability*	
	Expectation value, E	*Value at the pth percentile*					$P(N = 0)$	$P(N>E)$
		$p = 1$	$p = 50$	$p = 90$	$p = 99$	$p = 99.9$		
Early								
Death	0	0	0	0	0	0	100	0
Prodromal vomiting	0							
Lung morbidity	0	0	0	0	0	0	100	0
Late								
Fatal cancer	$1.9\ 10^{0}$	$1.5\ 10^{-1}$	$5.2\ 10^{-1}$	$5.8\ 10^{0}$	$1.4\ 10^{1}$	$1.8\ 10^{1}$	0	24
Non-fatal thyroid cancer	$1.3\ 10^{0}$	$1.1\ 10^{-1}$	$3.7\ 10^{-1}$	$4.3\ 10^{0}$	$1.0\ 10^{1}$	$1.4\ 10^{1}$	0	23
Non-fatal skin cancer	$1.6\ 10^{0}$	$1.2\ 10^{-1}$	$4.3\ 10^{-1}$	$5.2\ 10^{0}$	$1.3\ 10^{1}$	$1.7\ 10^{1}$	0	23
Non-fatal breast cancer	$3.3\ 10^{-1}$	$2.6\ 10^{-2}$	$9.1\ 10^{-2}$	$1.1\ 10^{0}$	$2.6\ 10^{0}$	$3.4\ 10^{0}$	0	23
Hereditary effects	$1.7\ 10^{0}$	$1.5\ 10^{-1}$	$5.0\ 10^{-1}$	$5.5\ 10^{0}$	$1.2\ 10^{1}$	$1.6\ 10^{1}$	0	24

NOTE [a] The numbers of health effects in the table are conditional upon the occurrence of the release. The frequency with which any given number of health effects may be exceeded can be obtained as f (100-p)/100 where f is the frequency of occurrence of the release (y^{-1}) and p is the percentile of the distribution.

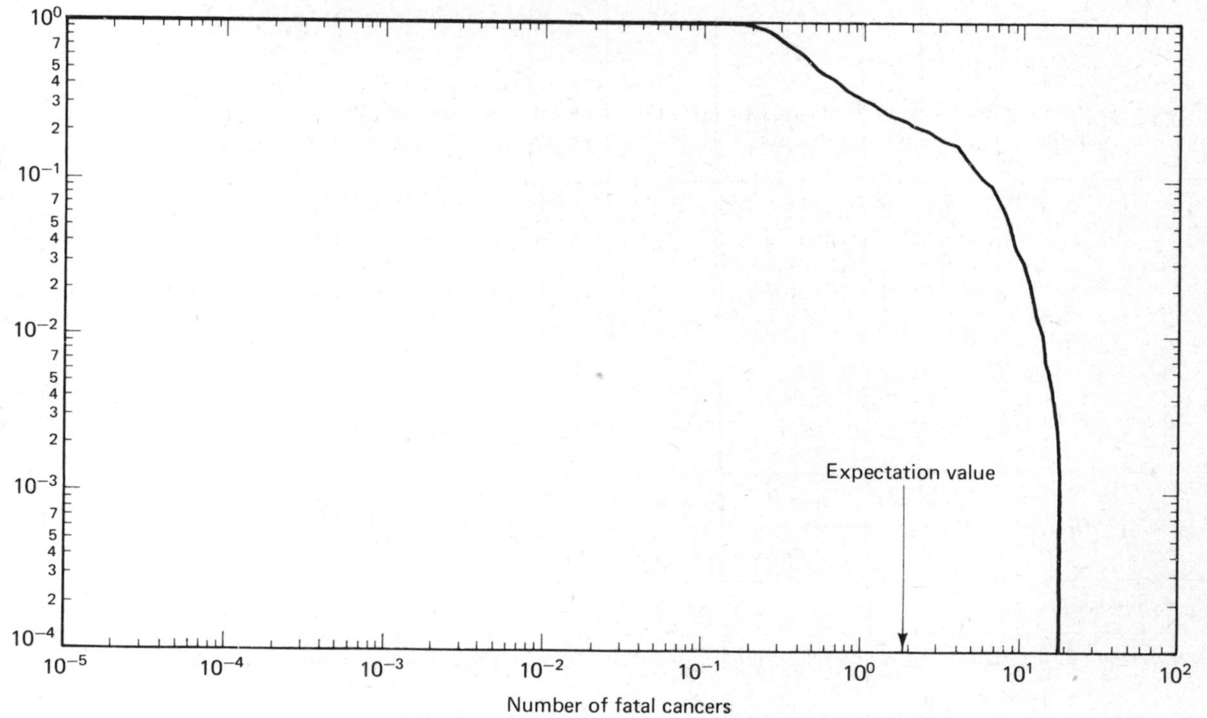

FIGURE 15.6 *Probability distribution for fatal cancers in the UK conditional upon the release*

conditional upon the release, is extremely small, the average being about two and even at a probability of one in a thousand, consequent upon the release, the number of fatal cancers is eighteen in a population of 55 million.

The individual risk of fatal cancer, conditional upon release, is given as a function of distance in Table 15.4. The individual risk is an average risk to an individual within a given distance band from the release location. It is a value calculated taking account of all meteorological conditions assuming that the release has occurred. The risks at the specified distances have been averaged over all directions and over the spectrum of meteorological conditions. In any particular direction the individual risk may differ from the average value, but in general only by a small factor. Further, the risks apply to an average individual in the population. In practice, there will be a distribution of individual risk at any distance. Factors such as age, sex, health and housing will have an effect on an individual's risk. Risk factors close to the release

TABLE 15.4 *Average individual risk of fatal cancer, conditional upon the release as a function of distance*

Distance from the release (km)	*Individual risk*
0.1	2.5 10^{-4}
0.2	8.2 10^{-5}
0.5	1.9 10^{-5}
1.0	6.2 10^{-6}
2.0	2.1 10^{-6}
5.0	4.7 10^{-7}
7.5	2.4 10^{-7}
10.0	1.5 10^{-7}
15.0	8.0 10^{-8}
20.0	5.0 10^{-8}
35.0	2.0 10^{-8}
50.0	1.2 10^{-8}
75.0	5.0 10^{-9}
100.0	2.9 10^{-9}
150.0	1.4 10^{-9}
200.0	9.0 10^{-10}
350.0	4.0 10^{-10}
500.0	2.3 10^{-10}
750.0	1.2 10^{-10}

NOTE The risks presented are conditional upon the occurrence of the release. The actual risk can be obtained as the product of these conditional risks and the frequency of occurrence of the release.

point are about $2.5\ 10^{-4}$, about one-thousandth of the natural risk of cancer.

It is of some interest to note that for the closest individuals the same risk arises from some 5 years exposure to natural background radiation and it must also be remembered that the accident risk is received only if the accident actually occurs.

CONCLUSIONS

Radiological analysis of a severe transport flask accident followed by fire indicates that the consequences in terms of both the risk to individuals and the total health risk to the population as a whole are very small. The people closest to the severe accident would have a risk equal to no more than a few years' worth of natural background radiation; at a greater distance, say 10 kilometres, the radiation dose received is roughly equivalent to that derived from one trans-Atlantic jet flight. This severe accident is therefore of no great radiological concern.

NOTES

1. G. N. Kelly *et al.*, 'An estimate of the radiological consequences of notional accidental releases of radioactivity from a fast breeder reactor', NRPB-R53 (1977). M. J. Crick and G. S. Linsley, 'An assessment of the radiological impact of the Windscale reactor fire, October 1957', NRPB-R135 (1982).
2. R. H. Clarke and G. N. Kelly, 'MARC – the NRPB methodology for assessing radiological consequences of accidental releases of activity', NRPB-R127 (1981).
3. G. N. Kelly and R. H. Clarke, 'An assessment of the radiological consequences of releases from degraded core accidents for the Sizewell PWR', Chilton, NRPB-R137 (1982) (London: HMSO); G. N. Kelly, C. R. Hemming, D. Charles, J. A. Jones, L. Ferguson, S. M. Haywood and R. H. Clarke, 'Degraded core accidents for the Sizewell PWR: a sensitivity analysis of the radiological consequences', NRPB-R142 (1982) (London: HMSO).
4. K. B. Shaw, J. H. Mairs, D. Charles and G. N. Kelly, 'The radiological consequences of postulated accidental releases of radioactivity during the transportation of irradiated fuel through Greater London', NRPB-R147 (1983).
5. See note 4.
6. H. Elder, 'An analysis of the risk of transporting spent nuclear fuel by train', Battelle Memorial Institute, USA, PN1-2682 (1981).

7. M. Broomfield, J. R. Simmonds and T. Chapman, 'POP-MARC and AG-MARC: population and agricultural distributions for use in the methodology for assessing the radiological consequences of accidental releases', Chilton, NRPB-M75 (1982).
8. See note 3.
9. In Kelly and Clarke, 'Radiological consequences'.
10. NRPB, 'Emergency Reference Levels: criteria of limiting doses to the public in the event of accidental exposure to radiation', NRPB ERL2 (1981) (London: HMSO).

16 Radiological Protection in Transport

ROBERT BARKER

Since 1959 the International Atomic Energy Agency (IAEA) has been the leader in establishing international regulations for radiation safety in transportation of all radioactive materials. I will discuss the safety requirements in the Agency's regulations for the transport of irradiated fuel and the procedures by which they are made applicable to such shipments. I will also review the record of safety in transport and some predictions of assessments in support of the conclusion that the safety measures being taken in transport of irradiated fuel are adequate.

TRANSPORTATION

Radioactive materials in transport are of many different types, forms and quantities and present a wide range of hazards. These hazards arise from two characteristics especially associated with radioactive materials: emission of ionizing radiation and, for certain types, the ability to sustain a nuclear chain reaction.

Most radioactive materials are transported in routine commerce by regular means. In most cases, packages of radioactive material are handled by carrier personnel who are not monitored or trained as radiation workers and who have no knowledge of radiation protection practices. The transport environment is predictable but uncontrolled.

Therefore radiation protection in transport is achieved primarily by requirements imposed by the shipper in packaging his materials for shipment and requiring the carrier to exercise only simple controls over the packages submitted to him. In principle, the safety standards require protective packaging and simple transport controls so radiation adds very little to the hazards normally associated with transportation.

Millions of shipments of radioactive materials have been carried safely with only a very few incidents involving releases or excessive radiation exposure. None are known to have caused serious injury from radiation. Nevertheless, each incident that does occur in transportation is considered by some to indicate a lack of control. However, the number of shipments is increasing steadily. More people are becoming involved. Inspite of every practical and reasonable precaution being taken, the occurrence of some incidents is unavoidable – absolute safety in any operation is not possible. As the number of shipments increase, the number of incidents may be expected to increase.

The Agency's Regulations for the Safe Transport of Radioactive Material, Safety Series No. 6,[1] provide practical, enforceable and adequate standards for ensuring safe and expeditious movement of all types of radioactive material, including irradiated fuel, by all modes of transport.

The Agency's Transport Regulations are primarily performance standards for shippers – packaging requirements and package test standards, labelling, *and* contamination and radiation level limits. Detailed carrier requirements are left to national authorities and other international organizations. All types of shipments are covered – from ores and luminous dial watches to irradiated nuclear fuel and high-level wastes.

OBJECTIVES OF THE TRANSPORT REGULATIONS

The Agency's Transport Regulations are designed to protect plant employees, transport workers and the general public from radiation under normal conditions and to limit the contents of each package or ensure that the packaging is designed and constructed to prevent loss of material or shielding so that exposures to people will not be serious, even under accident conditions.

The basic requirements for safety in transportation of radioactive materials are:

(1) adequate containment of the radioactive material;
(2) adequate control of radiation emitted from the material;

(3) safe dissipation of heat generated in the process of absorbing the radiation; and
(4) prevention of criticality if the material is fissionable.

CONTAINMENT REQUIREMENTS

For establishing the containment requirements, all radioactive materials are divided into four general classes:

(1) small concentrations and limited quantities which are exempt from special packaging;
(2) low-level solids and low specific activity materials;
(3) type A quantities which require packaging that will retain the contents and shielding under conditions likely to be encountered in normal transport, including minor mishaps such as rough handling, exposure to rain, compression and penetration by other goods; and
(4) quantities exceeding Type A quantities, called Type B quantities, which require packaging capable of retaining the contents and most of the shielding under conditions likely to be encountered in transport, including accidents and fires.

Compliance with design standards set out in the Regulations and implementation of quality assurance programmes, including proof-testing and independent reviews, provide a high degree of assurance that the packages will maintain integrity over their lifetime under normal and accident conditions.

ACCIDENT DAMAGE TESTS

The Regulations prescribe tests for demonstrating the ability of a package to withstand accident conditions in transport. These include impact, puncture, fire and water immersion, taking account of stresses arising out of accidents which might occur in all modes of transport (see Figure 16.1). A Type B package is expected to be designed so that safe recovery of the package would be feasible using normal emergency procedures after being involved in an accident; however, it is not required that the package be suitable for reuse after an accident.

I. Mechanical tests

A. 9-metre free fall onto flat, horizontal unyielding surface

B. 1-metre free fall onto solid mild steel bar 15 cm diameter and at least 20 cm long

II. Thermal test

Exposure to 800 °C for 30 minutes (emissivity 0.9, package absorptivity 0.8)

No artificial cooling for three hours

III. Immersion test – (fissile class I and II packages only)

Immersed in 15-metre water for not less than eight hours

FIGURE 16.1 *Accident damage tests*

Because of the importance of the packaging in assuring safety in transport, I shall discuss in some detail the development of the accident damage tests for Type B packages, which includes packaging for irradiated fuel.

The concept of Type A and Type B packaging was introduced in the first edition of the Agency's recommended Regulations for the Safe Transport of Radioactive Materials published in 1961. The packaging was 'to prevent any loss or dispersal of the radioactive contents and to retain the shielding efficiency'; Type A packaging 'under conditions normally incident to transport and under conditions incident to minor accidents', and Type B packaging under normal conditions and minor accidents and 'for the maximum credible accident relevant to the mode of transport'. The design of Type B packaging was to be approved by the competent authority of the country in which the shipment originated. The application was to 'indicate the nature of the accident postulated'. Some specific guidance was provided for nuclear criticality safety of fissile materials.

Applying the concept of a 'maximum credible accident' immediately raised difficulties. It was defined as the worst accident or combination of accident elements that is to be taken into account in a packaging design, having regard to the nature of the contents, the mode of transport and, in some cases, the route and operating procedures to be used. However, no matter how severe a test was selected, it was possible to foresee a more severe accident condition which could cause the package to fail. It became obvious that to achieve acceptance and uniformity of standards throughout the world, more precise package test standards were required.

In 1962, the Agency began a series of meetings to revise the Regulations and develop specific standards for the design and testing of packages.

In 1963, in preparation for the panel on design and testing of packaging, Alan Fairbairn and Tobey Messenger of the United Kingdom Atomic Energy Authority prepared a comprehensive report on environmental tests of packaging[2] based on a thorough analysis of information available on normal and accident conditions in transport.

That report, R-19, augmented by practical studies in various countries including USA, France and Italy, formed a basis of the work of a panel of experts convened by the Agency to develop packaging test standards. During its first meeting, the Packaging Panel agreed on three broad objectives:

(1) The testing requirements were 'to be *comparable* with conditions experienced in the transport environment, but *not* to represent or reproduce every conceivable condition or even any particular condition'. It was considered highly desirable that the tests be shown to be adequate for all modes of transport, but equal levels of protection for all modes were not considered necessary or, in fact, achievable. R-19 states, 'If a transport package were to be designed to withstand every conceivable accident resulting from man-made and natural forces, it would be non-transportable'. It was recognized that no tests can exactly reproduce actual transport conditions. It was the aim of the tests to simulate the damage to a package which would result from its exposure to the transport environment. And the test procedures should provide means of showing, at a reasonable cost, whether a design of package would be likely to withstand most accident conditions.

(2) As a second objective, the testing requirements were 'to be *practical* in that the number of tests be kept to a reasonable minimum and their nature be such that, as far as possible, they might be carried out using readily available facilities and equipment'. This meant that the tests must be simple, capable of being carried out in the laboratory or the field.

(3) And as a third objective, the test requirements were 'to *promote* the development of alternative methods, other than the actual submission of samples or prototypes to the tests, for determining

whether packaging could satisfy the relevant tests'. For this purpose, the tests were to be so specified that the results would be reproducible and, where feasible, compliance with the test requirements could be assessed by calculational methods.

It may be of interest to note that R-19 specified eighteen tests for Type A packaging; hence the name 'testomania', which is sometimes given to the document.

ACCEPTANCE CRITERIA

Any test programme consists of two parts: the physical test (i.e. action applied to a specimen) and the acceptance criteria (i.e. the test on the test). In the 1961 Regulations, the acceptance criteria were the same for Type B packaging under accident conditions as for Type A packaging under normal conditions of transport – namely, 'to prevent loss or dispersal of contents and retain shielding efficiency'. With the elaboration of more specific accident damage tests in the 1967 Regulations, somc relaxation was made in the acceptance criteria for Type B packages to allow some loss of shielding under accident conditions.

It was assumed that Type A packaging could fail in a major accident. Accordingly, the contents of Type A packages were limited to amounts which, if released, would not result in serious exposures, particularly if normal precautions for handling damaged goods were taken.

In practice, it was believed that the average person would take normal precautions if a package, identified as containing radioactive materials, were involved in an accident or damaged to an extent that could cause a release from a Type A package or loss of shielding for a Type B package. Under the 1967 and the present regulations, neither Type A nor Type B packaging is required to withstand all accident conditions to the extent that they can continue to be shipped without attention. Rather, if involved in an accident, the shipper and competent authority are to be notified and recovery operations carried out with the help of qualified persons.

However, under conditions normally encountered in transport including minor accident, such as rough handling, being dropped off the back of a truck and getting rained on, the average person would expect the package not to be adversely affected and would handle it in a routine fashion.

Therefore, Type A and Type B packaging were required to be adequate to prevent loss or dispersal of the contents and to retain shielding efficiency under tests simulating conditions likely to arise in normal transport and minor accidents. Those tests include the water spray test, 1.2-metre free-drop test on a flat, horizontal unyielding surface, compression test and penetration test.

To provide adequate shielding and containment, irradiated fuel casks are always very heavy – usually weighing several tons, and of high density. When such an object falls, even for a very short distance, the results can not be called a minor accident. In recognition of this fact, an adjustment in the height of free-fall is given for heavier packages.

After much discussion and study, the Packaging Panel agreed to develop three types of tests for simulating damage in transport accidents – mechanical, thermal and water immersion.

Mechanical tests

Various types of accidents and the probability of their occurrence are discussed in R-19. For the mechanical tests, tables of impact velocity versus height of drop were presented and the different types of tests discussed, including those for fireproof safes and 15-foot drop for solid irradiated fuel casks which the US Atomic Energy Commission had proposed in 1962. The test recommended in R-19 was a 30-foot free-fall onto a 1-foot wide, rolled-steel box beam.

In 1966, Alan Fairbairn, UKAEA, and Thurber George, Bureau of Explosives of the Association of American Railroads, published a discussion of the practical basis for the 30-foot drop height.[3]

> The choice of 30 ft. for the impact part of the mechanical test results from practical judgments; first, that in the course of transport, Type B packages are unlikely to suffer higher drops on to very hard targets such as dock wharves; and, second, that a part of the impact during collisions at high speeds will be absorbed by the vehicles. For example, an express goods train may suffer a crash at 60 mph. A commercial aircraft may crash at 250 mph, nevertheless information from such accidents (and they reference R-19) indicates that much of the energy of the impact is absorbed by damage to the rail vehicle or aircraft and, as a result, it was considered most unlikely that a package being carried would be subject to an impact significantly exceeding that in the 30 ft. drop onto the very hard target.

The test recommended in R-19 combined tests for 'impact and shear' for any package more than 1 foot long by using the 1-foot-wide beam as a target. For longer packages, the beam produced a 'back-breaking' action that was considered to be extreme. After much study and debate, the present two-stage mechanical test was devised: (1) impact – a 9-metre free fall onto a flat, horizontal unyielding surface; and (2) punch or shear – a 1-metre free fall onto a solid mild steel bar 15 centimetres in diameter and at least 20 centimetres long.

The impact test required a drop onto an unyielding target so specified that the package must absorb all of the impact energy. This simplifies calculations, makes it possible to duplicate results and permits extrapolation of results from one design to another with similar features. An unyielding target was described as a block of concrete of mass at least ten times that of the package with a thick steel plate on top. For large, 'hard' containers weighing more than a few hundred pounds, nature offers very few surfaces that can approach that degree of resistance to displacement or deformation. Most natural targets move, bend, break up or otherwise yield and absorb energy.

The test also requires that the package fall onto the target so as to suffer maximum damage. Usually that means with the centre of gravity above the point of impact. Under actual accident conditions such circumstances are unlikely to be experienced. The direction of travel is unlikely to be at right angles to the target and any deviation from 90° reduces the energy by rotation or translation. Also, the vehicle intervenes and absorbs much of the impact. This was demonstrated graphically in a series of full-scale tests conducted at Sandia Laboratories. In a 135-km/hr impact of a semi-trailer truck onto an immovable target carrying a 20-ton cask, the truck was totally demolished and the cask essentially undamaged.[4]

Punch test

The second part of the mechanical test, the punch test, is to assess resistance to shear and, for long packages, to 'breaking'. The punch, a solid mild steel bar 15 centimetres in diameter, was a practical simulation of a piece of railroad track acting as a penetrator. The 20-centimetre length allows for buckling if the package is heavy and 'hard'; railroad track also would be expected to buckle. The punch must be longer than 20 centimetres if penetration occurs and a longer length would do more damage; greater than 20 centimetres in length would not significantly affect the buckling.

Thermal test

The thermal test adopted in 1964 was expressed as exposure to the heat input from a radiation environment of 800°C for 30 minutes with an emissivity of 0.9, assuming the surface of the package had an absorption coefficient of 0.8.

Consideration was given to the fact that the fire resistance of safes and walls and fire doors is determined by a furnace test usually from 1 to 4 hours in duration following a standard time-temperature curve. Safes are subjected to a water spray immediately after, to assess the effect of thermal shock and a drop test. This simulates conditions in a building fire with water available for fighting the fire.

In practice, the conditions in a fire in transport are quite different from a building fire. A petroleum fire, in a transportation accident, heats up very quickly, burns uniformly and can produce blackening of the package surfaces. Also, in transport, fire-fighting equipment may not be readily available in remote areas.

The test requirements allow use of a furnace or open field burning and can be assessed by calculation. Also the specimen is not to be cooled artificially until 3 hours after the test or until the internal temperature begins to fall. This was to allow any combustibles to continue burning and to determine the effects of heat absorbed by heavy shielding and other parts of the package.

Subsequent research[5] indicates that credible conditions in transport are unlikely to subject packages to more severe thermal environments. Fires occur in only about 1 per cent of all accidents and even then radioactive packages are unlikely to be enveloped in the flames. It is of interest to recall that R-19 suggested the 'maximum credible accident' for air transport was represented by a crash to the ground followed by a 30-minute liquid fuel fire with a mean effective temperature of 800°C.

Water immersion tests

Finally, I would like to talk briefly about the two water immersion requirements which were specified in the 1967 Regulations.

(1) For Type B packaging, the containment vessel must remain leakproof at a depth of 15 metres in water. That was a design requirement and not part of the accident damage tests. In practice, it was thought unlikely that a package would be involved in an impact and a fire and then dumped in the water. The depth of

water was chosen because of concern for possible loss overboard of a package in Naples harbour.

(2) Included in the tests for demonstrating the ability of Fissile Class I and Fissile Class II package designs to withstand accident conditions in the 1964 Regulations was the following: immersion in water to a depth of at least 0.9 metre with each joint tested for a period equal to 24 hours divided by the number of joints. In the 1967 edition, the testing time was set at not less than 8 hours. In the discussion, it was suggested that if inleakage occurred so that a fissile material package went critical under a few metres of water, one should leave the package in place and install a pipeline to draw off the heat for use in the nearest city.

Water inleakage for other than fissile materials was considered but, in practice, it was considered to be very unlikely that inleakage, dissolving of contents and leakage of the dissolved contents out of the package would occur.

The present requirements are that all Type B packages be designed to withstand immersion in at least 15 metres of water for not less than 8 hours without excessive loss of contents and certain fissile material packages be designed to withstand at least 0.9 metre of water for 8 hours without inleakage.

DESIGN APPROVAL

For larger quantities of material, i.e. above Type A quantities, and for fissile materials, certification of approval of the specific packaging design by the competent authority is required. The underlying aim is to provide for scrutiny of the design by persons independent of the consignor, carrier or consignee to ensure that the relevant regulatory provisions have been and, indeed, are seen to be met.

As part of the Agency's programme for distribution of information, copies of certificates of design approvals are collected, computerized and lists issued of valid certificates with the names of the countries approving and those revalidating other countries' approvals.

The IAEA Regulations are designed to facilitate international movements by providing standards for unilateral approvals, i.e. approvals by the competent authority of the country of origin of the shipments. Such designs, which are referred to as Type B(U) packages, are intended to be accepted in other countries without further review.

Provision is also made for multilateral approvals, i.e. approval of the competent authorities of the countries through or into which the shipment goes. Multilateral approvals are required for movements under 'special arrangements'; that is, movements for which special precautions or special administrative or operational controls must be observed during transport to ensure a level of safety equivalent to that provided by the other types of packages.

NOTIFICATION

There is also another division of materials, large quantities not related to packaging requirements but to a requirement for the shipper to notify each competent authority involved of each shipment. Shipments of irradiated fuel fall into this category.

EXTERNAL RADIATION LIMITS

Radiation protection in the transport of radioactive materials involves both limiting the release of the material itself and control over radiation emitted by the material. Many radioactive materials emit penetrating radiation which is only partially shielded by the packaging.

For routine transport, radiation emitted from a package is limited to 200 mrem/hr (2 mSv/h) at the surface and 10 mrem/hr (0.1 mSv/h) at 1 metre from the surface. These limits are intended to ensure that the average doses to persons handling such packages do not exceed a fraction of the individual dose equivalent limits. The handling times for each package, in most cases, are small so the resultant total doses are also small. In specific cases, persons who are required to handle large numbers of packages of radioactive material are trained and monitored as radiation workers. For exclusive use shipments, and most irradiated fuels are shipped under those conditions, the radiation levels shall not exceed 200 mrem/hr at the accessible surface of the shipment, 10 mrem/hr at 2 metres from the external surface of the vehicle, and 2 mrem/hr at any normally occupied position in the vehicle. Under exclusive use conditions, exposures of handlers are controlled because all handling must be carried out by or in accordance with specific instructions of the consignee or consignor who are knowledgeable in radiation protection practices.

A simple system was developed in the late 1940s so that, in

transport, carrier personnel would only be required to read and add numbers and not be required to follow any complicated radiation protection procedures. Each package is assigned a number based on the radiation level at 1 metre from that package – 0.1 mSv/h (1 mrem/h) at 1 metre is a Transport Index of *1*. The transport index number is shown on the label on each package. For packages in storage or on the transport vehicle, carrier personnel are required only to add those transport numbers to limit the radiation levels from accumulations of packages and by the use of tables of segregation distances versus numbers of transport indexes, limit exposures to people and property (especially undeveloped film). The tables of segregation distances are specified by the national authorities or international transport organizations.

The TI system is also applied to fissile material packages in transport to prevent the accumulation of an unsafe number in one place or on one vehicle. As for other hazardous goods, identification of radioactive materials in transport is provided by distinctive labels on each package and placards on trucks and rail cars. Of course, for the details one must refer to the Regulations.

The nature and extent of the measures taken to determine the resultant radiation doses to the members of the general public and transport workers and to further control their exposures are related to the magnitude and likelihood of their exposures. Where it can be shown or has been determined that the doses are very small (for example, very few shipments are made in an area), no detailed monitoring or assessment of the radiation doses is necessary. This is the condition in most cases for members of the general public. Where the doses are small, for occupationally exposed workers from one-tenth to one-third of the individual annual dose limit, environmental monitoring and assessment of the radiation levels in the work areas should be conducted. For workers, where the doses are likely to exceed one-third of the individual annual dose limit, personal radiation exposure monitoring programmes (such as personal dosimetry and special health supervision) are required.

TRANSPORT RISK ASSESSMENT

Several countries have assessed the radiation exposures to the workers and the public from transport both by calculation and by actual measurements. Some of the results were reported in Mr Rosen's paper. A simplified system for assessing risk from transport has been developed for the Agency by Sweden and the USA, with assistance from other states. The Agency has encouraged each Member State to calculate the impact of transportation in the State. Individual results will be reviewed to improve the system, if necessary. Results can be used to respond to public concerns, to evaluate the adequacy of the standards and in making determinations whether exposures in transport are significant and within prescribed limits.

The acceptability of risk is a function of the perception of the hazard, which is often quite different from the reality of the hazard. The public reacts more from fear and emotion than logic and information. Lack of a balanced public information programme contributes to this. We have enough facts and figures on transportation risks from the studies already completed to allow a more objective evaluation.

CONCLUDING REMARKS

The Agency's recommended Regulations for the Safe Transport of Radioactive Materials have been adopted in many parts of the world and are applicable to all international shipments. As a result of following those standards, a very high degree of safety has been obtained. In the area of accident protection, the standards have led the way for package standards for hazardous goods. That record is directly attributable to very practical forward thinking attitudes of the experts who have participated in the development and evolution of the standards.

NOTES

1. Regulations for the Safe Transport of Radioactive Materials, Safety Series No. 6, International Atomic Energy Agency. The latest version is the 1973 Revised Edition (As Amended) published in 1979.
2. 'Interim recommendations for the application of environmental tests to the approval of packaging', AHSB(S)R-19, 1963.
3. R. Gibson (ed.), *The Safe Transport of Radioactive Materials* (London: Pergamon Press, 1966).
4. Proceedings of the Fifth International Symposium – Packaging and Transportation of Radioactive Materials, 7–12 May 1978, Las Vegas, Nevada, USA, Volume 1 (pp. 463–76).
5. *Severities of Transportation Accidents*: vol. I, *Summary*: vol. II, *Cargo Aircraft*: vol. III, *Motor Carriers*; vol. IV, *Train* – SLA-74-0001, Clarke, Foley, Hartman and Larson, Sandia Laboratories (July 1976).

17 Design and Safety of Flasks

ROBERT M. JEFFERSON

From the outset the transportation of radioactive materials has been recognized as an activity involving possible hazards to those moving the material and to the general public as well. This is particularly true in the case of transporting spent fuel. For that reason, those involved in developing the regulations concerning the packaging and transportation of radioactive materials in general and spent fuel in particular have continuously directed their attention to ensure that this activity was conducted in as safe a manner as possible.

The current international regulations, in large measure adopted by each of the countries participating in nuclear matters, reflect several decades of experience and are founded on the following principles:

(1) The packaging used to contain the radioactive material must provide all of the containment and shielding necessary to assure public and industrial safety. This minimizes dependence upon special vehicles, special routing, special crews, speed controls, or other administratively controlled factors during transportation. Since 'administrative' controls in transport are likely to be more subject to the vagaries of human behaviour than those at the reactor where packages are prepared for shipment, this principle can be seen as reducing dependence upon uncontrolled variables that could affect safety.

(2) The packaging must provide a level of protection proportional to the hazard of the material involved. Thus when shipping spent fuel very stringent packaging requirements are involved. These requirements demand that, unlike any other hazardous commodity, the packagings used to ship spent fuel must maintain their

protective features (containment, shielding, heat management, and criticality control) both under all conceivable normal environments and under severe accident conditions.

(3) The flasks will be designed to meet performance criteria which are clearly defined in terms of engineering test conditions.

The regulations which result include engineering criteria, explicitly stated in engineering terms, from which a designer can unequivocally determine his design objectives. Based on these design objectives, the engineer can use a variety of well-developed analytical and design tools to define a shipping cask for spent fuel. These analytical tools include structural, heat transfer, shielding and criticality codes of varying complexity. Stresses and deformations of structures are calculated using methods that range from simple one-dimensional models to the modern three-dimensional structural codes which have proven to be quite precise by comparison to actual test data. There are a variety of thermal codes which enable the designer to predict the behaviour of the cask system under both internal heat loads and external heat loads that might be encountered under normal and accident conditions. Similarly, shielding is an important part of the cask design and the engineer has available to him a variety of well-developed and validated shielding codes for determining those parameters. Finally, the analytical techniques in determining the criticality conditions of the fuel contained within the flask are well-developed and have been evaluated for their accuracy and recently benchmarked in an OECD study.

The final design must reflect all of these factors as in any design situation, but shielding considerations and the requirement to maintain shielding in accident environments usually dominate the other design factors involved. With short-cooled fuel, the next most critical design factor usually involves the thermal load of the fuel and the flask's response to external thermal environments. In the United States very little, if any, short-cooled fuel is being transported, thus downgrading the importance of thermal designs. The result of the shielding and thermal considerations is that most flasks are considerably over-designed structurally. This naturally resulting conservatism results in flasks which can survive accident conditions far in excess of those required for certification.

The failure of such engineered systems as spent fuel flasks is not a step function (i.e. instantaneous) but occurs progressively as the accident environment grows increasingly severe. The result of this gradual behaviour is that even after achieving some severity limit (e.g. the test levels specified in the regulations) which causes no degradation of performance, increasing accident severity does not instantaneously cause total failure. This range of behaviour is shown in Figure 17.1 where the increasing level of the threat environment is shown to cause a progressively larger and larger release fraction.

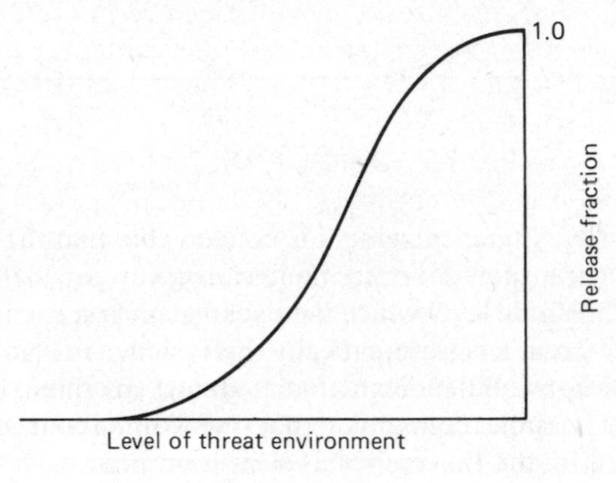

FIGURE 17.1

A similar concept applies to the level of threat involved. By plotting the probabilities of accidents versus the level of threat that the accident would produce it can be seen in Figure 17.2 that as the level of threat environment increases, the probability of an accident providing that level decreases.

The question facing most decision-makers is whether or not the two

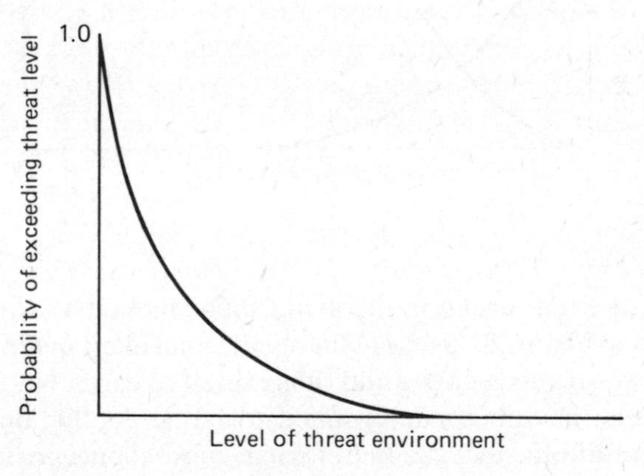

FIGURE 17.2

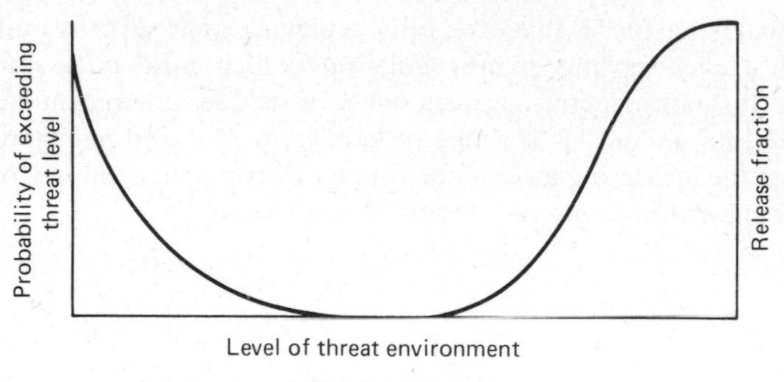

FIGURE 17.3

curves overlap. At one extreme it is conceivable that the threat level created by a transportation accident can never get high enough to exceed the threshold level which the package must see before it begins to release material. Diagrammatically this is shown in Figure 17.3. The opposite concept would indicate that at almost any threat level there is some release possible, demanding that one worry about the triangular area enclosed by the two curves as shown on Figure 17.4. In reality, neither of these extremes is the case.

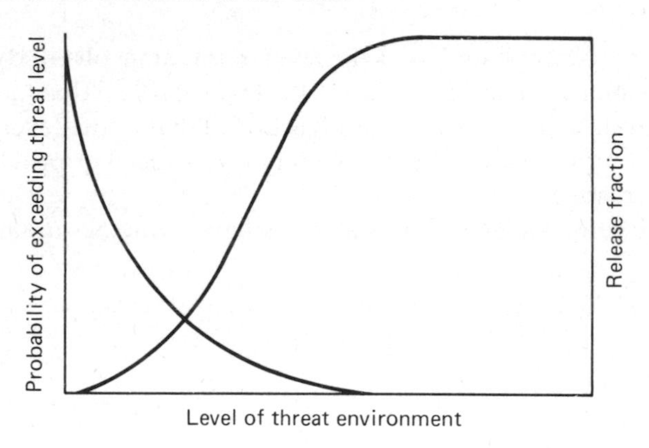

FIGURE 17.4

Instead, the real situation involving spent fuel flasks is more likely that shown in Figure 17.5 where there is a small overlap indicating that a few very severe accidents could cause small releases. Note that some small releases have been determined to be acceptable under severe accident conditions since the radiological consequences of such a small release would be consistent with these resulting from releases allowed to take place from other (lower quantity) radioactive packagings subjected to much less damaging conditions encountered during transport. In other words, there is a parity established in the regulations which do not require absolute containment under all conceivable conditions.

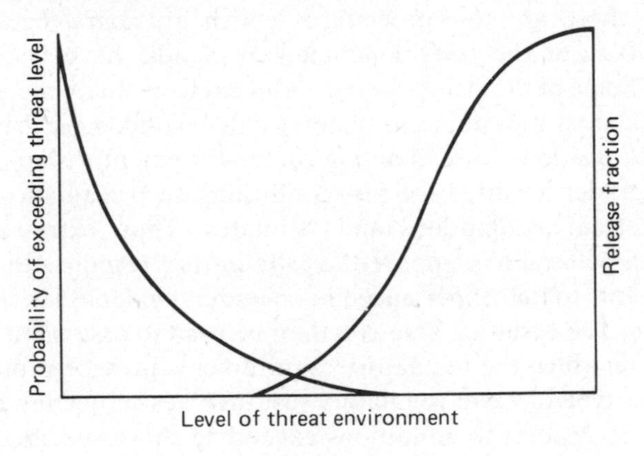

FIGURE 17.5

In order to assign values to the series of curves just shown, a variety of techniques are employed. These include gathering data on accident frequency and accident severities using the analytical tools previously spoken of to evaluate progressively more severe environments, and in some cases testing the systems under a variety of conditions.

The use of testing in engineering terms and, specifically, testing of spent fuel shipping systems can be divided into three categories. Some test procedures are conducted to evaluate the operability or operational ease of the system under consideration. This operational testing, usually conducted using mock-ups, is pursued to make the system safer and more efficient to operate for those who must work around the flask and to improve such things as the lifetime, maintenance and dependability of the flask. In some measure, operational testing utilizes experience with existing and previous systems to continually upgrade newer designs. It is, therefore, not necessary to perform operational tests on every system prior to its placement into service.

A second category of testing might be called regulatory testing. Some packaging systems are subject to the regulatory requirements in order to prove that they meet or exceed those engineering criteria laid down in the regulations. Although rarely used to qualify shipping flasks, testing could be used to prove the flask's ability to survive one or more of the environments when something about the flask design makes it difficult to bound the problem calculationally. When used on flasks, such regulatory testing can be viewed as a means of validating analytical calculations in a manner that is understandable to all involved.

Finally there are test procedures which are extra-regulatory in nature. Many of the tests conducted by Sandia have been in this category. Some of these tests were conducted to evaluate the adequacy of the regulatory requirements under accident conditions. While there are means available for estimating the environment a shipping flask might see under specified accident conditions, such evaluations always involve certain assumptions and estimates. Thus, extra-regulatory testing can be used to compare the estimates of response to accident environments to that experienced in an understandable set of realistic conditions. The resulting data can then be used to assess the accident severities to which the regulatory requirements provide protection.

Another type of extra-regulatory test involves subjecting full-scale or sub-scale systems to conditions exceeding those specified for the regulations in order to define the margin of safety incorporated into the flask design. In the case of flasks where shielding is the dominant design consideration, the actual margin of safety can be quite large. Further, while it is relatively straightforward to analytically determine the toe of the curve in Figure 17.1, it is more difficult to calculate the precise shape of the curve. Extra-regulatory testing, involving a series of increasingly severe tests, help to define the shape of that curve.

Sandia National Laboratories has conducted many such tests, under a variety of accident conditions involving rail and highway crashes, drop tests and explosions. For example, two identical 16,000 lb flasks were dropped, one from 2000 feet on to a clay and sand material (with properties similar to unreinforced concrete) and the other from 30 feet on to a fully unyielding surface; the impact velocities were 235 m.p.h. and 30 m.p.h. respectively. In both tests, the integrity of the flasks remained unimpaired, although the flask dropped from 2000 feet was less damaged than its twin dropped 30 feet onto an unyielding surface. In addition, Sandia has simulated a wide variety of sabotage modes, including a large quantity of explosive to compromise a truck type shipping flask. The final test in this series was conducted on a full-scale flask (in an enclosed chamber, etc.) so that all the effects could be measured. From this particular test it is possible to conclude that had such an act of sabotage occurred in mid-Manhattan during the rush-hour, the explosion itself would probably have killed around 150 people but the radiation release would have been quite small. Certainly it would have caused no early deaths and over the next 30 years it might be expected to produce two-tenths of one cancer in a population that normally expects a quarter of a million cancers.

Several conclusions can be drawn from this discussion of the regulatory and engineering concepts involved in flask design and use. First of all, the flasks are designed using concepts that have been subjected to a wide range of testing to assure that these techniques are indeed valid predictors of performance. Because these techniques have been shown to be valid and because their use is carefully evaluated by the designer and the regulatory review process, great confidence can be placed in the safety of the resultant designs. These factors, coupled with the fact that the regulatory requirements stated in engineering terms exceed the environments actually experienced adds further to the confidence that can be placed in these systems.

In transporting spent nuclear fuel, no one questions the fact that the material itself is dangerous if uncontrolled, nor that accidents may occur. The aim is to ensure that when an accident occurs there will be no radiological consequences to the public, even if it were severe and occurred in a densely populated area.

The past 30 years of international experience has shown that the regulations and the regulators have produced a level of safety which not only meets these two requirements, but which is far superior to that relating to the transport of any other hazardous material.

18 Shipping Flasks in Severe Rail Accidents

MARVIN RESNIKOFF

The Council on Economic Priorities (CEP) is an independent research organisation in the US. We have recently issued a report (*The Next Nuclear Gamble*) on the transport and storage of irradiated fuel which concludes that, whilst no transport accident leading to a major release of radioactivity has occurred in the US, the transport of irradiated fuel as currently practised in the US is unsafe. We do not think the flask (or 'cask' in the US) design is sufficient to withstand realistic highway and rail accidents.

Let me start by describing the current nuclear waste situation in the US. Around seventy-five power reactors are operating, but there is no commercial reprocessing plant in operation (a plant in West Valley, New York, opened in 1966, but was closed in 1972). All irradiated fuel in the US is being kept in 'pools' at reactor sites. This method is reaching its limits. From 1984 onwards reactor sites will run out of temporary storage space, but there is nowhere to take the spent fuel for reprocessing or permanent storage. To meet this problem Congress passed the 1982 Nuclear Waste Policy Act which establishes a limited-capacity, last-resort, temporary storage facility.

Over the past three years the number of flask shipments in the US has averaged only ninety-six a year. By the year 2000 the figure could be orders of magnitude bigger due to the currently growing backlog of spent fuel in temporary storage; furthermore, the bulk of such traffic in the US would continue to go by road rather than rail. However, US utilities are beginning to make provision for dry storage at reactor sites – in which case they will not need to use the temporary storage facility which the federal Government is making available. The CEP favours dry storage because it allows the fuel to decay to lower radiation levels and reduces the number of miles travelled, since the fuel has to move only once – directly to the final repository, as opposed to twice (initially to an intermediate storage facility, then to the final repository); it will also permit the development of safer transportation systems.

HISTORICAL BACKGROUND

The shipping flask performance standards, the 30-foot drop test, the 40-inch puncture test, and the 800°C fire test for 30 minutes, had their origins in England in the early 1960s. The IAEA (IAEA, 1961), while arguing that insufficient information was available to quantitatively describe the accident environment, nevertheless proposed tests which allegedly met highway and rail conditions that existed in the UK. The document is replete with qualitative assertions such as 'vehicles on the road travel at any speed up to, say, 60 m.p.h. In the majority of accidents some slowing down takes place in the odd second of warning' Presumably, these were the speeds on highways in the UK in the 1950s. The author describes a cask falling from a bridge or down an embankment, but concludes, without analysis, that glancing blows followed by rolling of the cask mitigate the accident's severity. The author concludes, 'Weighing up all the above information, it is felt that a direct drop of 30 feet on a hard surface is a reasonable test.' This 30-foot test exists to this day.

While stating that temperatures up to 1010°C might be obtained, noting the increase in the use of trains carrying only inflammable material, and the possibility of the standard fire reaching 1010°C in the second hour, the author nevertheless concludes that 'the test proposed of 800°C for half an hour is felt to be as searching as the British standard fire for one hour.' No proof was given. The test was to be conducted by placing the cask into a 800°C oven for half an hour, causing shock heating at the beginning of the test, whereas the British standard fire begins at room temperature. The IAEA panel rejected the British standard fire of one hour because the conditions in a transport accident would be much different. 'A petroleum fire, in a transportation accident, heats up very quickly, burns uniformly and can produce blackening of the package surfaces. Also, in transport, fire fighting equipment may not be readily available in remote areas' (Barker, 1981). No explanation was given as to why this shock heating was more severe than a 1-hour fire. It seemed to us that the total heat input to the cask is of greater concern than heat shock. The 800°C temperature test as described in 1961 is still in use today.

Regarding torch fires, in which a burning jet is aimed at the cask, the author concluded that 'torching can occur to produce high local temperatures, but the overall effect is small.' A torching fire standard was considered by the IAEA panel in 1963, but was rejected. Economics were a factor in the consideration. According to a member of the IAEA Cask Standards Panel:

> a torching flame is so hot that one can't design to protect against it without great effort or cost. The probability of this affecting a container is greatly reduced by the limited likelihood of such a flame being produced accidentally in a spot to impinge on a radioactive material container. Hence, it was thought that this type of accident condition, on the balance of judgment, was not of sufficient probability to justify it being considered. (Barker, 1965)

Torch tests, as one of the cask certification requirements, are not required even though the likelihood of torch-type fire has been greatly increased by the nature and quantities of chemicals in common use today.

No puncture test was suggested by the panel in 1961, but the 30-foot drop test actually employed in the UK was onto a 1-foot-wide steel slab which exerted very high shear forces, greater than 2000 g. The main discussion at the time among IAEA experts was whether to use the US drop test onto an unyielding surface from a 15-foot height *plus* the puncture test, or solely the 30-foot drop onto a 1-foot-wide bar suggested by the British. In the end a compromise was reached. The British 30-foot drop onto an unyielding surface was selected, plus the American puncture test. The US test was considered more reproducible than the British test. The puncture test – a 40-inch drop onto a 6-inch diameter spike – was 'a practical simulation of a piece of railroad track acting as a penetrator' (Barker, 1981). The bases for the tests actually employed in Great Britain in 1961 were published in 1963 (UKAEA, 1963). This 1963 paper became the basis for the adoption of international standards.

FIRE TEST

For rail shipments, it is not so much impact, but fire, puncture and crush environments which are of concern. Because of the presence of loaded railroad tank cars containing 10,000 to 40,000 gallons of combustibles, long-duration, high-temperature fires as a result of a train accident are quite possible. There are again limited data, but Sandia estimates (Sandia, 1978b) about 1 per cent of railroad collisions and derailments result in fire. As it has for truck accidents, Sandia has set up models to predict the temperature and duration of railroad fires. Some of the assumptions of the model, and the conclusions, are realistic and others are highly questionable. For truck accidents, aircraft fuel, JP-4, was used in the model, with an average temperature of 1010°C, ranging between 810°C and 1300°C. Sandia has ignored the higher temperature fires of hotter burning combustibles.

TABLE 18.1 *US production of some high-temperature chemicals (1960–80)*

Chemical	*Flame temp. –°F[a] Adiabatic*	*Production (million lbs) Year 1960*	*Year 1980*	*Increase %*
Acetone	3820	761	2076	273
Acrylonitrile	4430	229	1829	799
Benzene	4150	2267	14322	632
1-3 butadiene	4275	1883	2799	149
n-butane	4060	507	1506	297
1-butane	4175	—	126	—
Ethane	4040	734	7000	954
Propane	4050	2937	8712	297
Toluene	4220	1743	7279	418
Xylene (incl. o-xylene)	4205	1976	6895	349

NOTE (a) The adiabatic flame temperature (aft) is the maximum possible; for continuous furnaces, the gas temperatures are approximately 0.8 times the aft. From (ITC, 1961) and (ITC, 1981).

A major increase in the use and transportation of these and other chemicals has occurred since 1961, when the cask certification requirements were first conceived. Table 18.1 contains a representative sample of the major combustibles produced in the US and shows a ten-fold increase in the production of ethane, and dramatic increases in the production of others in the 20-year period between 1960–80. It seems clear, therefore, that the general fire test temperature should be raised substantially.

A fire at 1010°C for only 15 minutes could over-pressurize a cask and cause the rupture disk to fail 2.5 hours later (PNL, 1978). Battelle Labs believe that a reduction in risk level could be attained if corrective action were taken to cool the cask within 2 hours of the accident. This would entail having trained emergency personnel available at the accident scene within 2 hours (PNL, 1978). A hotter fire could cause the rupture disk to fail sooner.

PUNCTURE TEST

In the puncture test, a cask is dropped 40 inches onto a 6-inch-diameter steel plug. Though the diameter and geometry of the plug are reasonable, they are also arbitrary. As with the impact test, the cask itself is not dropped. Mathematical formulas model cask puncture and have been questioned from several standpoints by members of the industry. Most noteworthy and alarming were the results of a 1980 analysis of punctures of shielded radioactive materials containers performed at the Lawrence Livermore Laboratories (LLL). In assessing the manner by which cask designers substantiate puncture resistance in casks, the report stated: 'Existing test data is inadequate and analytical methods, largely empirical, are crude and unreliable. . . . The empirical formulas do not give designers the insight into puncture phenomena they need to produce a rational, safe design' (NRC, 1980b). According to LLL, the puncture formulas are 'crude and unreliable'. These strong criticisms have been independently supported by a former NRC engineer (Shieh, 1980). Yet, since no casks are physically tested, the cask designer must rely on these calculations to verify that a cask can satisfy the tests.

The results raised a new issue related to the temperature at which the puncture occurs. The present rule (49 CFR 173.398(c)(2)) assumed normal outdoor temperatures prevail at the time of the drop. The 30-minute fire follows the 40-inch drop test in the test sequence. The LLL tests showed that significantly less force could cause a puncture if the surface of the cask had been heated to 400°F (NRC, 1980a). In other words, heated surfaces could be punctured more easily than cool surfaces. It follows that a cask could more readily be pierced during a fire than before a fire, as the present test assumes.

CRUSH TEST

Very large crush forces can occur if an irradiated fuel cask is crushed underneath derailed cars. The structural members of a railroad car and the massive weight involved imply application of considerable force, estimated by Sandia as unlikely to exceed 50 tons (with a probability of one such event per billion car-miles) (Sandia, 1978b). The Sandia report fails to evaluate the crush force applied to a cask that is pinned between railroad cars in an accident. A more recent NRC-sponsored study (NRC, 1980c) shows that the crush forces exerted on a cask that is pinned between railroad cars can be as much as 550 tons, a factor of 2.6 times the force resulting from a 30-m.p.h. impact, which is all the cask manufacture standards require. This could produce cask deformations up to 22 inches, which the rail shipping casks may be unable to withstand (NRC, 1980c). While the probability of such crush forces was unspecified, the NRC considers it to be low.

A crush test to simulate the immense forces involved with train car pile-ups and with collisions between barges and larger ships is not presently required. IAEA regulations should be amended to include a crush test.

Crush tests have been 'under consideration' by DOT and NRC for five years (NRC, 1977), are supported by many members of the industry (Sandia, 1979), and 'would increase the current level of protection for both crush and impact' (NRC, 1980c), but are still in the talking stage. One NRC consultant suggests a test in which a weight four to twelve times heavier than the cask would be dropped from a height of 30 feet onto the cask as it lay on an unyielding surface (NRC, 1980c). The weight would have to be 400 to 1200 tons to adequately test a train cask, and performance of the test would present unique engineering problems.

No test is required to assess the effect of a torch fire on a cask, as might occur if a train with both liquid gas and irradiated fuel were derailed. If the tanker is pierced and the leaking gas ignites as it shoots out, a high-temperature flame directed at a small section of the cask could be created. While casks containing lead shielding are designed to allow the lead to expand during a fire, the space needed in a torch fire may be insufficient unless a large amount of the lead melts. If one spot on the cask is heated to a very high temperature very rapidly, the lead will expand at that spot, pressing against the outer steel shell with tremendous force. Such a problem was indicated in another recent study (Sandia, 1980d) in which it was found that tests caused failures of the cask seal or shell 'due to local stresses (when) lead and/or gamma shield material could melt locally, away from expansion volumes, and the outer shell could rupture' . . .. In contrast to the absence of torch test standards for casks, a torch test is presently required of some types of railroad tank cars, during which a 2200°F flame blasts against the car at 40 m.p.h. for 30 minutes (49 CFR 197.105–4(c)). Some preliminary tests are now being conducted at Sandia, but none of the casks in use today are designed to withstand a torch fire.

Since the CEP study was published a report by an NRC contractor, REA (REA, 1982), has confirmed our criticisms of these standards. The contractor examined 500 real truck and rail accidents and concluded that the standards should be strengthened to accommodate these accidents. For example, they recommended that truck and rail casks withstand 1- and 2-hour fires at a temperature of 1600°F, not 1475°F; they also recommend greatly increased resistance to puncture, impact and crush forces. Table 18.2 compares the present NRC testing standards with the REA-recommended standards.

TABLE 18.2 *Comparison of present nuclear regulatory commission cask performance standards to NRC contractor proposed standards*

Testing standard	*Present NRC standard*	*REA proposed standard*	
		rail	*highway*
1. Crush	none	100 ton, 3 inch wide	30 ton, 4 inch wide
2. Impact	(a) 40-inch drop onto spike	sill of 200 ton locomotive with 2300 ton force	
	(b) 30-foot drop onto unyielding surface	80 foot drop onto concrete	90 foot drop onto concrete
3. Torch	none	1 hour, 6 foot diameter, 12 feet from cask	½ hour, 6 foot diameter, 12 feet from cask
4. Fire	½ hour engulfing fire at 1475°F	2 hour fire at 1600°F	1 hour fire at 1600°F

VALVE DESIGN STANDARDS INSUFFICIENT

To prevent the cask seal from rupturing under pressure, cask designers build in a pressure-relief valve that vents coolant. Except for Regulatory Guide 7.6, a revised version of Section III of the American Society of Mechanical Engineers (ASME) Boiler and Pressure Code, which is used for the design of *stationary* nuclear power plant components, no specific NRC guidance or design criteria exist for the design and manufacture of shipping cask valves. The adequacy of ASME standards for cask design has never been determined. Since the ASME Code is written for stationary systems, it does not specify the deceleration forces for which the valves must be designed. Sudden impacts may cause the cask valve to open. It has been noted by a citizens group (Sierra Club, 1980) and by industry personnel as far back as 1970, that 'once this (pressure) valve is actuated, it is extremely difficult to reseat' (ORNL, 1970).

In June 1981, five years after the casks were put into service, the valve manufacturer for the four General Electric IF-300 shipping casks informed GE that the valves failed during operational testing due to a 'generic problem' affecting all IF-300 casks. The IF-300 is the cask most frequently used for rail transport in the US. While the valve would open under pressure, it would not close again, thereby creating an avenue for the continued release of radioactive material. All IF-300 casks have now been restricted to shipping older irradiated fuel in a dry state, where relief of pressure will not be necessary (GE, 1981). It turns out that 'dry' does not mean non-wet. Common English breaks down here. Because of the location of the valves, the IF-300 cask has 1 cubic foot of water remaining, easily enough to rupture the valves in a severe fire.

The NRC was unable to furnish any evidence that it had inspected the IF-300 casks during their manufacture. The tests conducted on shipping casks by Sandia, in which trucks crashed into barriers at 84 m.p.h., never adequately tested the valves. The casks at Sandia were not pressurized to the levels attained during normal operation, which are far below accident pressure conditions.

Another problem with cask valves and seals involves their performance in a fire. Because the valves and seals are made of a Teflon-like coating, the failure threshold for closure drain valve and vent valve seals is 30 minutes in a fire of 1010°C. The failure threshold for a rupture disk is a mere 15 minutes at the same temperature (PNL, 1978). 'Data on Teflon valve seals indicate that failure would occur if the *cask* temperature [not fire temperature] exceeds 278°C' (PNL, 1978). Even the all-metal seals may not hold (AGNS, 1978).

CEP has concluded that valve design criteria and performance are sorely deficient. ASME valve standards must be formulated, and valves tested under accident conditions.

WELD TEST QUALITY CONTROL TOO LENIENT

Computerized certification tests are based on the assumption that the cask is perfectly constructed. Tests which check casks for welding imperfections, however, have been criticized in the literature. Tests usually include physical and X-ray inspection of welds and are referred to as non-destructive tests (NDT) because they do not involve damage to the item being inspected. Published in 1978 (after the fabrication of almost all existing casks), a study by Sandia Laboratories noted that:

> it is difficult to understand the lack of imposition of state-of-the-art NDT methods on critical cask elements such as closure weldments and tie-downs on some of the LWR casks. In most instances, the cask design agency and the cask fabricator had access to knowledgeable in-house NDT personnel and equipment that they used routinely to process ASME code vessels [e.g. high-pressure tanks]. (Sandia, 1978a)

These tests were not performed on casks because the NRC regulations did not require them. Further, once the cask is constructed, the ultrasonic tests can no longer be performed. Since some of the current casks did not have ultrasonic tests performed 'on critical weldments following initial fabrication', it is impossible to know now if the welds were done properly. The same Sandia study analysed the two main types of weld tests with the following conclusion: Ultrasonic weld testing does not work well with austenitic steel (the primary steel used in many irradiated fuel casks): 'False defect indications and masking of real defects in the weldments . . . have raised real doubts as to the credibility of an ultrasonic method for such weldments.' Two-view radiography 'is questionable as to its validity and efficiency when compared with ultrasonics.' One cask expert estimates that the smallest crack detectable by this method is 3/16ths of an inch, and therefore a cracked weld could go undetected (Shappert, 1980).

In the course of CEP's study, it was learned from the author of the welding study that the older NAC-1 cask welds were not properly examined for defects (Ballard, 1982). Thus it appears that at least some welds were not thoroughly checked, that they cannot now be checked, and that the testing methods are, in any case, open to question. Casks of uncertain construction may be travelling the nation's highways. Whether this critique of weld tests applies to other commonly used casks, such as the GE IF-300 and NLI-1/2, has not been established.

CASK FABRICATION

Irradiated fuel casks are certified for compliance with federal regulations by NRC or DOE, depending on the use of the cask. If inspections of cask manufacturers are carried out by the NRC, the inspection reports are publicly available in the US. After studying inspection reports for casks in current use, we have concluded that:

(1) Both agencies have monitored cask fabrication on the basis of papers filed by the cask manufacturer and have rarely examined the fabrication while it was in progress.
(2) The documentation for casks that have been on the road for many years either is not available or has been lost or destroyed.
(3) Manufacturers have occasionally failed to construct casks according to drawings or specifications. The NRC's response has been to change the licence to allow the alteration after the fact instead of requiring compliance with the original design.
(4) Casks have been manufactured or altered improperly and then used for extended periods without awareness by any of the federal agencies. Discovery of the alteration has been made by purchasers or by a laboratory testing a cask, sometimes after it had already been taken out of service due to obsolescence.
(5) The response time of the NRC to serious questions about cask capability is far too slow.

It is worth noting that the record-keeping on irradiated fuel casks is occasionally sketchy. Some reports are missing from NRC files. Documentation that could influence these conclusions may have been improperly withheld from public disclosure.

RADIATION RELEASE IN AN ACCIDENT

Although there is clearly a relationship between accident severity and radiation release, it is difficult to quantify. According to one Government contractor, 'accidents are unique events and cannot be experimentally duplicated, so engineering judgment was required to arrive at realistic release estimates' (PNL, 1978).

The irradiated-fuel shipping cask used in the US is a massive steel and lead cylinder weighing 21 to 100 tons or more, with stainless steel walls 0.75 to 1.5 inches thick and lead walls 4 to 6 inches thick. Clearly, an improbably high-speed accident would be needed to crack open such a structure. Valve failure or seal leaks are more likely. But even these have not been quantified by the NRC. In fact, the basic information needed to predict the amount of radioactivity release is not available. One must know what percentage of the fuel rods is damaged due to impact or elevated temperatures, how much crud is loosened, what percentage of the radioactivity is released from the fuel rods through damaged cladding to the cask coolant, and what percentage of radioactivity is ultimately released from the cask.

Basic data are lacking on each of these points:

(1) Damage of fuel assemblies due to impact. An NRC contractor has said, 'Little work has been done in the area of transmission of impact through the cask inner structure to the fuel assemblies within. A better understanding of this area will also be required to obtain a basis for determining adequate but not overly conservative criteria for design adequacy substantiation' (NRC, 1979).
(2) Cladding failure. The same NRC contractor says that 'detailed data on the shape and size of cladding failure and on the degree of fuel fragmentation do not exist' (NRC, 1979).
(3) Deposition of radioactivity within the cask. 'Review of the existing literature disclosed no reports of integral studies of radionuclide transport and deposition in spent fuel shipping casks under accident or non-accident conditions. Integral tests refer to experimental work in which actual or simulated cask geometrics were used, and fission product leakage was characterized for a range of thermal, fluid flow, and leakage conditions. Therefore, there is no cask deposition information' (NRC, 1979).

Thus it is virtually impossible to correlate accident severity with radiation release.

Certain higher probability events which have not been quantified do not require high-velocity accidents. These could be caused by an unseated valve or a damaged seal releasing radioactive steam to the environment (from casks with water coolant or from 'dry' casks with residual water). The jolt or deceleration which would cause a valve to open has not yet been quantified. Under conditions in which coolant contaminated by radioactive crud was released but fuel cladding did not fail, the NRC has estimated damage as high as $4 billion and thousands of latent cancer fatalities, in a densely populated area (Eisenberg, 1980). The NRC calculations assume explosive damage to the cask by saboteurs but they could apply to the release of contaminated water coolant under any circumstances, including accidents.

According to the NRC, successful sabotage could release up to 154 curies of cobalt-60, arising from one crud-contaminated fuel assembly (NRC, 1980b). In an accident, the cask would have to be breached or a valve unseated, and the irradiated fuel cladding would have to crack or shatter in the same impact and/or fire in order for greater amounts of radioactivity to be released. However, impact tests have not been conducted with irradiated zircaloy cladding. The Sandia crash tests were conducted only with irradiated stainless-steel-clad fuel rods. This lack of data concerned one NRC contractor who stated that the fact the cladding did not fail in the Sandia tests is 'not very comforting in view of the large differences in strength and ductility between irradiated stainless steel and irradiated zircaloy cladding' (NRC, 1979). Battelle Labs has concluded that: under an impact of the cask at 41.4 m.p.h., the rods would fail at a single point, releasing all radioactive gases and allowing the fuel pellet to contact the water coolant. Under a side impact at 28 m.p.h., there would be possible multiple cladding failures on each assembly rod' (PNL, 1978). Battelle has also concluded that 'the neutron shield would fail during any significant impact accident which would result in damage to the cask' (PNL, 1978). Without a neutron shield, neutron radiation doses to emergency personnel could rise by a factor of ten (NRC, 1976).

An important factor in determining the amount of radioactivity that will be released in an accident is the amount of each radionuclide located in the gap between the pellet and the cladding where it is more readily available for release. The radioactivity in the gap arises from vaporization of the radionuclides during reactor operation. Additional volatile radioactivity, particularly caesium and tellurium, will be released from the fuel pellet itself at elevated temperatures in an accident. Following an impact in which a valve is unseated or a seal leaks, a certain percentage of radioactivity will be immediately released to the cask coolant if the cladding is damaged and then to the environment during the cask blowdown. (Blowdown is the initial relief of pressure to the environment through an opening such as open valve or cask seal.) If a long-duration, high-temperature fire occurred or if the fuel were highly irradiated in the reactor and shipped shortly after removal, continual volatilization caused by oxidation of the uranium fuel pellets and release to the gap of volatile radionuclides would occur (NRC, 1979). Train derailment and fire involving tank cars is the most likely scenario in which this could happen.

Calculations for a high-impact collision, but only 1200°F internal cavity temperature, predict up to 2.7 per cent release of caesium and tellurium (NRC, 1979). In some cases, however, the internal temperature could rise to 2000°F. The NRC calculation of caesium release also assumes that 10 per cent of the fuel rods are damaged in the collision and that 27 per cent of the caesium and tellurium is in the gap (NRC, 1979). The estimate of 27 per cent was assumed in a model in the Rasmussen Report (NRC, 1975). Lower estimates of caesium release, generally less than 1 per cent of the contained caesium, also appear in the literature (NRC, 1977, 1980b).

Based on the amount of radioactivity released from the fuel assemblies to the reactor coolant at Three Mile Island, 50 per cent, CEP projects that much more radioactivity will be in the gap and be released in an accident. The exact physical and chemical mechanisms which accounted for this high caesium release at TMI have not yet been explained.

As a result of the uncertainties in the potential release modes and quantities of radioactive materials that could escape from the cask, CEP concludes that the existing probability calculations do not properly describe the likelihood of radiological releases in the most severe accident, nor the distribution of releases in less severe accidents.

ACCIDENT IMPACT

The fine radioactive particles and gases that are released from an irradiated-fuel shipping cask would be carried downwind, the radiation cloud spreading out by means of turbulent diffusion as it moved. People downwind would inhale radioactivity from the passing cloud and be exposed to 'cloudshine' (gamma ray exposure from the passing cloud). People and animals would be exposed to whole body radiation.

As the cloud moved downwind, it would be depleted as the radioactive particles settled on buildings, pavement, soil and water. This would subject people and animals to long-term or chronic exposure caused by 'groundshine' (gamma ray exposure from the settled caesium and other radionuclides). The radioactivity clinging to buildings and pavements would cause whole body radiation doses that could cause latent cancer fatalities, birth defects and other genetic effects. Inhalation of resuspended radioactivity is also possible, as from dusty dirt roads or ploughed fields. Over the long term, ingestion of radioactivity from the water or food supply is also possible. The season during which a potential accident takes place and precipitation at the time of the accident are other variables that can affect the distribution and uptake of radiation.

The most important pathways for radioactivity to reach humans is by inhalation of the passing radioactive cloud and by 'groundshine', which yields a chronic whole body radiation dose. Of course, if persons were evacuated from the contaminated area and the area were cleaned up, this latter exposure would be reduced, but then the economic consequences of the accident would become very high. In the CEP study we examined the economic consequences and projections of the NRC in some detail. We found that the NRC seriously underestimated the economic effects of an accident.

It is a complex task to calculate the cost of an event that has never occurred. In the Rasmussen Report on reactor accidents (NRC, 1975), the costs are heavily skewed: almost all the studied evacuations were in non-urban areas, where costs are lowest, population density is low and the proportion of private vehicles is high, thereby facilitating low-cost evacuation. The values developed in such incidents are not applicable to urban areas.

Similarly naive is the assumption of rapid decontamination or accommodation to evacuation by the populace. WASH-1400 assumes 'resettlement' of the affected population in 90 days, with new homes and jobs, and businesses reopened in 6 months following a disaster (NRC, 1975). No consideration was given to businesses going bankrupt due to loss of stock, equipment or records, dispersal of trained employees, difficulties in developing jobs for a large influx of people, site-specific businesses now lacking a function or a market, and lack of housing, warehouse, factory and office space.

A general error that pervades the cost accounting of an accident involves the use of *per capita* calculations of losses and expenses. For example, 'total relocation cost' is developed by adding income lost to moving cost *for the residents of the area affected*. Such calculations neglect to include the loss of income to commuters whose jobs are lost when their place of employment is contaminated, or the disruption of a company when its headquarters has been 'quarantined'. (Approximately 2 million commuters enter New York City each day.) The NRC method underestimates the costs of an accident in the Central Business District, where fewer persons reside.

When focusing on the clean-up of buildings, WASH-1400 looks almost exclusively at the exterior spaces, even in cities, despite the likelihood of internal contamination of buildings by infiltration and through ventilation systems. No cost of the clean-up of the inside of a building is detailed and no estimate for the loss of such valuable material as corporate records is indicated. While levels of internal contamination would be lower than those outdoors, they could be difficult to clean up because of the large number of surfaces that could be involved.

A more recent analysis by the NRC did expand the WASH-1400 analysis for urban areas by developing an equation for inhalation dose rates for people inside buildings. However, it made the assumption that particles not entrained by a building's air filters would settle in its ductwork and then merely doubled the clean-up charge if the ductwork required cleaning (NRC, 1980b). Most ductwork is concealed and not readily handled without extensive removal of suspended ceilings, disturbance of interior walls (which may also be contaminated) and potential dispersal of the contaminants inside the ductwork. Contaminated ducting would probably be simply sealed in plastic shrouds and disposed of, as a less expensive option than cleaning. Massive replacement of such systems, especially in an urban area, would take many times longer than the ten days envisioned by Sandia's parameters. Most modern office buildings cannot operate without mechanical ventilation systems and would therefore remain closed until those systems were replaced.

In residences with open windows, personal property would have to be scanned for contamination and removed if 'hot'. To merely search all apartments near an accident would require personnel and equipment in quantities beyond the available resources.

Sandia's need to simplify its model resulted in no calculation of contamination below ground level. Thus Sandia did not realistically consider contamination of subways and tunnels. Decontamination of a subway system would be a herculean task and certainly impossible in ten days, the time period contemplated in the Sandia report (NRC, 1980b).

Despite these underestimates, the Sandia report (NRC, 1980b) did calculate economic damages up to $4 billion (Eisenberg, 1980). Our re-analysis of Sandia's calculations, using real property values, puts the estimated damages in the tens of billions of dollars. The details of a transportation accident, the levels of radioactivity from the scene of the accident outward, and the economic consequences of an accident are discussed more fully in our study.

REFERENCES

AGNS (1978) (Allied-General Nuclear Services) *Current Status and Future Considerations for a Transportation System for Spent Fuel and Radioactive Waste*, OWI/SUB-77/42513, Barnwell, SC (February 1978).

APL (1982) (Applied Physics Laboratory, Johns Hopkins University, Laurel, Md.) letter from R. Fristrom to N. Bernstein, CEP, (3 June 1982).

Ballard, D. W. (1982) Personal communication with N. Bernstein, CEP, (16 August 1982).

Barker, R. F. (1965) *Proceedings of the International Symposium for Packaging and Transportation of Radioactive Materials*, in question and answer period of paper by G. J. Appleton and J. Y. Servant, 'Packaging standards' (1965).

Barker, R. F. (1981) 'How some of the package tests were conceived', Seminar on Transport Package Test Standards, Tokyo, Japan, (1 October 1981).

Eisenberg, N. (1980) 'Transport of radionuclides in urban environs', presented at the 6th International Symposium on the Packaging and Transportation of Radioactive Materials, Berlin, West Germany (November 1980).

GE (1981) (General Electric) letter to R. Odegaarden, US Nuclear Regulatory Commission, Washington DC (June 1981).

IAEA (1961) (International Atomic Energy Agency) Regulations for Safe Transport of Radioactive Materials, Safety Series No. 6, Vienna (1961), article by A. Grange, p. 83.

ITC (1961) (International Trade Commission) *Synthetic Organic Chemicals, U.S. Production Sales 1960*, Washington, DC (1961).

ITC (1981) (International Trade Commission) *Synthetic Organic Chemicals, U.S. Production Sales 1980*, Washington, DC (1981).

NRC (1975) (Nuclear Regulatory Commission) *Reactor Safety Study*, WASH-1400, Washington, DC (October 1975).

NRC (1976) *Final Generic Environmental Statement on the Use of Recycle Plutonium in Mixed Oxide Fuel in Light Water Cooled Reactors*, NUREG-0002, Washington, DC (August 1976).

NRC (1977) *Transportation of Radioactive Materials by Air and Other Modes*, NUREG-0170, Washington, DC (December 1977).

NRC (1979) *A Scoping Study of Spent Fuel Transportation Accidents*, NUREG/CR-0811, Washington, DC (June 1979).

NRC (1980a) *Puncture of Shielded Radioactive Material Shipping Container*, NUREG/CR-0930, Washington, DC (April 1980).

NRC (1980b) *Transportation of Radionuclides in Urban Environs: Draft Environmental Assessment*, NUREG/CR-0743, TRUE Study, Washington, DC (July 1980).

NRC (1980c) *Potential Crush Loading of Radioactive Material Packages in*

Highway, Rail and Marine Accidents, NUREG/CR-1588, Washington, DC (October 1980).

ORNL (1970) (Oak Ridge National Laboratory) *Siting of Fuel Reprocessing Plants and Waste Management Facilities*, ORNL-4451, Oak Ridge, Tennessee (July 1970).

Perry, R. H. (1973) *Chemical Engineers' Handbook, Fifth Edition*, (New York: McGraw-Hill, 1973).

PNL (1978) (Pacific Northwest Laboratory) *An Assessment of the Risk of Transporting Spent Nuclear Fuel by Truck*, PNL-2588, Richland, Washington (November 1978).

REA (1982) (Ridihalgh, Eggers & Associates Inc) *Final Report on Severe Rail and Truck Accidents: Toward a Definition of Bounding Environments for Transportation Packages*, prepared for the Nuclear Regulatory Commision (December 1982).

Sandia (1978a) Sandia National Laboratories, *Non-Destructive Testing Evaluation of LWR Spent Fuel Shipping Casks*, SAND78-0309, Albuquerque, New Mexico (February 1978).

Sandia (1978b) *Severities of Transportation Accidents Involving Large Packages*, SAND77-0001, Albuquerque, New Mexico (May 1978).

Sandia (1979) *Proceedings of the Nuclear Materials Transportation Program Development Seminar*, SAND79-2262, Albuquerque, New Mexico (April 1979).

Sandia (1980d) *Assessment of Accident Thermal Testing and Analysis Procedures*, SAND80-0060C, Albuquerque, New Mexico (April 1980).

Shappert, L. B. (1980) 'U.S. cask requirements and industry capability survey', presented at the 6th International Symposium on the Packaging and Transportation of Radioactive Materials, Berlin, West Germany (November 1980).

Shieh, R. C. (1980) 'Recent developments in puncture analysis techniques of nuclear shipping container components', presented at the 6th International Symposium on the packaging and Transportation of Radioactive Materials, Berlin, West Germany (November 1980).

Sierra Club (1980) (Sierra Club Radioactive Waste Campaign) 'Shipping casks: are they safe?', Buffalo, N.Y., 1980.

UKAEA (1963) (United Kingdom Atomic Energy Agency) 'Interim recommendations for the application of environmental tests to the approval of packaging', AHSB(s)R-19, England (1963).

19 Fire Tests and their Relevance

HARBANS L. MALHOTRA

I shall be the exception in that I shall not concentrate on fire tests specifically for nuclear flasks. I shall attempt to provide important background about the nature of fire tests in general and to do that I need to sketch some history.

BRIEF HISTORY

Fire tests, in some shape or form, seem to have been conducted ever since concern grew over damage caused by fires and the need to design structures to minimize the undesirable consequences. The past concern had been with fire damage in buildings, particularly when conflagrations such as the great fire of London in 1666 wiped out large parts of the city. Experience on fire behaviour was gained at that time by examining the fire damage. However, in 1773 David Hartley, to prove the effectiveness of a particular form of construction, bought a house on Putney Heath and set fire to it. The techniques were improved over the next hundred years or so. When the Denver Equitable Building Insurance Company wanted to assess a floor construction, it built a special chamber to conduct a test in 1890. The art was brought to much greater refinement by the British Fire Protection Committee in 1895 by building special furnace chambers at Regents Park and later on in Westbourne Park to undertake a programme of systematic testing. Using producer gas and boosting the heating where necessary with wood fuel, it was possible to subject walls, floors and doors to standardized heating conditions of 816°C (1500°F) or 982°C (1800°F) for up to 4 hours. Testing at constant temperatures was not unique to BFPC as in the USA a specification issued in 1907 required a temperature of 926°C (1700°F) to be maintained for 4 hours as this was claimed to approximate the heat of a burning building.

The first standard specification on fire resistance tests was published in the UK in 1932 as BS 476, and gave a grading system. Other countries notably the USA, Germany and Sweden, issued similar specifications in the same era. One main difference between these specifications and earlier test methods was the use of a transient heating curve rising rapidly to over 900°C in 60 minutes and then more slowly to above 1200°C in 6 hours. The type of heating regime, known as the standard temperature time curve, has now acquired an international recognition by being specified in the International Standards Organization (ISO) specification no. 834 and is being put forward by the EEC in document 1202 as a basis for harmonizing fire test procedures in the member countries. This so called fire curve has been used for the last 50 years to provide information on the behaviour of building components in fires and has sometimes been mistakenly assumed to represent the pattern of a building fire.

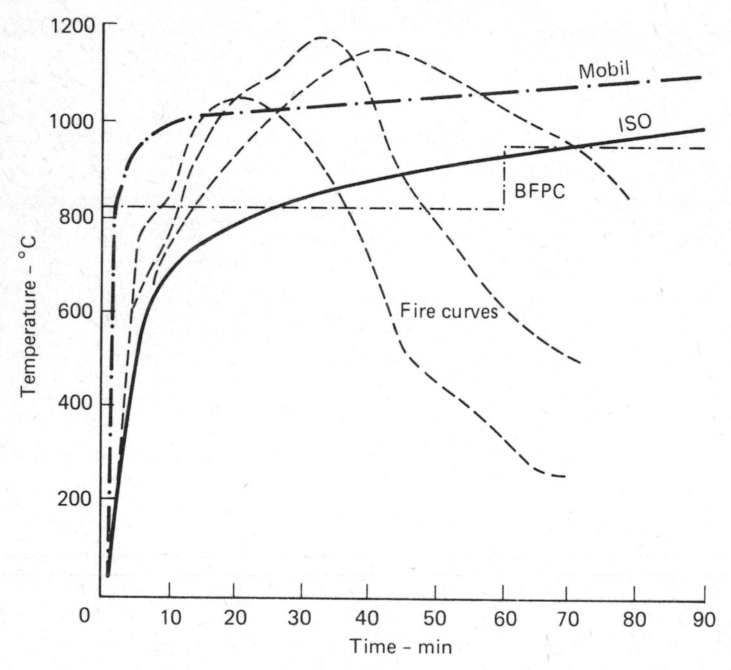

FIGURE 19.1 *Temperature/time curves*

Figure 19.1 shows the ISO fire curve, which is the one in common use internationally, including the UK; it also shows the 'Mobil' curve, which is used in the oil industry in relation to hydrocarbon fires (the main difference being that the temperature rises very rapidly to nearly 1000°C – much faster than the normal building fire curve). The curve marked 'BFPC' shows the recommendation of the British Fire Protection Committee in 1892 – a constant temperature of about 800°C, rising to around 900°C for up to 3 or 4 hours. In relation to modern thinking the BFPC curve is outdated. The important point is that fire is a dynamic situation involving temperatures which vary through time according to the combustion conditions and properties of the materials concerned.

THE CURRENT POSITION

Fire tests have become an integral part of fire safety and regulatory systems. Levels of safety are specified for different buildings in terms of fire resistance times and compliance is automatically assumed when appropriate test evidence can be provided. This is rather a unique position for a test to have and differences in performance of a few minutes can affect the acceptability of a product. The major activity over the last two or three decades in fire resistance testing has been to improve and refine the test procedure to decrease variability and improve reproducibility. No fundamental conceptual change has occurred. However, recently questions have been raised about the meaning of tests and their validity. Studies of actual incidents as well as experimental fires have shown that fire tests are unable to reproduce the complex phenomenon of a real fire where fuel, ventilation and the environmental conditions interact in a random manner generating virtually different fire patterns each time. The temperature/time relation of the standard curve occurs but rarely and many factors which influence the behaviour of composite systems cannot be easily reproduced in a laboratory test.

This has created a dilemma for the practitioners who have been relying absolutely on fire tests. In some instances attempts have been made by the industry to introduce new fire curves to assist with their particular problems. The oil industry is an interesting example as it has produced special temperature/time relationships to represent the severer heat output from a hydrocarbon fire. Consequently the laboratories have been busy adjusting their apparatus to see if it can be run at higher levels to meet these new criteria. Unfortunately, the approaches used to resolve some of these newer problems have not always followed a rational or a logical path by analysing the problems before seeking a resolution. The issues of validation have not been adequately dealt with and, therefore, an assumption cannot be made that the newer tests are nearer to reality than the earlier ones. More recent discussions are encouraging as the scientists and the applicators, i.e. safety authorities, have become aware of the need to consider the total problem and not fire tests in isolation.

TYPES OF TESTS

Before dealing with a more rational system, it is worth considering the general nature of fire tests currently in use. They can be divided into four different types on the basis of their design and use (see Figure 19.2).

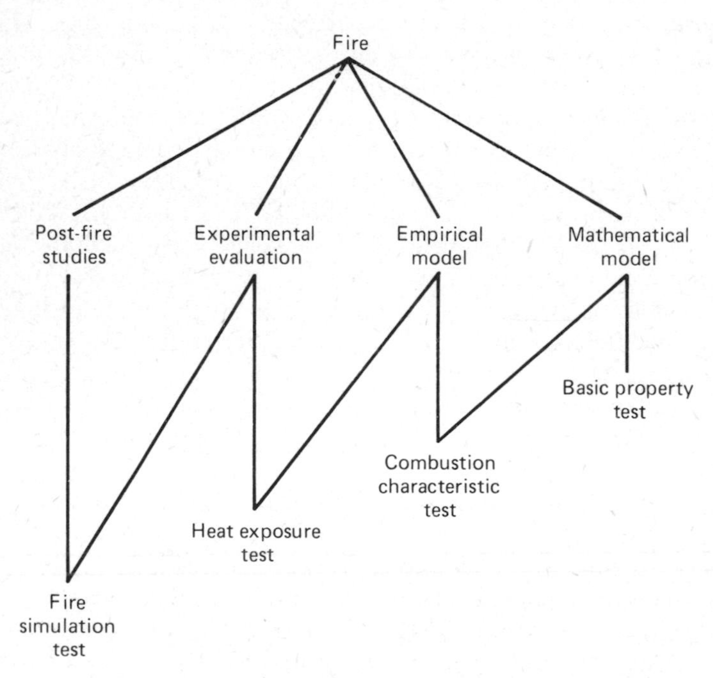

FIGURE 19.2 *Different types of fire tests in relation to fire*

(1) *Basic property tests* establish thermal or other properties having influence on the behaviour of a material or product in a fire.
(2) *Combustion characteristic tests* establish under specified conditions the burning behaviour or the decomposition characteristics of a material or product.
(3) *Heat exposure tests* are applied to components and sub-systems and assess under a particular condition of fire simulation the behaviour of a component or a building element.
(4) *Fire simulation tests* are usually *ad hoc* procedures which reproduce a particular fire scenario in order to examine the fire behaviour of a complete structure or a part under conditions approaching a real-life fire. This type of categorization has been accepted by an ISO Committee on the coordination of fire tests and whilst in the UK its counterpart has taken a slightly different route, the basic concepts are similar as indicated in a recently published document, BS 6336 'Guide for the development and presentation of fire tests and for their use in hazard assessment'. Another more basic document will emerge shortly from the committee responsible for the majority of the British Standard fire tests to provide a guide to the principles and application of fire testing. It warns that fire tests alone cannot measure fire hazard nor can the results of a test guarantee a particular degree of safety. In connection with fire resistance tests in particular it insists that no direct relationship should be assumed between performance in the test and in real fires, nor does the period of fire resistance indicate that in a fire of an equal duration similar performance will be obtained.

For our present purposes perhaps the fire simulation tests are of greater interest. These are designed to reproduce in a realistic manner, by choosing a suitable fuel and controlling its burning state, a selected fire scenario and subject a construction to the resulting heating conditions.

These provide a link between standardized laboratory procedures and observed fire procedure. However, they require careful design and planning, and expertise in the interpretation of data. Where fire knowledge is lacking, the tests can be misused and the result misapplied in ignorance.

VALIDATION OF FIRE TESTS

Without validation a fire test cannot provide confidence that is needed in its reliability and its appropriateness for the purpose for which it has been specified. Reliability of the test is concerned primarily with the soundness of design, i.e. mechanical features and the arrangement of the test. The mechanical reliability is the easiest to deal with as it requires the application of general apparatus design principles and should include the ease of operation and acceptable levels of repeatability and reproducibility. Systematic variability should be eliminated and random variability reduced as far as possible. The test arrangement is a translation of the test concept into practical terms and attention should be paid to the representative nature of the heating system, the sample, the method of exposure and the measurements to be made. The appropriateness of the test takes into account the test concept, the type of data obtained and the ability of the data to predict behaviour under conditions of use and exposure which are of interest. In other words, the test must be able to provide meaningful information. Ideally, a scenario should be prepared for the fire aspect on which information is required. This may be the burning characteristic of an individual material or a component or the behaviour pattern of a complex structure or some other system. Decisions have to be taken on the fire phase which is to be considered. Is it the early growth from the start of ignition, or its development to the flashover point, or a fully developed fire within a building, or a liquid fire in the open, or some other special hazard such as liquid fuel burning on water? It is possible to quantify the heat exposure conditions. Very few fires produce constant temperatures; they are usually characterized by transient conditions. Description of a fire by temperature is an inadequate specification; fires transfer their heat by convection and radiation, hence a precise description requires both these factors to be specified. One of the criticisms of the standard fire resistance test is its incomplete definition by the standard temperature/time curve which allows real differences to exist between equipment in different countries.

Once the fire has been defined, the next factor to consider is the manner of exposure of the construction, manner and duration of application of the heat. The time so arrived at is not necessarily the real duration of a fire, but the time for which a known type can burn at a specified burning rate. Perhaps the most difficult aspect of validation procedure is to develop suitable criteria which enable the test data to be used in a meaningful way. This requires a knowledge of the behaviour pattern in a fire and then developing precise relationships between the test and the fire environment. Many tests have been wrongly developed and used in the past because this aspect of the validation process was ignored. A good example of this is the range of tests developed to measure self-extinguishing properties of materials.

Appropriate criteria can be developed by undertaking an analysis of the fire, by studying past incidents and, when necessary, by undertaking well-planned fire-simulation tests. By collecting and analysing data, it should be possible to build a model of the situation for which information is needed to develop criteria expressed as precise phenomenon or technical measurements which are indicators of performance. For example, enough experience exists with fire-resistance tests to show that in building fire structural stability and fire containment as two objectives can be defined by criteria related to loadbearing capacity, resistance to passage of flames and heat transfer. The same criteria may be inappropriate if the fire were to occur in the open or in an underground tunnel.

FIRE SAFETY SYSTEM

The main objective of fire safety is to provide adequate protection against hazards that may be created by the occurrence of a fire. As mentioned previously, passing a test does not and cannot guarantee safety. The whole fire problem has to be studied and its various components considered and a fire-protection strategy developed. The essential components of such a strategy are studies of statistical data, defining levels of protection needed or demanded, evaluating the

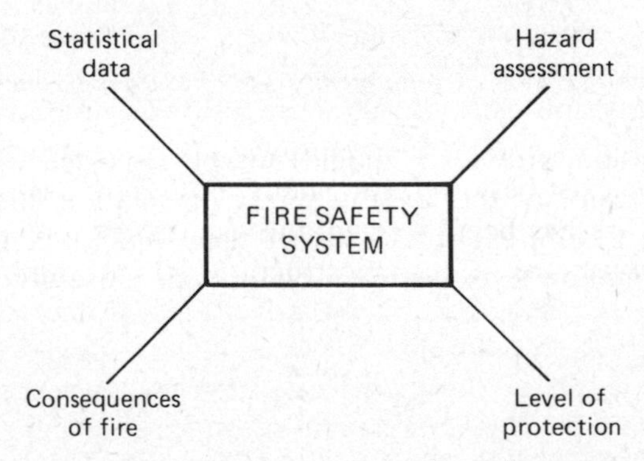

FIGURE 19.3 *Components of a fire safety system*

hazard and considering the feasibility of various protective measures (see Figure 19.3). Hazard assessment is one of the subjects worthy of greater study than it has received so far. Fire hazard may be defined as the potential of a fire to cause harm to people or property in its vicinity. Hazard assessment is an evaluation of this potential and is an essential component of any fire safety system. Hazard assessment takes into account the risk of a fire starting, the nature of materials likely to be exposed to the fire, the method of their use and application, the consequence of such an exposure on their properties and the potential damage that might occur (see Figure 19.4). The burning of a building in isolation in the middle of an open area or wilderness may be of little consequence, but the same fire in the middle of a city or in a high-rise building can have much greater potential for harm.

Fire tests can be used in a number of ways in a hazard assessment scheme: they can provide a measure of fire risk on the basis of the nature of materials, they can establish the likely behaviour of materials on the basis of their use and they can indicate the likely damage when they are used and exposed to a fire.

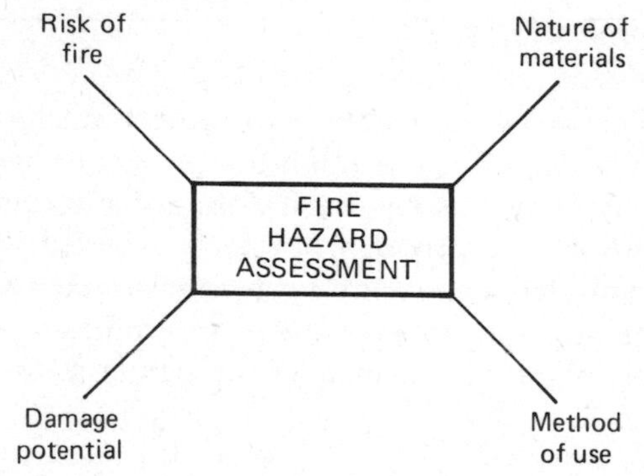

FIGURE 19.4 *Components of fire hazard assessment*

In conclusion, fire tests should not be regarded as capable of indicating hazard or the level of safety in isolation, they should be regarded as a component of a hazard assessment scheme and a fire safety system. Fire tests used sensibly provide a useful tool for safety experts.

20 Impact Modelling and Reduced-Scale Tests

BRIAN EVASON

We have been told that the transportation flasks have been designed with large factors of safety against failure due to foreseeable impacts, and of how this has been confirmed by experiment. We have also heard that credible doubts have been raised as to the validity of the tests at a reduced scale. I am sure that you wish me to comment on the question of whether or not the flasks are proof against any conceivable collision. That I cannot do as I have had no involvement, nor sufficiently detailed information, to justify such comment.

What I hope to do is to explain, or clarify, the principles, capabilities and limitations of reduced-scale modelling of structural behaviour in impact situations. The purpose of this paper is to provide background technical information.

IMPACTS

It might be as well to start by discussing the salient features of an impact, or collision between two bodies. In this case, as is usual for accidents in the sphere of civil and structural engineering, we are concerned with the impact of a moving body, referred to as the missile, on a stationary body referred to as the target.

I apologise for the emotive terminology but there are not necessarily any militaristic connotations. It is unfortunate that the predominant use that man has found for impacts is belligerent. As a result most impact testing was carried out for military ends and it is only in recent years that tests have been carried out for specifically civilian purposes.

My own interest in impacts is normally with the behaviour of the target since I am concerned with the havoc that might be wreaked by errant ships or bursting boilers. This is immaterial to the principles since it is of fundamental importance to remember that impact behaviour is dictated by the interaction between the two bodies, and little useful information can be gleaned from the characteristics of one only. A falling trapeze artist would prefer to land in a safety net although the parameters describing his (the missile's) flight are unaffected by the surface he will meet, until contact is established.

Intuitively it can be realised that weight, or rather mass and velocity, of the missile are important factors which will dictate impact loading. Equally obviously, strength parameters of each body, dictating the rate of deceleration of the missile, can be expected to be important. These and several other parameters must be taken into account in the analytical or experimental investigation of an impact.

For normal working loads, under which a structure must remain essentially elastic in its response, analytical techniques are found to give an accurate assessment of performance (see Figure 20.1). However, as response deviates further and further from this elastic behaviour, and permanent deformation becomes significant, and as structural com-

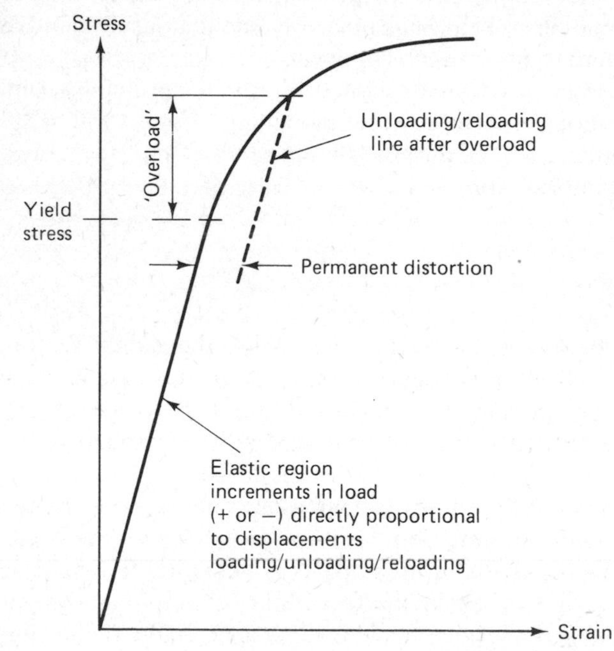

FIGURE 20.1 *Typical uniaxial stress/strain relationship for structural materials*

plexity increases, it is found that the simplifying assumptions, necessarily introduced into the analysis, cause progressively greater deviations between calculated and observed behaviour. When these difficulties are exacerbated by the introduction of time varying loads, the designer's problem approaches the intractable.

Analytical solutions of impact conditions tend to be realistic only for slow (low deceleration, essentially elastic response) collisions, or for very simple structures (or those which can reasonably be represented as such). For the more complex problems the designer must make recourse to experiment, either to calibrate or validate this calculation (possibly using existing empirical evidence), or to obtain *ad hoc* design data. For such experiment to be useful the experiment must echo local and global deformations of both missile and target zone, and, generally, give information on loads transmitted to foundations or remote parts of the target.

Logically the best experimental evidence would be obtained by fabricating exact duplicates of both missile and target and observing the exact postulated impact. This is not necessarily practicable, desirable, or even possible.

As specific examples of impact cases that are actually considered, this could involve the provision of a North Sea oil-rig and ship, or a nuclear power station and an aeroplane. In fact, since a single test is generally inadequate to gauge consistency or effects of variations of detail, it would require several sets.

The problem can be reduced by fabricating only those parts in the region of impact, shorn of trivial details, and simpler structures to simulate the support or dynamic mass and stiffness characteristics of the target and missiles. This full-scale modelling requires comparable care to that needed for reduced-scale modelling. It does not alter the weight of the missile to be projected under strict control of velocity, aspect and flight path. The resulting waste products and the risk to personnel and nearby structures, when handling excessive loads, must also be considered.

REDUCED-SCALE EXPERIMENTS

In many situations it is appropriate to fabricate and test reduced-scale models to represent the significant parameters with components which can more readily be handled, controlled and observed. In general the parameter of greatest interest is the degree of damage. This is not readily quantifiable and it is desirable that the damage in the actual structures, or prototypes, is duplicated in the model. It is then possible to see which parts are broken, severely distorted or undamaged and judge whether this level is acceptable.

The idea of using reduced-scale models to simulate structural behaviour is not new. It is traceable back to Roman times when Vitruvius concluded it did not work. Da Vinci attempted to disprove this conclusion but failed. The reason for these failures was essentially that they were trying to model gravitational or weight effects without similarly scaling gravitational acceleration; they were unaware of the need to do so. This illustrates a main danger of any form of modelling, the failure to include all relevant parameters. Galileo is credited with initiating the basic ideas of dimensional analysis, the primary tool for defining scaling laws. These ideas were improved and used by such distinguished researchers as Newton and Lord Rayleigh, leading to numerous advances in many and varied branches of science. But it was not until 1915 when Buckingham outlined his π theorem that the approach was put on a formal basis. Coincidentally, in the same year, Hopkinson propounded his scaling rules for blast loading to the Ordnance Board. This is believed to be the first scaling law pertinent to the response of structures to dynamic loads.

DIMENSIONAL ANALYSIS

The basic principle of dimensional analysis is that equations relating physical properties are dimensionally consistent. A length measured in pounds per square foot is meaningless – unless some calibration constant is included with the description. It is in fact quite common to measure hydraulic pressure in feet – actually in feet head of water. There is a precise, dimensionally consistent relationship between hydrostatic pressure and depth of water.

The first stage of the examination of any physical system is to prepare the general equations of motion, or energy balance, for the relevant conditions. The relationship must include all the relevant parameters including dimensional constants such as pressure per foot head of water, or gravitational acceleration.

If the ensuing relationship is capable of exact mathematical solution the prediction will be reliable. As this is very rarely the case even in static conditions, the next alternative is to obtain a solution by numerical approximation. This method, particularly since the advent of the computer, allows many more realistic solutions to be obtained. When conditions of load, structure and material become more complex the approximations in such computer simulations can get out of hand.

It is then left to attempt a solution by experiment. It may not be acceptable to model the exact conditions but it is not necessary. All that is required is a model which is governed by the same defining equations. This, of course, is not as easy as it may sound.

All the physical quantities must be associated with units, or dimensions. Some dimensions are considered to be fundamental, the others derived. For example, taking length and time as fundamental, the units of velocity can be derived from the immutable relationship that velocity is the rate of change of displacement. Velocity, therefore, has the units of length per time (m.p.h., metres/second). We could, of course, also say that length has the units of velocity times time.

The choice of which units to take as fundamental is arbitrary and they are chosen for convenience; that is, simplicity in application. In a purely mechanical or structural system three fundamental units are normally adequate. Those most commonly used are force, length and time in static conditions or mass, length and time in dynamic conditions. Newton's laws of motion provide the relationships between the alternatives. The law of similitude, or dimensional homogeneity, formalised by Buckingham, states that any physical relationship can be expressed as a function of a set of independent dimensionless products composed of the relevant physical parameters.

When the associated manipulation is carried out on the defining equations of the system, a number of dimensionless groups will be obtained, depending on the number of physical quantities necessary to describe the behaviour. By varying one dimensionless group, and keeping the others constant, a series of experiments will establish the dependence on that group. With a large enough series of tests it may be possible to establish the dependence on each physical parameter. If the results of tests with variation of a group remain constant then that group, and usually some physical quality, can be eliminated from the relationship as not affecting the behaviour.

MODELLING

In order for an experiment to represent the behaviour of the prototype, those dimensionless groups that are relevant must be invariant between prototype and model. These requirements remain whether the model is at full or reduced scale. If one variable is scaled by a model scale factor, then factors are dictated for the scales of other variables or groups of variables. The details, apparent exceptions and peculiarities of this analysis are a fascinating but lengthy study. This is, however, the basic principle of the model definition.

Consider the blast modelling proposed by Hopkinson. From a theoretical study and background experimental experience, fifteen parameters are considered relevant to the evaluation of blast pressure at a target (see Figure 20.2). Some, such as the energy of the explosive charge (directly proportional to its mass), and distance between charge and target, are immediately obvious.

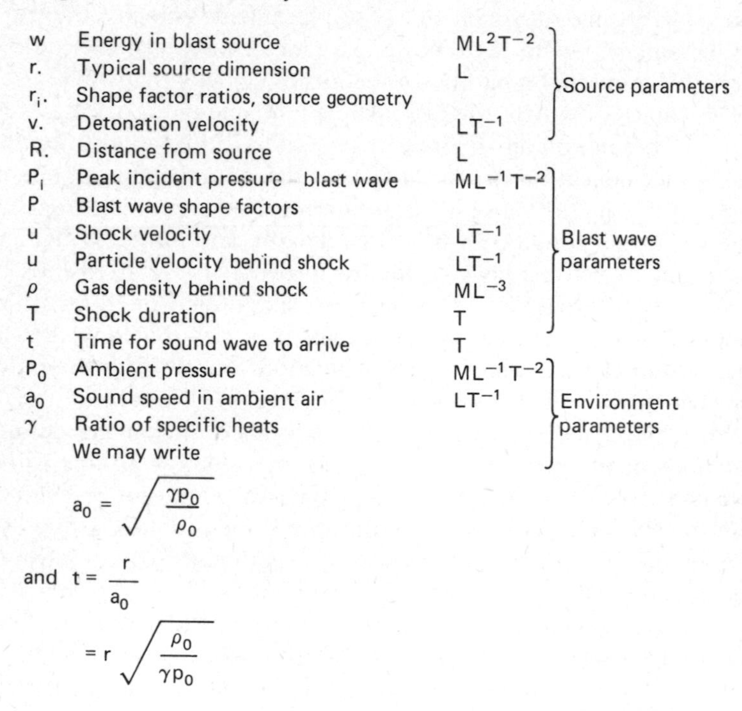

w	Energy in blast source	ML^2T^{-2}	Source parameters
r.	Typical source dimension	L	
r_i.	Shape factor ratios, source geometry		
v.	Detonation velocity	LT^{-1}	
R.	Distance from source	L	
P_I	Peak incident pressure – blast wave	$ML^{-1}T^{-2}$	Blast wave parameters
P	Blast wave shape factors		
u	Shock velocity	LT^{-1}	
u	Particle velocity behind shock	LT^{-1}	
ρ	Gas density behind shock	ML^{-3}	
T	Shock duration	T	
t	Time for sound wave to arrive	T	
P_0	Ambient pressure	$ML^{-1}T^{-2}$	Environment parameters
a_0	Sound speed in ambient air	LT^{-1}	
γ	Ratio of specific heats		

We may write

$$a_0 = \sqrt{\frac{\gamma p_0}{\rho_0}}$$

and $$t = \frac{r}{a_0}$$

$$= r\sqrt{\frac{\rho_0}{\gamma p_0}}$$

FIGURE 20.2 *Parameters for evaluation of blast pressure*

The dimensional analysis leads to the dimensionless groups shown in the left-hand column of Figure 20.3. These represent a completely generalised modelling law until restrictions are introduced.

Hopkinson's similarity rules for blast loading state, essentially, that using the same explosive in the same air conditions, similar blast waves

		Effect of Hopkinson restriction
$\pi'_1 = r_i$	$\lambda_{ri} = 1$	Geometric similarity of whole experiment
$\pi_2 = \frac{R}{r}$	$\lambda_R = \lambda_r = \lambda_L$	
$\pi_3 = U\sqrt{\frac{\rho_0}{\gamma p_0}}$	$\lambda_U = \lambda a_0 = 1$	Invariance of velocities
$\pi_4 = \frac{U}{u}$	$\lambda_U = \lambda_u = 1$	
$\pi_5 = \frac{pi}{po}$	$\lambda_{pi} = \lambda_{po} = 1$	Invariance of pressure and pulse shape
$\pi_6 = p$	$\lambda_p = 1$	
$\pi_7 = \frac{pi\, r^3}{w}$	$\lambda_w = \lambda_r^3$	Scaling of blast energy
$\pi_8 = \frac{\rho u^2}{pi}$	$\lambda_p = 1$	Invariance of shock density
$\pi_9 = \frac{t a_0}{r}$	$\lambda_t = \lambda_r = \lambda_L$	Equivalence of time and space scaling
$\pi_{10} = \frac{v}{u}$	$\lambda_u = \lambda_v = 1$	Equivalence of velocities
$\pi_{11} = \frac{\tau}{r}\sqrt{\frac{\rho_0}{\gamma p_0}}$	$\lambda_\tau = \lambda_r = \lambda_L$	Time and space scaling
$\pi_{12} = \gamma$	$\lambda_\gamma = 1$	Atmospheric invariance

FIGURE 20.3 *Effect of Hopkinson restriction*

TABLE 20.1 *List of parameters for impact studies*

Parameter	*Value*	*Dimensions*
Missile diameter	d	L
Shape factor ratios	r_i	
Angle of attack	β	
Density of missile material	P_m	ML^{-3}
Density of target material	P_t	ML^{-3}
Impact velocity	v	LT^{-1}
Target thickness	h	L
Missile temperature	θ_m	θ
Target temperature	θ_t	θ
Specific heat of missile material	C_m	$L^2\,\theta^{-1}T^{-2}$
Specific heat of target material	C_t	$L^2\,\theta^{-1}T^{-2}$
Heat of fusion of missile material	n_m	L^2T^{-2}
Heat of fusion of target material	n_t	L^2T^{-2}
Ultimate strength of missile material	σ	$ML^{-1}T^{-2}$
Ultimate strength of target material	S	$ML^{-1}T^{-2}$
Other stress ratios	σ_i	
Strain	ϵ	

TABLE 20.2 *Scaling relationships associated with impact studies*

Parameter	*Dimensions*	*Full-scale value*	*General model value*	*Model value using Replica modelling*	*Typical quantities*
Length	L	d	$s_L d$	sd	Missile dimensions
					Target dimensions
Mass*	M	m	$s_m m^*$	$s^3 m$	Missile mass
					Target weight
Time	T	t	$s_t t$	st	Duration of load
					Natural period of target
Density	ML^{-3}	P	$s_\rho P$	P	Densities of materials in structure and missile
Strain	—	E	$s_E E$	E	
Stress	$ML^{-1}T^{-2}$	σ	$s_\sigma \sigma$	σ	Includes pressure and material moduli
Velocity*	LT^{-1}	v	$s_v v^*$	v	Missile velocity
					Stress wave velocities
					Ejecta velocity
Acceleration	LT^{-2}	a	$s_a a$	$s^{-1}a$	(Except gravity)
Rate of Strain	T^{-1}	r	$s_r r$	$s^{-1}r$	
Force*	MLT^{-2}	w	$s_w w^*$	$s^2 w$	Applied load
					Structural reactions
Energy*	ML^2T^{-2}	E	$s_E E^*$	$s^3 E$	Kinetic energy of missile

NOTES The scale factor s is defined as $s = f_m/f_p$.
Parameters marked * are *not* independent quantities.

with identical peak pressures are produced at scaled distances from similarly scaled charges of the same geometry. The familiar cube root term is introduced if the size of the charge is referred to by mass instead of characteristic length.

Assuming that peak pressure is the property requiring evaluation it can be directly measured in reduced-scale experiments, using the same materials, provided that geometrical similarity is observed. This concept is known as Replica Scaling.

If this rule is applied to the general modelling law it is found that the other dimensionless groups remain invariant. Provided that all the relevant parameters have been included, or that any omitted also fit the scaling laws (in their normalised form), the experiment will yield the correct pressure irrespective of whether the scaling is one-to-one or one million-to-one.

This model can be extended to include normalised target response since similarity of loading has been achieved, in intensity and

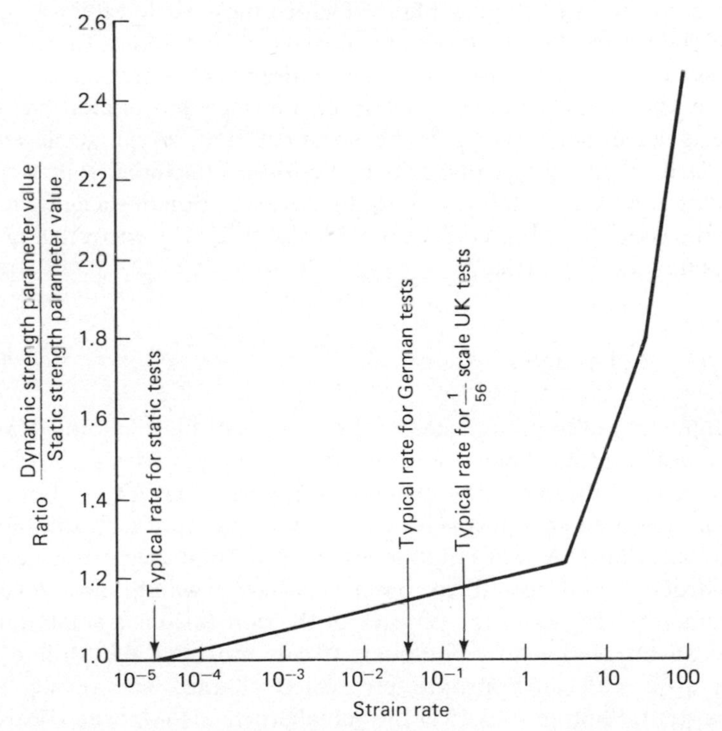

FIGURE 20.4 *Effect of strain rate on compressive strength of concrete*

distribution and the relationship between stress history and strain is uniquely defined by the nature of the material, which is the same in both cases. Impacts are a similar type of problem except that the impulse or energy is carried by a solid rather than the air, and the transfer zone is localised to a smaller contiguous area.

However, we might expect, or know from experience, that some of the kinetic energy of the missile is dissipated as heat, originating at the contact area. In order to maintain similarity between model and prototype conditions, parameters to reflect the generation and transmission of heat must be included. Replacing blast wave parameters by missile parameters a comparable set of pertinent physical properties is obtained.

If the restrictions of Replica Scaling are applied to the ensuing dimensionless groups, it is found that the temperature effects do not scale, and thermodynamic similarity is not achieved. In the absence of a better scaling law, impacts have been investigated using Replica Scaling, anticipating scale effects or deviations with scale due to this defect in the law. At very high velocities these scale effects become very significant. For the types of material used in building this requires speeds of several hundreds or even thousands of m.p.h.

For the types of impact met in civilian accident conditions heat effects have been found to be so small that, in practical terms, similarity is not jeopardised. It is therefore concluded that Replica Scaling can be acceptably used for the investigation of such impacts as are the concern of this conference. This sounds very encouraging, but does it work in practice?

BLAST LOADING

Within certain limitations it does. In the case of blast loading it works very well, and has been confirmed over a range of scales. The most spectacular example is the examination carried out for the bouncing bomb designed by Barnes-Wallis for the RAF attack on the Mohne and Eder dams. A series of models at about one-twentieth scale were constructed and tested to optimise charge weight and location parameters. To verify the efficacy of the conclusions a small dam in Wales, very similar to a one-quarter scale model of the Mohne dam was used with an appropriately scaled charge. The results were consistent. Photographs after the actual attack also showed consistent damage.

Tests at one-twentieth, one-quarter and full scale show, by their agreement, a major validation for Replica Scaling, for the already-checked Hopkinson rules, and more importantly its value in reproducing structural response. There are, however, limitations in the range of validity which can give rise to scale effects. One in particular is that, due to the nature of the chemical reaction, the explosive energy in very small charges is no longer directly proportional to the charge mass.

LIMITATIONS IN MODELLING IMPACTS

In the impact case the position is less obvious and what appears to be scale effects can assume major importance. The main factor is found in the basic definition of Replica Scaling, that the same materials are used. What is meant by the 'same material' has not yet been defined. The actual requirement is that the relationship between stress and strain for the prototype stress history is duplicated in the model having the model stress history, which, though similar, is faster in objective time.

At first sight this does not appear to present difficulties. However, it is known that the experimentally recorded material properties of nominally identical material samples depend on the rate at which load is applied. Also there is a finite limit beyond which samples of nominally identical material, of different size, exhibit different properties.

Suppose a one-tenth scale model is required to examine the behaviour of a 6-inch slab of standard concrete with 1-inch maximum diameter aggregate. Clearly it is not only unreasonable but actually impossible to fabricate the model from the same mix of concrete. A mix of micro-concrete which echoes the stress-strain relationship would be used.

A similar situation exists with steel but is less obvious and often unimportant for the physical sizes in the model and prototype. Of greater concern is the fabrication detail. For a welded structure (unless annealed or specially treated) properties will not scale in the heat-affected zone. Welds must therefore be fabricated to give the appropriate strength behaviour rather than geometric exactitude. Since they do not significantly change the mass or stiffness distribution, realistic models can be obtained. This may also apply to other forms of joint, change of section or discontinuity. While these problems can realistically be catered for they do tend to limit the scale factor that can be used for structures with complex fabrication.

The enhancement of strength with increased rate of loading is generally considered to be a function of the increased rate of strain (Figure 20.4). Strain itself is invariant between model and prototype but strain rate varies with scale and its effect must be assessed.

This therefore implies that the model strength will be greater than the full-scale prototype. This is not so serious as it sounds, partly because typical strain rates in normal civilian accidents lie in the shallow first portion of the curve where the increase is only a few per cent for an order of magnitude scale change, and partly because the local strength increase demands a greater decelerating force and higher imposed stress. Provided that the object is to discover extent of damage the effect is partly self-compensating although not strictly obeying similarity requirements. If the effect is partly due to rate of deformation, rather than rate of strain, this effect may scale.

Problems can also arise because of difficulties in modelling tolerances and imperfection sensitivity. An important fact to remember is that gravity does not scale. For impact tests dead weight is generally unimportant to the structural behaviour. However, once material is broken off, e.g. scab or spall, it is ejected from the target appropriately, with the correct velocity, but its trajectory is then governed by gravity. The result is that debris will scatter further in a reduced-scale model. If the position at which it comes to rest is important this can be calculated, or re-scaled from the model position. This is a tedious process but it can be achieved. It is not usually of interest, since in most applications it is the amount broken off rather than where it lands that is important. This also applies to the blast condition.

Since gravity does not scale it may immediately suggest doubts as to the validity of a drop test to model a horizontal impact. It is, of course, incorrect but it is conservative since an additional component of load has been added, albeit incorrectly. Modelling a vertical case, the effect does not scale but even at the low speed of 30 m.p.h. (9 metre drop) it means that 9 per cent of the total load is incorrectly scaled. The error will be a proportion of this 9 per cent which will increase as the model scale is reduced. Within experimental tolerances the effect is trivial except for very low speeds and very large-scale factors.

Specific tests were carried out in the UK to examine the effect of moderate velocity horizontal impacts of solid steel billets and of deformable pipe missiles against micro-concrete targets. In a collaboration they were replicated in Germany at a scale 5.6 times larger, against normal concrete targets of the same strength as the micro-concrete. Initially some inconsistencies were apparent but these were traced to unplanned variations in parameters. The tests then indicated reasonably similar damage (i.e. similar within the range of repeatability of either tests). It should be remembered that the results of two nominally identical tests are unlikely to be identical; this is referred to as experimental tolerance.

The deformable missile tests, at over 400 m.p.h., agreed with analytical prediction. The programme used did not incorporate strain-rate corrections. The same programme was able to predict the extent of unseen damage in a ship collision at less than 10 m.p.h. The observed conclusion that strain-rate effects do not greatly affect the behaviour appear to be further validated by these analyses.

CONCLUSION

It is therefore clear that analytical methods can be used for impact conditions within computational capability. Beyond this range damage levels can be realistically reproduced by reduced-scale models provided that sufficient care is taken in the model details. A single test yields information about the actual impact that is modelled. Care is required in extrapolating test information at reduced or full scale for variations in impact conditions. There is no technical reason why an impact test of shipping flasks containing reactor spent fuel should not be carried out at a reduced scale, provided that the scale factor is not excessive and sufficient care is taken in the model details.

21 Public Perceptions of Risk

BRIAN WYNNE

INTRODUCTION

Whether or not it is true that 'transportation is the most vulnerable part of the entire fuel cycle from the point of view of accident or sabotage,'[1] it is certainly the part of the nuclear fuel cycle which is the most vulnerable to public hostility. Transport from reactors to reprocessing is regarded as a main artery of the fuel cycle. Whereas immediate conflicts over reactors or other plant are at least localised, the transport of spent fuel is somewhat like a highly mobile siting issue. The analogy is emphasised by reminding ourselves that a single typical flask might have 0.5–1.5 million curies of radioactivity – as one critic put it, like 'iodine-free reactors; unguarded and mobile'.[2] The dislocation between meanings which different social groups give to the issue is reflected in the natural language they use to express their experience of it. Spent nuclear fuel is widely and unambiguously treated as 'nuclear waste' in popular perception, yet it is seen as a valuable resource to the industry. Indeed, one suspects that the common term 'spent' fuel has been expunged because of its negative connotations.

The importance to the state as sponsor of nuclear energy of keeping the arteries open has been underlined by the conflicts between local and central government over the right to control nuclear fuel shipments through urban districts. When Ipswich Council attempted in 1980 to create a bye-law to ban such shipments, the Government refused to allow the bye-law any legal status. Various other local authority resolutions banning 'nuclear waste' from their territory have been similarly neutralised, highlighting the general issue of principle about the central restriction of local democracy. This is one of the most potent dimensions of the issue of irradiated fuel transport.

In this general controversy there has been virtually no distinction recognised either between nuclear waste and recyclable nuclear fuel, or between civil and military shipments. Opposition to civil shipments has been closely intertwined with the broader re-emergence of opposition to nuclear weapons generally. The real confusion and official prevarication over possible military uses of plutonium from civil reactor fuels has only added to this embroilment.

As an exercise in risk interpretation, cursory oversight shows that the issue of irradiated fuel transport is far from self-contained, even to within the broader reaches of the full nuclear fuel cycle. Like the long-standing controversy over fluoridation of water supplies but more so, one sees a linkage of apparent conflict between 'objective' and 'perceived' risks of an action with an underlying conflict about the legitimate scope of centralised authority.

In this paper I want to take up the general issue of how to interpret the gaping conflict between the apparently 'objective' and 'perceived' risks of transport of irradiated fuel and of nuclear power in general. In the view I will advance the 'objective risk' framing of an issue is not so much an attempt to identify and rank the risks as to establish a particular framing of the problem to which it is thought public attention should be given, and thus of the problems to which public attention should not be given. My overall aim will be to argue that most approaches to this question – which arises in a wide range of policy issues – are too rooted in the specifics of the technology giving rise to the physical risks, and too 'psychologistic' – they pay inadequate attention to institutional relationships and processes which, when properly treated, imply a radically different view of the problems involved.

RISK DEFINITIONS

Others have drawn attention to the problems in defining precisely what the risks are to the public and transport workers; they have also given examples of the inconsistencies and ambiguities in testing and regulatory practices which are assumed to foster public hostility.[3] I do not know whether there is any large-scale public hostility to nuclear fuel transport as such. This conference has defined (urban) transportation as one separate risk setting for nuclear energy and much effort has been devoted to calculating the discrete risks for this specific setting. Much effort of a different kind has also been devoted to identifying perceptions of risk in similar discrete categories.[4] Each shares the assumption that public reactions are related to these categories of risk – transport, reactor accident, disposal, etc. – and evaluation of their rationality or otherwise is implicitly based on their relation to an identified physical risk factor which gives shape to that category.

It is important to note, and it should give us pause to consider how meaningful are our basic categories for social decision-making, that these basic classifications arise from technological and/or regulatory categories which may have little to do with the ways in which real people experience events and processes. A corresponding set of assumptions about these risk conflicts involves an approach which has been enshrined in the influential but, I believe, fundamentally misleading framework of 'decision analysis'.[5] The relevant aspect of this approach here is that it defines risk decisions according to some taken-for-granted framework, or problem setting, e.g. whether or not to wear seatbelts, or to take out insurance against floods, and then evaluates people's perceptions according to how far they help optimise their position with respect to that one problem. This also leads naturally to the 'objective–perceived' risk dichotomy.

The risk, and social response to it, is assumed to be generated by some given 'intrusion' into a situation; the risk decision and associated options are defined according to that factor. Risk perceptions are then given their multiple attributes also in relation to that factor. Thus, for example, perceptions of some aspect of the array of physical nuclear risks are understood in terms of their deviation from the so-called 'objective' risk, by sub-theories, such as the extra load given to involuntariness; emotional association with images of Hiroshima; the imperceptible nature of radiation; the perceived benefits relating to that factor, etc.[6] Even when these are defended as 'rational' biasing factors, or at least neutrally described, they remain cognitive–emotional categories defined in relation to the physical risk generator.

More progressive decision analysis or psychological approaches have introduced qualifications: (1) that the experts' assessments of the 'objective' risks can be systematically biased.[7] However, this insight has so far only obscured the deeper point that the ensuing expert arguments and uncertainties over the 'objective' risk only consolidate as natural the common basic framing which identified the targets of the risk-identification process in the first place; (2) they have acknowledged the sociologists' criticism that risks as defined by regulatory bureaucracies searching for an artificially precise uni-dimensional scale in which to set an acceptable level may be a meaningless category.[8] They note that 'from a decision making perspective, it makes no sense to speak about acceptable risks. One adopts (or accepts) *options*, not risks. The option may entail some level of risk, but there is no way in which that risk can be separated from its other features'.[9]

However, this sense of enlightenment is tempered by the more fundamental problem this approach does not address, namely that it persists in defining 'options' (and implicitly evaluating rationalities of perception and response-optimisation) in terms of a decision problem assumed to be common to all actors. They are all assumed to adopt a deliberate, active response by selecting amongst options defined in relation to that factor which (amongst other things) embodies some risks.

The approach which I shall propose as a way of interpreting risk conflicts generally, and that over irradiated fuel transportation in particular, takes a different starting point. The essence of this alternative approach is that the most general problem confronting people and organisations is to maintain a sense of order and security: this means a concern to maintain social credibility, some autonomy and trust in surrounding actors. This general problem pervades all our social dealings and all our more specific problems to a greater or lesser extent.[10] Every 'problem' in this sense is actually a mosaic of several connected problems of different levels of visibility and precision. The connections and levels will differ for different actors, who concretely experience an issue in very different ways.

Nuclear fuel transport is a different issue for a railway worker than for a nuclear engineer or say a shoe-factory worker who may never have heard of irradiated nuclear fuel. This is not merely a point about cognition. How a railway worker relates in real life to a nuclear fuel consignment as a 'problem' may have direct empirical implications for other consignments and work relations he has to handle routinely; how a nuclear engineer relates to it embodies a different set of consequences for his or her professional status, social credibility. Others who have no active interest will have an 'attitude' that is only the accidental by-product of other commitments which have an incidental impact on the 'problem'. However, even these can be constraints, since these incidental commitments, e.g. to the passive idea that the local marshalling yard is a benign evironment, can be the source of sudden active interest if attention patterns are somehow changed. What we call an issue is in reality a common context defined by some influential interest group as a 'problem to be managed'. But that common setting is the crossroads for multiple experiences and issues.

It is a common limitation of policy approaches, and of policy research that interests impinging on an 'issue' as defined from the policy-making position are defined by their relation to that issue (if they are even recognised at all) rather than the other way about.[11] This lack of sociological perspective contributes to unrealistic policy formulation in ignorance of problems in later implementation and practical social viability. From this 'verstehen' perspective[12] social institutions, be they formal ones or informal ones like friendship networks, are the origins of meanings, problem-definitions and perceptions: 'risks' are part of a structure of meaning based in the security of those institutional settings in which people find themselves. Such settings interconnect the personal scale and the scale for example of organisations or political cultures. These cultural patterns are being constantly reproduced by people through routine social interaction, in their constant cultivation of a sense of security and autonomy.

Institutions are about the maintenance of social order and trust, whatever their more specific agenda. A sense of threat, crystallised onto discretely identified 'risks' may then arise when these institutional processes break down or change into unfamiliar modes for whatever reason. A significant risk may be identified in the form of some external physical factor as the symbol of this breakdown. Conversely, the presence of some otherwise 'significant' physical risk factor may occasion no sense of threat or risk if those institutional processes are operating more or less satisfactorily. Thus Rayner's fieldwork in hospitals specialising in radiological work, with regular use (and abuse) of radioactive materials, showed a virtually complete lack of concern for the heated controversy in society at large, about the effects of low doses of radiation.[13] Risks were expressly recognised, so it was not a case of psychological suppression. But these risks were defined according to social relations in the institutional setting, for example, whether a plumber trusted technicians not to throw radioactive materials down the ordinary drains rather than the special 'active' materials drains.

In a complex industrial society, characterised by high levels of differentiation, even fragmentation, these patterns of familiar cultural experience and commitment may vary a lot. Anthropologists who have studied risk controversies have identified a few basic systematic patterns, each with its own characteristic, underlying diverse settings and interactions. This is not the place for an exposition of this body of work, which has anyway been clearly described elsewhere.[14] The main thrust is that social interactions can be distilled into basic styles demarcated by orthogonal axes indexing degree of 'grid' (roughly, how much an individual is constrained by others) and 'group' (how strongly an individual's significant relations are socially bounded from others). These styles dictate forms of social experience, e.g. of control, cause–effect, unity, etc., which parallel experience of nature and the physical environment.[15] Different rationalities derive from these different social roots.

Despite the normal crop of questions for further development, this cultural approach to the interpretation of risk conflicts offers a much richer and less loaded understanding of conflicting perceptions and their roots than do the essentially 'psychologistic' ones which dominate present thinking on the issue. One can see how the definition of meaning of a physical risk issue will vary irreducibly between groups, reflecting these different kinds of social experience. It is not a case of identifying cognitive factors causing deviation from some 'objective' risk, but of understanding fundamentally different problem configurations, organically rooted in different empirical social experiences and agenda.

PERCEIVED RISKS

One of the associated tenets of the 'objective–perceived' risks approach is that uncertainties in the scientific understanding of the objective risks are the origin of conflicting perceived risks – what regulators or industry representatives regard as marginal scientific differences over risks are believed to allow strong emotional forces enough room for extreme polarisation of perceived risks. What is thought at any point in time to be a scientifically given amount of uncertainty is thus a resource for emotional licence: this not only polarises social perceptions but tends to pull scientific debate away from an otherwise natural consensus point. Some irreducible uncertainties are recognised especially in evaluation of dose-effect questions involving latent periods, chronic doses, statistical effects and multiple causes. But even apart from this category of scientific uncertainties, research has shown how even in areas with no external implications or interest scientists themselves are inconsistent about the extent or significance of uncertainty in their own specialty.[16] In other cases they have been shown to relate their evaluations of the scale and significance of scientific uncertainty to underlying behavioural judgements. Campbell discovered that scientists' judgements of the scale of uncertainty in the environmental questions surrounding the proposed Mackenzie Valley pipeline in northern Canada, were systematically influenced by their tacit behavioural judgements about some key social questions, such as whether the oil industry could really be regulated to build only at the correct times of the year, and whether one pipeline would lead to a whole 'corridor' of associated developments.[17]

In the Windscale inquiry in Britain, similar dislocations in the basic meaning of the issue – the problem to be addressed – were created by different empirical predictions of the social behaviour of key interests.[18] Thus the authorities, including Mr Justice Parker, the inspector, took the question as the immediate 'factual' one of the identifiable 'objective' risks and benefits of a single new plant for oxide fuel reprocessing, THORP. The regulatory establishment or those culturally associated with it could understand and trust the social system which would make subsequent decisions about proposed follow-on decisions as essentially separate issues to be divorced from the present one. However, to those outside this esoteric cultural milieu, the present decision had rationally to be evaluated in terms not only of any risks identifiable *sui generis*, but also in terms of the further power and autonomy it gave to that elite, historically remote decision-making culture.

Given the uncertainties surrounding the immediate 'factual' dimension, the institutional behaviour and trust dimension naturally came more into prominence (though reciprocal influence also affects the level of perceived factual uncertainty). If past experience indicated that the authorities had acted by assuming that every single step was only a stage in the continuous growth of nuclear power it was rational of people to formulate an open-ended problem, and to condense all possible consequences of all possible future follow-on developments onto the single THORP decision. Instead of recognising and negotiating with this rational product of concrete social relations to the technology and its controlling institutions however, Parker abruptly dismissed the extensive definition as emotive nonsense – one more example of the perceptional bias from objectivity brought about by the usual catalogue of factors (psychological dread, hysterical media, etc.) carried by this traditional approach.

As I have commented in a lengthier discussion of this issue,[19] historical causation and the bounding of issues into 'decision problems' will inevitably be objectively different from different positions in the web of social causation. Objectively different behavioural experiences and predictions will prevail; democratic processes would allow for some accommodation between these in the usual negotiation and formulation of a public issue. Imposition by fiat of the narrower definition of the problem and thus of the framework of the 'objective risk' discovery question, was simply an authoritarian way of disagreeing with and obliterating rational, empirically grounded behavioural predictions. The 'objective–perceived' risk approach in its general form is identical.

This is not in any way an evaluation of the specific contents of those experiences and inarticulate rationalities hidden by the pejorative language of 'perceived risk' defined in relation to an imposed problem as psychological categories. For example, one does not have to support or deny the argument that the growth of civil nuclear power may increase the chances of nuclear war by horizontal weapons proliferation, to see that the supposedly psychological factor in nuclear risk perception – association with Hiroshima – embodies an empirically grounded behavioural judgement. The judgement is about whether human institutions can realistically control the spread of nuclear weapons under conditions of extreme international competition in civil nuclear commerce, with ambiguous connections between the two sectors. Whatever the answer, this is not a merely psychological factor, but is rooted in social relations of the technology.

One behavioural prediction – that the institutions cannot control the problem – is dressed in such a way as to be presented as an emotive psychological reaction: the opposite prediction no more justified by any empirical evidence or greater behavioural expertise, is buried in the public language controlled by the decision-making elite as a taken-for-granted, natural setting for defining other, 'objective' risks.

THE SOCIAL CONDITIONS OF RISK – TECHNOLOGY AS SOCIAL PROCESS

The preceding argument has included the point that the 'objective–perceived' risk dichotomy is false, for one reason because 'perceived' risks embody relevant empirical behavioural experience and predictions which reflect a different domain of objective risks; the further reason is that in addition to any physical uncertainties, 'objective' risk definitions often themselves embody tacit behavioural assumptions. As I have argued elsewhere in relation to three examples, nuclear power, computer software systems and the safety of pesticides,[20] the structure of control often places in key positions the very people – scientists – least fitted even to recognise that they have incorporated social judgements in their 'objective' statements, let alone to make realistic ones. The conflict over 2.4.5-T between the expert Pesticides Advisory Committee (PAC) and the National Union of Agricultural and Allied Workers (NUAAW) for example, has been a conflict about the relative significance of certain social conditions of production and use of the pesticide. The PAC took it that the prescribed conditions of use were fulfilled when they asserted the safety of 2.4.5.-T. Yet these behavioural predictions were woefully unrealistic in the empirical knowledge of the NUAAW, thus calling into question the whole universe of reassurance cultivated by the PAC in initially ignoring the key social conditions as a trivial matter of implementation.

Parallel examples are not hard to find in the nuclear issue, whether relating to risk, downright viability, or both. As a typical example, we could examine Sir Alan Cottrell's view of the safety of PWR pressure vessels and Sir Walter Marshall's interpretation of that view.[21] Although Cottrell expresses himself reassured by later work of the Marshall Committee on this problem since its strongly criticised first Report in 1976, the conditions which Cottrell attaches to this view – about the extreme rigour of construction and inspection needed for safety – are arguably unattainable, according to one's behavioural predictions. By being under-emphasised these behavioural judgements are tacitly incorporated into the scope of authority of the expert's pronouncement, yet there is no reason why this scope should cover such behavioural predictions. Thus what is widely quoted as the top expert's endorsement of the safety of PWR pressure vessels could just as easily be interpreted as the opposite.

The nuclear fuel cycle taken as a whole is replete with such dependencies upon behavioural disciplines previously taken for granted or at least optimistically assessed. This is no different in principle from other technologies, except for the greater elaborateness of nuclear technology and its image of exemption from such mundane realities. It is a complex *social* process. If even loyal professional insiders can get it wrong, then it must be widely evident that there are large opportunities for error by those who play a key role but who are not even trained and socialized into the nuclear industry – such as ordinary railway personnel with only occasional intersection of their concerns with the nuclear industry.

These are areas of experience which ordinary people can understand and relate to, more realistically perhaps than technical experts, and the more the latter are shown to have neglected these dimensions, the less trustworthy they are naturally seen to be. Once the precarious conditions of public trust in the responsible institutions have been breached, then it becomes natural for judgement of risk to be based on the worst possible physical case, since there is no longer any credence given to the institutional behaviour that might restrict that physical possibility to an extremely low probability. In these circumstances if flasks containing millions of curies of activity are going through residential areas it is little use arguing about how well the flasks are tested and how long it would take for fuel vaporisation. It is not that technical conflict over these questions creates social mistrust in the authorities, but more the latter creates the former.

Ironically the reaction to the growing realisation of the importance of unrecognised thus relatively unregulated behavioural areas in risk management leads to programmes for greater behavioural control which can itself evoke a negative reaction.[22] This is part of a general syndrome to which the nuclear fuel cycle may be particularly vulnerable, but by no means uniquely so. All areas of public life depend on much greater informal trust and accommodation between groups and individuals than is usually recognised. As the general conditions of trust in collective institutions break down, for example by the further spread of contractual relationships as is widely recognised, these collective interactions to which we are committed need to be regulated by formal 'coercive' means to replace the dissolving informal accommodations. Yet once this process starts it reveals a virtually inexhaustible and impossibly large field of interactions to 'externally' regulate.[23] The dimensions of this process are indicated in the contrasting approaches to regulation in Britain and the USA.

In the USA, mistrust is institutionalised in the political culture even in the constitution. Traditional, informalised political authority is always highly problematic, so that institutions have to resort to formalised approaches – legal procedures, elaborate reference to 'scientific' criteria and facts – to a degree which British authorities regard as neurotic. They tend to take for granted the authority of traditional institutions which are therefore freer to exercise informal discretion in interpreting statutes and regulatory criteria from case to case. If this substratum of authority in the political culture is undermined, however, then reasons for decisions are demanded in public, and formal responses may have to be given. At this point the lack of unambiguous technical rationale for decisions (as exposed for example by the cross-examination of energy or traffic forecasts) may become acutely problematic. The importance for policy viability of sustaining public trust in decision-making institutions is far greater in Britain than it is in the USA, and thus places more of a long-term burden upon the former.

CONCLUSIONS

It is usual to assume that concern about the risk from some activity corresponds amongst other things with the degree to which it is regulated. Yet despite the discovery of areas of intrinsic difficulty in regulation in the nuclear fuel cycle, in comparison with many other risky activities it is inherently easy to regulate. Compared with the transport and disposal of hazardous chemicals, for example, where the number and variety of producers, transporters and other handlers is vast and largely unknown, nuclear fuel transport is highly centralised, completely known (at least to the regulators) and small-scale. The resources devoted to the nuclear case are colossal in comparison. In the USA, for example, over 200,000 million ton-miles of hazardous materials, in nearly 1.5 million vehicles are 'regulated' by about 250 inspectors.[24] Yet despite the furore over the USA Environmental Protection Agency's Superfund for hazardous waste dumps, hazardous waste transport has not yet evoked the same level of general hostility and concern as nuclear fuel transport.

Indeed one could say that over the full range of activities, 'unregulability' is normal; nuclear power is one of the more regulable ones. As argued earlier the question is why so many risky activities, badly controlled, are more or less accepted without much sense of risk; put this way the question highlights the positive role played by institutions in constantly reproducing and repairing social trust and reassurance. Risk is defined not so much by the penetration of some identifiable risk-generating factor, but by the mere absence of those familiar and traditional institutional supports which can be so strong as to allow people to live happily in the presence of 'objectively' large risks. A 'problem' such as the 'objective' risk of some activity, central to a given decision-maker (or decision-analyst) may be an incidental item in a very different world of meaning to some other person or group involved.

Unlike abstract policy formulations, real political issues are never unicentric and self-contained. One of the basic tensions in politics is the regular tendency for specialist, single-interest professional groups to distort the negative and flexible political interpretations of issues with their own artificially self-contained meanings of key prescriptive terms, such as 'risk' or (as in the race–IQ controversy) 'intelligence'.[25] Intentionally or not, this uncritically imposes single meanings whose artificial precision and objectivity conceals their questionable social relevance. It expresses a political bias by obliterating a whole set of normative questions to do with social relationships, historical commitments and their entailments, etc.

The conflict about the risks of urban transport of irradiated fuel is only the symptom of a deeper issue. Perhaps technical fixes could be found which could avoid the symptoms, for example by shipping all fuel to Windscale by sea and avoiding virtually all land transport. But this would only shift the expression of the symptoms to some other point. After more than a quarter of a century, nuclear technology, including its institutions, is still not integrated into popular culture – largely through the unanticipated consequences of its own previous means of gaining authority. We should not be surprised if there is tension when, inevitably, the magical expectations confront the mundane realities, for example of flasks containing spent nuclear fuel lying apparently neglected in railway sidings easily accessible to children.

Nuclear power was launched and is still driven by a kind of apocalyptic, synoptic vision and commitment which accommodates very badly with the realities of its implementation, which are much more akin to the policy model of disjointed incrementalism – 'a series of (relatively) incremental, remedial choices among a narrow range of options, reconciling only a narrow range of interests. The strategy works to the extent that interests and alternatives ignored at one step can assert themselves and demand attention at a subsequent step'.[26]

The perceived risks of irradiated nuclear fuel transport are only a response to the objective reality of the integration of this activity with the rest of that impatient, culturally alien commitment. No guarantee should be expected that yesterday's practices have not already set in train today's and tomorrow's responses; nor that the results of their now demanding attention will not be paralysingly disjointed. The only feasible route is to recognise the social realities underlying people's cognitions, and take the risks of, first, respecting them for what they are; and, second, negotiating with them.

NOTES

1. L. R. Solon, 'Some public health aspects in the transportation of radioactive materials involving the city of New York', quoted in I. Welsh, *Don't Take the A-train; A Critical Examination of Nuclear Waste Transport* (Edinburgh: SCRAM, 1981), p. 4.
2. C. Wakstein, *The Ecologist*, 21 (April/May 1980), 131.
3. See the other papers in this Conference.
4. As summarised for example by Terence Lee, in this Conference. See also, B. Fischhoff *et al.*, 'The "public" vs. the "expert": perceived vs. actual disagreement about nuclear power,' in V. Covello *et al.*, (eds), *The Analysis of Actual vs. Perceived Risk* (New York; Plenum Press, 1982).
5. For recent reviews, see B. Fischhoff, S. Derby, R. Keeney and P. Slovic, *Acceptable Risk* (New York: Cambridge University Press, 1981); and O. Svenson and B. Fischhoff, 'Active response to environmental decision making', Swedish Research Council for Humanities and Social Sciences (1983).
6. Ibid., notes 4 and 5. For a critique, see H. Otway and K. Thomas, 'Reflections on risk perception and policy', *Risk Analysis*, 2 (1982), 147–59.
7. Fischhoff *et al.*, 'The "public" vs. the "expert" '.
8. Otway and Thomas, 'Reflections'. Brian Wynne, 'Technology, risk and acceptance: on the social treatment of uncertainty', in J. Conrad (ed.), *Society, Technology and Risk Assessment* (New York: Academic Press, 1980).
9. Svenson and Fischhoff, 'Active response', p. 143.
10. The sociological writings of Goffman have been especially clear in describing these multiple dimensions of social interaction. See e.g. E. Goffman, *Relations in Public* (Harmondsworth: Penguin, 1972).
11. This is a problem for example in the way in which multiple actors are conceived in the decision process for siting liquid energy gas terminals, in the IIASA comparative study by H. Kunreuther, J. Linnerooth *et al.*, 'Risk analysis and decision processes: the siting of LEG Facilities in Four Countries', Springer Verlag, Berlin (forthcoming).
12. 'Verstehen' – understanding from the actor's point of view, or attempting to get inside their meaning structure – is a principle of social research identified with Max Weber.
13. S. Rayner, 'Effects of workplace organisation on the perception of occupational hazards', paper to IIASA seminar (March 1983), IIASA mimeo.
14. M. Douglas and A. Wildavsky, *Risk and Culture* (Berkeley: University of California Press, 1982); M. Thompson, 'Among the Energy Tribes', IIASA WP-82-59, Laxenburg, Austria, 1982.
15. For collections of readings showing the relevance of this perspective for

scientific knowledge, see S. B. Barnes and S. A. Shapin (eds), *Natural Order* (London: Sage, 1980); S. B. Barnes and D. O. Edge (eds), *Science in Context: Selected Readings in the Sociology of Science* (London: Open University Press, 1982).

16. T. Pinch, 'The sun-set: the presentation of certainty in scientific life', *Social Studies of Science*, 11 (1981), 142–56.
17. B. Campbell, 'Disputes among experts: the debates over biology in the Mackenzie Valley Pipeline Inquiry', Ph.D. thesis, McMaster University, Canada (1982).
18. Brian Wynne, *Rationality and Ritual: the Windscale Inquiry and Nuclear Decisions in Britain* (London: British Society for the History of Science, 1982).
19. Ibid., pp. 163–4.
20. Brian Wynne, 'Redefining the issues of risk and public acceptance: the social viability of technology', *Futures*, 15 (1983), 13–32.
21. Correspondence reprinted in *Atom*, 310 (August 1982).
22. Again this is a cycle of general relevance. See e.g. B. Wynne and H. Otway, 'Information technology, power and managers', in N. Bjorn-Andersen *et al.* (eds), *The Information Society: For Richer or For Poorer*, North-Holland, 1982).
23. There is a suggestive analogy here with the condensed (informal) and elaborated (formal) codes of linguistic interaction analysed by Bernstein, *Class, Codes and Control* (Oxford University Press, 1972).
24. C. Diver, 'A theory of regulatory enforcement', *Public Policy*, 28 (1980), 257–99.
25. Y. Ezrahi, 'The authority of science and politics', in E. Mendelsohn and A. Thackray (eds), *Science and Values* (New York: Humanities Press, 1974).
26. Diver, 'Regulatory enforcement', 280, note 24.

22 The Psychology of Nuclear Anxiety

TERENCE R. LEE, J. BROWN AND J. HENDERSON

The public, according to Conrad, suffers from 'a diffuse and ill-articulate sense of confusion, anxiety, alienation and ambivalence towards expertise'. Although modern industrialised societies are safer than any previous ones, there is burgeoning interest in safety and, in particular, *risk assessment*. Scientific risk assessment can be divided into two main approaches: (1) the objective, statistical extrapolation from past experience, and (2) the study of public perceptions, attitudes, etc., towards hazards. The majority of papers at this Conference deal with the former, but the present paper concentrates on the latter.

The need for risk perception research is widely acknowledged, arising as it does from the sometimes huge discrepancies between the risk assessments made by scientists and those made by ordinary people. This is nowhere more evident than in the case of nuclear power. It is easy to find cases to support the assertion that the public's view usually prevails because it is translated into decision-making through elected representatives. One example is in the pharmaceutical industry, where 37 per cent of costs currently go on safety and where the total sum expended on safety is approximately 2500 times greater than in agriculture.

The shortcomings of the public's approach to risk are as follows:

(1) They fail to study the issues about which they are worried.
(2) They create a demand for sensation in the media and are then influenced by its expression.
(3) They use 'heuristic', short-cut devices for guessing probabilities, which produce bias.

(4) They do not synthesise the costs and benefits of hazards in the trade-off sense, but oscillate in their attitudes, thinking first of benefits and then on the next day of costs.

The shortcomings of the scientific statistical approach are:

(1) Scientists frequently fail to appreciate that hazards involve the probability of occurrence compounded by the severity of the consequences. The latter can only be assessed by the public, especially where moral values on attitudes or e.g. attitudes to technology are involved.
(2) The necessary experience/statistics are often not available.
(3) When available, the statistics deal usually with mortality and not with the more common morbidity, while the public is often concerned with the severely reduced quality of life following an accident and, at worst, with 'fates worse than death'.
(4) Experts are human and biased by self-interest and self-perpetuating institutional contexts.
(5) Extrapolation is always approximate.
(6) The scientist is not in a position to make judgements about which section of the population (or future generation) should carry the burden of a hazard; these are value-laden judgements.

It is clearly absurd for scientists to claim that the public is irrational, biased or in error in its risk assessments. The only sensible strategy is to recognise that the gap must be closed from both sides. Lord Flowers was, in my opinion, misguided in saying in his opening address that the scientists should look after the objective assessment of risk and the politicians should look after the public. This is a cynical view which implies that scientists are right and the public are wrong, but troublesome, and so in need of political handling. I suggest that physical scientists should deal with the estimation of physical aspects of risk while social scientists should provide understanding of the lawfulness underlying the public's attitudes and perceptions. Both should furnish the politician with the facts that will help him to make decisions. Decisions that involve values and the differential treatment of different sections of the public can only be made by its elected representatives. They are not directly amenable to any form of science.

Research is being carried out at Surrey University under the auspices of the CEGB, the DOE and the HSE (Health and Safety Executive). The main study is concerned with the processes of consultation with the public in relation to the provision and planning of nuclear power stations in the south-west of England. There is also research at an early stage into the public's attitudes towards the management of nuclear waste.

Is the public for or against the use of nuclear power for the generation of electricity? The assertion by Friends of the Earth that the British public is wholly 'against' is totally unjustified. Surveys in this country and other parts of the world show considerable variation, opinion is volatile, both as a result of events such as Three Mile Island and because of substantial variation in the precise meaning of the questions that are put in surveys. The most dependable recent national sample was carried out by Social and Community Planning Research (SCPR) recently under the auspices of the Health and Safety Executive. They put the following question:

> There are people nowadays who say we should not be building nuclear power stations because the dangers associated with them are too high a price to pay for electricity. Do you yourself agree or disagree with this view?
>
> Results: 35 per cent agree
> 50 per cent disagree
> 15 per cent don't know

Our own study in the south-west put a more general (and simpler) question:

> What is your opinion about using nuclear power to generate electricity?
>
> Results: 66 per cent in favour
> 22 per cent against
> 13 per cent uncommitted

It has to be noted that these surveys were carried out mainly in areas threatened with nuclear power stations.

However, an attitude is a very complex multi-dimensional phenomenon. It includes cognitions, i.e. thoughts and perceptions, about the object, and also 'feelings'. Secondary analysis of the SCPR Survey carried out at Surrey University suggests that although there are high proportions of the public who claim to 'think' about the six risks included in the survey very much smaller proportions conceded

TABLE 22.1 *Perceived risks of nuclear power*

	N	%
Radiation leak	556	59
Disposal of waste	251	27
Environmental impact	80	8.5
Transportation of fuel	58	6
Terrorism	44	5
Gaps in knowledge about nuclear technology	37	4
Social impact	31	3.3
Stealing plutonium for making illicit nuclear weapons	18	2

that they are 'worried' about them. The order of priority puts nuclear power second after smoking for concern or general preoccupation, but it is lowest in the list for worry/anxiety (it is important to note that in an earlier section of the questionnaire, respondents were asked to list their main worries and hardly anyone mentioned nuclear power).

As well as different modes of responding, attitudes include orientations on different facets of the environmental object, both risks and benefits. Tables 22.1 and 22.2 show the perceived risks of nuclear power and the perceived benefits. These data are derived from the south-west study.

TABLE 22.2 *Perceived benefits of nuclear power*

	N	%
Cheaper power	412	33
Conserves fossil fuel	172	18
Unlimited fuel	112	12
Less pollution	84	9
Brings employment	54	6
Freedom from Middle East	21	2

It is possible to focus down more closely on one of these facets, i.e. waste management. A small feasibility survey has been carried out in this area. When asked whether radioactive waste disposal was an issue that people worried about, 27 per cent responded positively. The remainder had either 'never thought about it', or 'thought about it but were not worried'. When asked to estimate how long (in years) they thought the waste to be potentially dangerous, 32 per cent said 'up to 100 years', 19 per cent said 'hundreds of years' and the remainder various figures in excess of this. When asked what aspects of radioactive waste disposal were of particular concern 31 per cent were concerned about 'disposal at land or sea', 20 per cent about 'transportation of fuel', 18 per cent about 'long-term effects', 17 per cent about 'pollution/leaks' and 13 per cent about 'strength of containers'. The large majority of our sample considerably overestimated the bulk quantity of nuclear waste produced in the UK, most of them by a factor of some thousands!

More general research by psychologists on the perception of risk suggests that risks are perceived as serious by the public in ways that can be partially predicted by a number of variables. These are:

(1) Involuntary/voluntary.
(2) Delayed/immediate in effect.
(3) Familiar/unfamiliar.
(4) Catastrophic.

Anxiety over nuclear energy is partly due to association with atomic bombs, which gives it an image of very high catastrophic potential, albeit low probability. Also, it is completely unfamiliar, ordinary people having no personal experience from which to project the future. It is also involuntary, people feeling impotent to control it in the face of massive and remote national and international forces.

It is hoped that this paper has demonstrated that public attitudes are complex and multi-dimensional but ultimately understandable. At present, the DoE spends £10 million per year on research into the technology of risk assessment but until recently nothing at all on research into public attitudes. This research can be quantitative and sophisticated in analytical terms. There should be more of it.

Attitudes are resistant to change but change is possible. It is necessary for national and local government institutions to take a much more positive line and to use modern, effective methods of evaluating their communications with people. We have to overcome a reluctance to educate which arises from a fear that we might be engaging in propaganda. On the contrary, effective and open communication of views is the essence not only of democracy but of human socialisation.

REFERENCES

The public's perception of risk and the question of irrationality. Proceedings of the Royal Society, London, 376, 5–16, 1981.

The Perception of Risk. The Perception and Assessment of Risk, Royal Society, London, 94–148, 1982.

23 The Economic Aspect

GORDON MacKERRON

The principal economic concern about nuclear power reactors has always been their capital cost. Currently, the capital cost of a reactor contributes some 70 per cent to the total cost of nuclear-generated electricity, even when somewhat speculative allowance is made for decommissioning and waste disposal costs (see Table 23.1). Given that nuclear capital costs have been rising in real terms by as much as 10–14 per cent *per annum* in the USA, this concern is clearly not going to disappear.

TABLE 23.1 *Components of estimated cost for nuclear and coal-fired power stations – UK 1983 (%)*

	Nuclear	*Coal*
Capital	69	26
Fuel	22	67
Other	9	7
	100	100

SOURCE CEGB *Analysis of Generation Costs* mimeo (1983), Illustration VII, p. 19.

In addition, however, there has been growing attention to the technical and economic questions at the so-called 'back-end' of the nuclear operation – all the activities that need to be undertaken after fuel has been irradiated in the reactor. Interest has concentrated on reprocessing of spent fuel, waste storage and disposal, and reactor decommissioning. In many of these areas, technology is not yet commercially established, so that the economics of these back-end operations are highly uncertain. The costs of back-end activities have been a particular focus of UK attention, because problems in the reprocessing of Magnox fuel elements have led to a virtual doubling in overall nuclear fuel costs (in real terms) in a matter of only 6 or 7 years.

Later I shall consider the ways in which economic forces are likely to influence the pace of future nuclear development in general. However, I begin by considering the economics associated with one of the back-end areas which is the main subject of this conference – irradiated fuel transport. Fuel transport is clearly an area which has political importance, but economically it is of profound insignificance. The economics of irradiated fuel transport are necessarily, therefore, very short.

IRRADIATED FUEL TRANSPORT COSTS IN THE UK – THE SIZEWELL PWR

Until recently, there was, to my knowledge, no detailed UK cost information available on this subject. However, the public inquiry into the CEGB's proposal to build a PWR at Sizewell has brought forth an enormous volume of new and valuable information, including detailed estimates on fuel transport costs. The basic data are extremely simple: for the Sizewell PWR, the Board would need to buy 1 Excellox (or equivalent) flask at £400,000 plus one railway wagon at similar cost. Ten railway trips a year will cost £70,000, and £10,000 a year would be spent on maintenance. This, over the project lifetime, amounts to £3.28 million, allowing for the 30 years of journeys needed. Once

TABLE 23.2 *The estimated costs of irradiated fuel transport for Sizewell B*

	£m
1 Excellox flask	0.40
1 Railway wagon	0.50
Journey costs (31 years)	2.17
Maintenance (31 years)	0.31
Total (at 1982 prices)	£3.38
At 5% annual discount rate, £3.28m amounts at 1982 prices to	£1.008

SOURCE CEGB, Sizewell B Power Station Public Inquiry, Proof 9 by J. K. Wright on *The Nuclear Fuel Cycle* (November 1982).

these costs, which do not start until the year 2007, are valued in 1982 terms (by the use of a 5 per cent per annum discount rate), the entire lifetime's cost of irradiated fuel transport for Sizewell will amount to almost exactly £1 million (Table 23.2). This tiny sum contributes 0.003 p/kWh to overall costs; this is less than 0.5 per cent of all fuel costs, or less than 0.1 per cent of all Sizewell's costs. Put more graphically, all other project costs are roughly 2000 times greater than irradiated fuel transport costs.

Sizewell's overall profitability is therefore almost totally insensitive to even quite large changes in irradiated fuel transport costs. To take an extreme case: if irradiated fuel transport costs rose by a multiple of ten, the net monetary benefit of Sizewell would be reduced by around 1 per cent.

IRRADIATED FUEL TRANSPORT COSTS – EXISTING UK REACTORS

The apparent importance of irradiated fuel transport for Sizewell is of course reduced by the fact that its costs are distant, and are discounted at 5 per cent per annum. It is also important to know what the CEGB is *currently* spending on fuel transport from existing nuclear power stations. I have not discovered any direct data on this, but it is not difficult to arrive at an approximate order of magnitude.

This rough calculation is based on the following crude assumptions: each of the nine existing nuclear stations (excluding Scotland) has its own waste flask and railway wagon; the average trip to Sellafield (Windscale) is of the same distance as from Sizewell; and as current nuclear capacity (4500 MW) is roughly four times the capacity of Sizewell, the volume of current waste arising will require four times as many trips as planned for Sizewell. Finally, it is assumed that the annual capital charges associated with flasks and railway wagons amount to 10 per cent of their initial capital cost. In general, these are conservative assumptions, that is, they will probably overstate the cost of existing irradiated fuel transport. The results are that total costs probably amount to no more than £1.1 million annually, of which £730,000 is capital charges, £280,000 is direct transport costs, and £90,000 is for maintenance.

The context of this figure is that the CEGB told the Monopolies Commission in 1980 that it spent some £226m annually on all nuclear fuel services. The likely current costs of irradiated fuel transport therefore amount to less than one-half of 1 per cent of all nuclear fuel costs.

The message of these cost figures for existing irradiated fuel transport and for Sizewell is simple and needs no labouring: irradiated fuel transport costs are by any criterion negligible, and even the largest likely change in them would have an insignificant impact on the overall economic status of nuclear power.

ECONOMIC APPRAISAL

The future volume of irradiated fuel transport needed will obviously be a direct function of the quantity of future installed nuclear capacity. The uncertainties surrounding the future international prospects for nuclear power are so great that it is worthless to attempt quantitative forecasting more than a decade into the future. Instead I shall concentrate on the factors that I believe will mainly determine the scale of future nuclear power.

Most of the factors are of a broadly or narrowly economic kind. This does not, of course, mean that economics in the narrow sense are the dominant consideration in decision-making about nuclear power. In the first place, it is clear that we owe the existence of electric power from nuclear fission largely to the development of nuclear weaponry. Many past decisions about civil nuclear power, and doubtless many in the future, will be heavily influenced by military considerations.

It is also clear that many people view nuclear power as a technology that will succeed because, historically, its time has come. The nineteenth century was based on coal, the twentieth on oil, and, surely, the twenty-first will be based on nuclear power? Fossil fuel is becoming scarce, so nuclear power will replace it on an inexorable, deterministic tide. With such a view, decision-making becomes a matter of high and long-term strategy, and the mundane economic details can be left to catch up later.

In the real world, these military and strategic matters are undoubtedly important, but I shall nevertheless assume from this point that the more every-day considerations of the relative economics of nuclear versus other options will, in the long term, be the prime determinant of the scale of future nuclear electricity.

PAST NUCLEAR ECONOMICS

In assessing the likely economic prospects of nuclear power in the future, it is useful to look first at past economic evaluations, and the extent to which they have been accurate. In the UK and the USA, proponents of nuclear power have been arguing consistently since the middle 1960s that nuclear electricity is competitive with, or cheaper than, the next best alternative sources. Has this view been vindicated?

The first thing to be said is that the effort devoted to evaluating the actual economic record of nuclear power has been minuscule compared to the effort spent in successive forecasts of future nuclear economics. It should not be thought that there is something unique about nuclear power in this respect: it is a permanent complaint about the practice of economic evaluation that it is hardly ever informed by even a cursory attempt to understand the successes and failures of past efforts. Nor is it particularly surprising that most of the evaluations have been conducted by parties with a strong interest in creating a favourable market environment for nuclear power. This again applies to some degree to all major public and private investments and accounts to some considerable degree for the often-observed 'appraisal optimism' that is a widespread feature of economic project evaluation.

The fact that the real volume of resources expended on nuclear projects has often greatly exceeded appraisal expectations is therefore no real surprise. In the UK this has taken extreme form in the case of the Advanced Gas-Cooled Reactors (AGRs), where actual costs have been roughly double – and construction times more than double – the anticipated values. Whether this kind of experience renders nuclear power uncompetitive depends entirely on whether the degree of optimism in nuclear appraisals has been significantly higher than in appraisals of the main competitor to nuclear electricity – since 1973 this has been almost entirely coal-fired electricity. Looking at capital costs alone, it seems generally true internationally that nuclear appraisals have been significantly more optimistic than those applying to coal-fired projects. Given the technological (and regulatory) immaturity of nuclear technology at the time of major commitment to nuclear projects in the late 1960s and early 1970s, this is again not particularly surprising. If capital costs were all that mattered, however, this would have meant a significant shift in the economics of electricity production away from nuclear and towards coal.

But the overall status of alternative generating options depends on fuel cost as well as capital cost, and here the experience has been that fossil fuel costs have gone up much more rapidly than expected. The oil price rises dating from 1973 and 1979 are very familiar, but rather less publicity has attended the increases in coal prices that followed. In the UK, for instance, the real price of coal rose by over 60 per cent between the end of 1973 and the end of 1975. With fuel costs accounting for around 70 per cent of all coal-fired costs, such increases in fossil fuel prices made significant dents in the competitiveness of coal-based power.

In general, therefore, the failure to control capital costs in nuclear technology has been roughly matched by a failure to control fuel costs in fossil-based technology, so that the comparative economics of the two sources are not greatly different now compared to 10 or 15 years ago.

There would, I believe, be fairly widespread agreement about the above analysis. There is a good deal more controversy about whether this leaves nuclear generally still ahead of coal in absolute terms, or whether the reverse is true. The first thing that needs to be said is that as world markets in steam coal are still small and rudimentary, and as coal costs and quality are highly variable across the world, the target for nuclear competitiveness is highly variable. It would, for instance, be hard to imagine nuclear power making any economic sense at a South African mine-mouth site and about as difficult to see coal-fired power as competitive in central France.

In general, however, it has been widely asserted in western Europe and the USA that nuclear power has, despite the vicissitudes of nuclear capital and coal costs, retained some economic edge over coal-fired power. In practice, I believe the picture to be much less clear-cut. I have already mentioned the case of central France, where nuclear will be cheaper under most imaginable circumstances. In the UK, however, the historical record is a mixed one. Until recently, the CEGB have argued that all their past nuclear investments – in Magnox and first-generation AGRs – have proved economically superior to all their past coal-fired investments. This argument was based on faulty economic logic (in which historic – low-price – capital costs were added to current – high price – fuel costs) and came under considerable external pressure. Very recently, the CEGB has conceded the force of some of these external arguments, and has recalculated the economics of its past investment decisions using a more consistent methodology. This yields a rather complex set of results. In their lifetime to date, the Magnox reactors have operated about 18 per cent more expensively than coal-plants contemporary to them: if coal prices keep rising steadily, this disadvantage may narrow to 7 per cent by the end of their lifetimes. More important is the economic competitiveness of the AGRs, as these are still a future supply option. Only Hinkley Point B (in England) has accumulated any commercial operating experience: so far this has been poor, and results in a 27 per cent economic disadvantage, over its lifetime to date, compared to the roughly contemporary coal-fired Drax I. In 1981/2, Hinkley and Drax I had virtually identical costs, and (again if coal prices rise steadily) the CEGB expects this to turn into a 12 per cent advantage to Hinkley over its expected lifetime. Hinkley is, it should be noted, much the cheapest and best of the first-generation AGRs, and the CEGB has acknowledged that the worst, Dungeness B, will always be much more expensive than coal-firing.

What does this amount to? It is clear that it does not unambiguously establish the superiority, historically, of either nuclear or coal-fired power in Britain. The uncompetitiveness of Magnox is not surprising, especially as the rationale behind its ordering was not primarily economic. The best of the AGRs promise to compete fairly well with coal-firing at current coal prices, while the intermediate cases (Heysham I and Hartlepool) will depend on continuing rises in coal prices to break even. The significance of these results should not be exaggerated as they have no direct bearing on the economic worth of investments currently under consideration. Their importance is mainly as a more objectively based contrast to the earlier official view, which has persisted over the past 20 years, that in all circumstances nuclear power was proving to be a better economic option than all others. This view of history, coupled with the expectation of further rises in real coal prices in the future, has lent weight to the kind of determinism about the inevitable success of nuclear power that can make a future strategy of investing solely in nuclear power seem inevitable and rational.

FUTURE NUCLEAR ECONOMICS

As I suggested earlier, no quantitative forecasts about the long-term expansion of nuclear power are sensible beyond a few years into the future. It is worth perhaps saying that in these next few years, the world's capacity of nuclear power stations will probably rise very substantially because of the large backlog of nuclear orders, particularly in the USA and France. From a current world capacity of 140 GWe (thousand megawatts of electrical capacity), we are likely to reach some 300 GWe by 1990 purely as a result of past investment decisions. This will represent a much larger proportion of the world's electric supply capacity than the current 10 per cent.

Beyond 1990, things are much less clear. There are four factors that will mainly determine nuclear capacity beyond that point (in each case some illustration is provided from the current public inquiry at Sizewell):

Electricity demand

As John Surrey points out (see Chapter 1), the single most important cause of the severe world-wide recession (France excepted) in nuclear power ordering has been reductions in electricity demand growth. In the 1960s, rates of growth of electricity demand of 6 per cent and 7 per cent were widely expected to persist indefinitely, and utility-ordering policies reflected this expectation. Since the early 1970s, growth has fallen dramatically, and not only in the UK; last year, in the USA, electricity demand actually fell for the first time in several decades. This has meant that many countries now have substantial surpluses of generating capacity either in place or on order, and limited need for new plant. In this situation, a drying up of nuclear orders is simply part of the general lack of need for new generating plant investment.

At Sizewell, it is important to notice that the economic case being made for the PWR does not rely at all on any prospective shortage of capacity. New capacity is not needed on the CEGB's own figures for 4 to 5 years after the scheduled completion of Sizewell, and in practice this period could well be extended. The case for Sizewell is made on the separate argument of 'net effective cost'. This says that for a system of a *given* size, costs will be lower if Sizewell is built than if nothing is done. This would happen if, as the CEGB argues, the total cost associated with building and running Sizewell is smaller than the fossil fuel costs saved as a consequence of its operation. In this framework, it is conceivable that a substantial number of new nuclear orders could be justified without any demand growth at all.

In the world context, future expectations of electricity demand growth are, however, clearly going to be critical to the future scale of nuclear ordering. If or when the world recession abates, this could have a large impact: it remains true that electricity demand grows as fast as, or faster than, national income. However, from the late 1990s onwards there will be another major source of demand for new plant, and that is replacement needs. An obvious consequence of the high levels of capacity constructed in the 1960s and early 1970s is that there will be a replacement need as it wears out, or becomes uneconomic. This will mean a high level of ordering by utilities from the late 1980s onwards, and if nuclear captures a significant share of this market, there could be very considerable expansion in nuclear capacity. In England and Wales, for example, the CEGB estimates that it will need to commission 35 GW of plant between 1997 and 2010 simply to keep its system at constant size.

Technological maturity

If nuclear power is to become a more important source of electricity, it will be necessary to achieve more control over nuclear power capital costs and more reliable operating performance. The explanation for past cost increases and variable performance is a complex one, and reflects not only escalating safety standards but also some serious technological problems in what is still a fairly young technology. The immaturity of the AGR design is a major cause of the poor UK experience in gas-cooled technology. Generic problems of stress corrosion and brittle fracture have had serious consequences for BWRs and PWRs respectively. At Sizewell, the chronic Westinghouse problem of designing reliable and safe steam generators is reflected in the fact that the 'F' type design to be used there is new and has not yet accumulated significant operating experience. The nuclear industry is of course only too well aware of these technological problems and has devoted a very substantial R & D effort to their solution in recent years. It is reasonable to expect that this effort will produce a technically mature product within a few years, provided that new safety concerns do not intervene.

What is less certain is that outstanding problems at the 'back-end' of the fuel cycle will be satisfactorily resolved. Decommissioning, reprocessing and waste disposal are all technologies that have evoked little sustained research effort until very recent years. In consequence, they are considerably less mature than reactor technology, and the acceptability of reactors is increasingly being linked to perceptions of better solutions to 'back-end' problems. The military and inter-generational problems that back-end issues raise – the latter because of the well-known long-lasting effects of radioactivity – mean that in this area, even more than in reactor technology, technical solutions to problems need to carry with them a degree of social acceptability if they are to be implemented. This takes us into issues of acceptance.

Social and political factors

In the past, opposition to nuclear power has, in many countries, been vocal and highly visible. As in many other areas of politics, noise and visibility have been symptoms of a *lack* of power and influence rather than of the reverse. The direct impact of opposition to nuclear power has been marginal and temporary. The indirect impact, via influence on the regulatory authorities and safety standards, is hard to assess but has probably been of some importance. It is necessary to stress, however, that much of the escalation in safety standards has been a necessary response to improved knowledge about nuclear plant characteristics. The main exception to this argument comes from US experience since the Three Mile Island accident in 1979, though it is important to point out that licensing of nuclear plants is now in normal operation again.

For the future, it is exceptionally difficult to predict what kind of political and social resistance there may be to nuclear power. Resistance is likely to focus on the back-end operations referred to in the previous section, because this is an area of serious past neglect and one where many strategic decisions are yet to be made. Whether opposition to nuclear power will make much impact is also hard to assess. It is noteworthy that none of the major political parties in the main nuclear-using countries (Sweden aside) adopts an anti-nuclear stance. Supporters and opponents of nuclear power each have a scenario in which public acceptability would be greatly changed. Supporters look to a time when there may be serious power shortages, at which point they expect public resistance to crumble, while opponents suggest that a major accident would stiffen public resistance. In the (hoped-for) absence of either eventuality, it appears to be the case that the current political controversies over nuclear weapons are having a spill-over effect on public perceptions about the acceptability of nuclear power.

At the Sizewell inquiry, analysis of any social and political impacts on nuclear development are noticeably (and understandably) absent. The CEGB clearly hopes that the PWR will raise no new issues of acceptability, though the very existence of the inquiry makes this unlikely. Whether opposition will have any impact on Sizewell is highly questionable, especially if one assumes the continuation of Conservative Governments.

Competition to nuclear power

Since 1973 oil-fired power has no longer been a viable investment option. Even if oil prices drop somewhat from their current reduced levels, this is unlikely to change. For base-load (continuous) generation, this leaves only coal-firing as a serious rival. (Unless electricity storage problems are much more satisfactorily resolved, renewables will not be a significant contender for some considerable time.)

On current resource estimates and current consumption rates, the world has some 300 years' worth of coal remaining. This is sufficiently large to suggest that coal-fired power could continue to be developed on a large scale for some time. Whether coal-firing will offer serious future economic competition to nuclear power depends crucially upon the future world price of coal, and of environmental standards. If coal prices in real terms were held steady at current levels, there is little doubt that in most of the world, coal-firing would prove more economical than nuclear, because the forces leading to increases in nuclear capital costs have not yet fully worked their way through to actual nuclear costs. In the Sizewell case, constant fossil fuel prices would render Sizewell uneconomic, and would mean, if the CEGB is right in all its other assumptions (some of which appear optimistic about nuclear power) that PWRs and coal-fired plant would roughly break even. In practice it is difficult to avoid the conclusion that world coal prices will rise in real terms: it is hard to see how greatly increased trade in steam coal could come about unless the profitability (and hence price) of coal could be raised. The extent of such rises, and their impact on the nuclear/coal balance, is much harder to foresee.

Finally, there are the environmental questions associated with coal-burning. Current standards vary widely between countries, with the UK having, among industrialised countries, rather limited controls. It is possible that environmental regulation may tighten substantially in this area, and this would be to the clear advantage of nuclear power. On the other hand, the political forces that might succeed in achieving this goal might also be turned on nuclear power, and cause tightening of nuclear regulations, though these are, by any standards, already very stringent.

SUMMARY AND CONCLUSIONS

The economics of irradiated fuel transport are simple, short and quite unimportant. The sums expended are trivial compared to the scale of nuclear projects and even if multiplied ten-fold the effect on the economic viability of nuclear power would still be negligible.

Internationally, the future volume of irradiated fuel will be a direct function of the level of nuclear capacity. The historical record is that nuclear power has not proved superior to alternatives in all circumstances, and in UK the overall viability of AGR technology is still dependent on increases in the future price of coal. The deterministic view – that the age of nuclear will inevitably succeed that of oil, as oil succeeded coal – should be resisted.

Important strategic and military considerations will doubtless influence future nuclear development, but electricity demand and the relative economics of nuclear and its main competitors will mainly determine the scale of future capacity. This capacity will more than double to 1990 on account of past investment decisions, but quantitative forecasts beyond that point are hazardous. Four factors are likely to have the main influence on future nuclear ordering: (1) expectation about electricity demand growth, plus the need to replace the heavy plant investment of the 1960s and early 1970s, will be a very important conditioning factor in setting the size of the overall market; (2) nuclear capital costs and operating performance will be important; improvement in turn depends on the technological maturity of the industry; at the back-end of the fuel cycle this will be more difficult to establish than in reactor technology; (3) the degree of public opposition to nuclear power could play an important role in the future, though its practical impact in the past has been quite small; (4) the economics of the main alternative, coal-firing, will be critical. In this context, the evolution of world coal prices will be vital (it is likely that the real price of coal will rise) as will the stringency of environmental regulation surrounding coal combustion.

The directions and relative strengths of these four factors are impossible to foresee, but in the absence of a major nuclear accident, it is quite possible that they may lead to a much larger volume of new nuclear orders than the world has seen in the past few years.

24 The Security Aspect

RICHARD CLUTTERBUCK

In broaching the question of the risks of deliberate criminal or politically motivated attack, theft or hijacking of irradiated fuel in transit, I have attempted to look through criminal and terrorist eyes – what would I do to achieve the aims I wished to achieve? The questions I then asked were: Does such material offer an attractive target to criminals or terrorists? If so, what would be their aims and how would they set about them? And how effective are the security arrangements in relation to these potential threats?

Whatever the longer-term aims of extreme political movements – whether they are nationalistic, religious, social or revolutionary – the choice of targets and form of attack are usually dictated by short-term aims. Thus the motive may be publicity, political blackmail (e.g. the release of prisoners or convicts), extortion of money to finance the movement, and political destabilisation by creating public alarm, discrediting the authorities and provoking them into repressive action.

POSSIBLE FORMS OF INTERFERENCE

In considering the possible forms of interference it is necessary to highlight some salient points about the methods of transporting irradiated fuel. Apart from a short road journey from the power station to the railhead, shipping flasks are moved almost exclusively by rail in trains dedicated only to these CEGB flasks and which therefore require little or no marshalling or shunting en route. Each flask is carried on a flatrol (FR) wagon and there are normally up to three such wagons to each train, with a brake van and sometimes empty barrier vehicles. The flasks are made of 14-inch steel or, for AGR fuel, of steel lined with lead. Each flask weighs about 50 tonnes and is designed to withstand the impact of a 30-foot drop onto an unyielding surface, being at the centre of an 800° fire for half an hour and submersion for 8 hours in 50 feet of water. Each flask contains about 2 tonnes of irradiated fuel immersed in water to keep it cool. The fuel rods continue to generate their own heat during transport and the level of radioactivity decreases with time. The flask dissipates this heat into the atmosphere by convection so that the temperature remains stable.

If the flask were to be ruptured and water leaked out, the temperature of the rods would rise but would again stabilize at a higher temperature. If the water leaked onto the ground it would cause contamination in the immediate vicinity but at a low level. If it seeped into a water or drainage system this contamination would be quickly dissipated to a level within the accepted background level of radioactivity present in the environment. The fuel rods themselves are not explosive and not combustible below temperatures of about 600°. The lid of the flask, which is bolted and sealed, weighs 9 tonnes. If this lid were removed and the fuel rods ejected, they would remain at a low temperature in either the air or water and radiation would not constitute a health hazard beyond 50 yards from any one rod. Such an area could be quickly cleared under standard emergency procedures. Should the exposed rods be subjected to an intense and prolonged fire, there would be an airborne radioactive release which would require further evacuation, again covered by the emergency procedures for fires involving toxic materials.

To sum up, to achieve contamination or radioactive release such as to cause a significant health hazard, terrorists would need to exceed the effects of the worst accident which the flasks are designed to withstand.

In addition to the fuel from CEGB power stations, consignments of spent nuclear fuel come in by sea for processing at Windscale. This fuel is almost all carried in light water reactor flasks which are cylindrical and weigh about 70 tonnes, heavier than the CEGB's 50-tonne flasks. The main variety, the Excellox flasks, comprise 3.5 inches of steel and 8.5 inches of lead. By virtue of their extra weight they would present still greater problems to terrorists. They are carried by rail, mainly on cross-country lines. If the CEGB opens a PWR station at Sizewell the spent fuel from this will also be transported in this type of flask.

Terrorist interference could take the form of hijack or attack while the shipping flask is halted, e.g. in a marshalling yard, or on the move, or interference at a site to which the flask had been hijacked or maliciously causing an accident to the train or road vehicle. Since irradiated fuel is not an attractive object for theft – it is not explosive, without a reprocessing plant it cannot be used for making nuclear weapons, it is not combustible at temperatures below 600°C, its direct radiation is not hazardous beyond 50 yards, and it will not effectively contaminate water supplies – the aim of such interference would therefore be to cause disruption and public alarm.

The 14-inch steel of the flask could be penetrated by some modern anti-tank weapons. This could cause a leakage of irradiated water, which would contaminate the ground in the immediate vicinity (or spread by surface water in the event of heavy rain) but the level of contamination would be unlikely to cause any significant health hazard. Such an attack would not result in explosion or fire.

It would be extremely difficult, if not impossible, for a terrorist to insert any large explosive charge into the flask through the very small hole made by the anti-tank weapon. Even if he did, the fuel itself is not explosive. The lid could be removed to enable the terrorist to insert an explosive charge large enough to throw the irradiated fuel elements out of the flask, but distribution would still be localised and the health hazard could be obviated by normal evacuation procedures for accidents involving toxic materials.

To create significant airborne radioactivity, terrorists would need to conduct a complex three-phase operation. Having hijacked the flask removed the lid and ejected the contents, they would need to subject them to intense fire (e.g. by bringing in large quantities of flammable liquids or liquefied gases) to raise the fuel rods to their combustible temperature of about 600°C. As soon as such a fire was detected the police and fire brigade would be summoned. The police would then organise the evacuation of the area subject to potential airborne radioactivity using existing emergency procedures for fires involving any toxic materials. The fire services are equipped with protective clothing and, once they had extinguished the fire and damped down the ash with foam, the danger of airborne radioactivity would cease after the settlement of any residual contaminated airborne dust.

Thereafter the fall-out area would have to remain closed until tests had been completed to detect whether any significant quantities of the more persistent isotopes were contained in radioactive dust which had settled on the ground or on trees, vegetation, buildings, etc. Techniques for detection, clearing and removal are now well tested and reliable so that the contaminated area would need to be closed for a limited period only but during that time considerable disruption might be caused if the site had been cleverly chosen.

The protection of flasks against accidents is such that even the most intense and prolonged fire would be unlikely to create airborne radioactivity as great as that discussed above, if at all. It is conceivable that this might be the objective of a maliciously caused accident, e.g. with a tanker lorry containing flammable liquids but, apart from the complexity of such an operation, the design of the flasks is such that release of radioactive material would be most unlikely. Once again, however, there might be considerable disruption and public alarm.

Irradiated fuels in transit do not offer any practical means of contaminating the water supplies. If the water from a punctured nuclear flask seeped into a main water or drainage system the level of contamination would rapidly dissipate to a level below that constituting a health hazard and this would be easy to monitor to ensure safety. The same would apply if terrorists were able to puncture and immerse the entire flask in a reservoir or to immerse fuel blown out of a flask by explosions. This water would keep the fuel cool and the radioactivity would again be so low that it would quickly dissipate to a safe level. Since there are vastly easier and more effective ways of contaminating water and drainage systems, it is most unlikely that terrorists would embark upon this method though they might try to gain publicity or create public alarm by claiming they could and would do so.

It is conceivable that terrorists could hijack a flask with a view to hiding it in order to maintain an indefinite threat of using it for some future purpose. Spent nuclear fuel and waste do *not* constitute raw materials from which a nuclear bomb could be made nor, as shown, can they effectively be used to contaminate water supplies so the threat could be publicly discounted. The theft itself, however, would present the terrorists with immense difficulties. Attempts to conceal the flask on the rail system, even on remote and disused sidings, would be easy to detect. Transfers of the 50-tonne flask to a low loader and transport by road to a secret hideout would again almost certainly be detected. Removal of the contents from the flask would create unacceptable hazards for the terrorists themselves.

If a trainload of flasks were seized and held by armed men on the railway system (e.g. in a marshalling yard) there would be immediate justification for an SAS attack to recapture them. As already indicated the load itself, unlike say liquefied gases or toxic materials, would not offer the terrorists any immediate and credible threat (e.g. of an explosion in a built-up area) to deter such recapture. Hostages, such as the train crew, held by the terrorists would be at no greater or lesser risk than those in any other SAS rescue (e.g. in the Iranian Embassy siege in 1980).

The flasks are normally only transported by road for short distances. The transporter could be hijacked and driven away but, since the alarm would at once be sounded, the chances of such a ponderous and recognisable vehicle reaching a secret hideout undetected would be small. Since the hijacking for similar purposes of any of the much more attractive and less conspicuous targets on the road (e.g. of liquefied gases or motor spirit) has not been attempted by terrorists, there seems very little likelihood of them selecting a nuclear flask for such an operation.

It would be possible for terrorists to declare that unless, say, certain prisoners were released or the movement of irradiated fuel through London ceased forthwith, they *would* attack a flask in transit. Public ignorance of the very limited effects of such an attack might cause alarm. The answer would be to inform the public of the facts and say that the authorities concerned had no intention of giving way to the threats.

Terrorists might falsely claim that they had hijacked a flask and that they would use it to release radioactivity unless certain conditions were met. This too could create alarm. The answer would be to announce categorically that all flasks in transit are monitored by computer and that none are missing.

SECURITY MEASURES

Spent nuclear fuel elements are stored in their power stations under water (in 'ponds') to absorb and dissipate the heat they generate until the rate of heat production and radiation has fallen to a permitted safe level. They are then loaded into flasks as described above. Transport by road on low loaders is kept to a minimum and distances are usually short. Speed is limited to 12 m.p.h. As already described these transporters, if hijacked, would be extremely difficult to get away and hide without detection.

CEGB nuclear flasks have, since 1982, been transported exclusively in dedicated trains carrying nothing else. Each train normally carries up to three flasks and about 500 CEGB flasks are moved each year. Some pass through or around London. Routes and times are varied and kept on a need-to-know basis. Since dedicated trains require little or no marshalling or shunting en route, halts are rare and of short duration. Nuclear flasks trains halted in marshalling yards must be placed apart from wagons containing other dangerous goods, specifically those containing explosive, flammable gases, highly flammable liquids, spontaneous combustibles and organic peroxides. Such loads are not normally carried on the same lines as nuclear flask trains. Movement of every such train, including ones carrying nuclear flasks, is subject to control by an on-line computer system covering the whole of British Rail. Its position can be checked on any of the computer terminals within a few minutes. It cannot leave any control point without specific clearance being sought on the computer for its route and next control point. If the clearance were withheld, only a conscious and deliberate rejection of this instruction, probably requiring a conspiracy of five or more people, could override this safeguard.

All public services, including British Rail, local authorities, medical, police and fire services, have standard emergency procedures to deal with accidents and natural disasters, including explosions, fires involving toxic materials, and contamination of ground, water and drainage. Many of these have been discussed earlier in their context. There is no hazard which could be created by interference with irradiated fuel in transit which would be as serious as the most severe of the other contingencies which these procedures are designed to meet. Specifically, in the case of accidental or deliberate damage to nuclear flasks, the standard procedures include the immediate summoning of police, fire and medical services and, should there be any risk of radiation, health physics and other experts as required will be called from any of twenty-five nuclear establishments (CEGB, AEA, BNFL, military establishments, etc.) spread all over the country. Regular exercises conducted without warning by CEGB and British Rail have resulted in police and fire brigades reaching the sites of simulated incidents within a few minutes and all the experts required including health physics teams in less than 2.5 hours. These exercises, and the handling of such real accidents as have occurred involving dangerous loads, leave no doubt as to the dedication and efficiency of the services involved in transporting such loads and in public safety.

Mr Justice Parker, who was in charge of the inquiry into the proposed THORP reprocessing plant at Windscale in 1977, concluded 'that the transport of spent fuel creates no significant risk and that such risks as may exist are less than those involved in the transport of other substances which cause no alarm to any substantial section of the public'. In response to a Parliamentary Question in 1979, the Department of Energy carried out a detailed 'Study of consequences of a terrorist attack on an irradiated fuel flask'. On 27 July 1982, the

Parliamentary Under-Secretary of State announced the results of the study which, he said, showed

> that even under the most adverse combination of circumstances these flasks would not give rise to any significant hazard to the local population. I am satisfied that the existing arrangements, which I shall continue to keep under review, are adequate to protect public safety against any consequences of such an attack, and that there are no grounds for altering them.

Under the Nuclear Material (Offences) Bill published by the Home Office on 21 January 1983, legislation will be introduced whereby the receipt or holding of or dealing with nuclear material with intent to commit a criminal offence will be punishable by prison sentences of up to 14 years. Under the same Bill, an offence would be committed if a person used nuclear material as part of a threat to commit a criminal act or 'if, in order to compel a state, international government organisation or person to do, or abstain from doing, any act, he threatens that he or any other person will obtain nuclear material' by means of such offences as theft, robbery or burglary. This Bill would also enable the Government to ratify the international Convention on the Physical Protection of Nuclear Material which has been signed by a large number of countries in east and west Europe, the Soviet Union, the Americas and the Third World. This Convention would amount to an extradition arrangement for such offences even with signatories with whom the UK had no other formal extradition agreement.

Because of the appalling and continuing effects of uncontrolled contamination and radioactivity arising from the nuclear bombs dropped on Hiroshima and Nagasaki in 1945, nuclear materials of all kinds arouse a degree of public emotion and alarm out of all proportion to the actual hazards involved in the use of nuclear fuel. This gives any person threatening to misuse them a disproportionate power of coercion. That is the primary justification for the Nuclear Materials (Offences) Bill. This power of coercion, however, would be decreased if there were better public understanding of the true level of hazard in using, storing or moving nuclear materials, particularly in comparison with other dangerous materials, and of the high degree of effectiveness of the arrangements for their security.

My own research confirms the views quoted above, namely that no changes are at present required in the existing security arrangements. However, I do feel that it would be beneficial to raise the level of public information both about the degree of hazard and the effectiveness of security arrangements, especially against terrorist attack. This would reduce any attractions for terrorists or others to interfere with nuclear materials.

ATTRACTIONS AND LIMITATIONS AS A TARGET

Shipping flasks are designed to protect the public from the most severe impact, explosion, fire or immersion that could arise from the worst conceivable accident. Irradiated fuels in transit present extraordinarily difficult targets for terrorists – far more difficult than other dangerous goods which are less well protected and could be attacked with far simpler and quicker operations and with less chance of detection and more chance of success. A terrorist organisation which assembled the formidable resources needed for a successful operation against a nuclear flask could use its resources to far greater effect if deployed against a more vulnerable target. Nevertheless a well-sited attack or hijack of a nuclear flask could cause serious disruption by closing a bottle neck or (with the case of a fire creating a fall-out of radioactive dust) evacuation of a large area for a considerable time while testing, clearing and removal is completed.

The public alarm with which nuclear material is regarded will ensure that the media will make the most of any incident or threat of an incident involving irradiated fuel. The more sensational the media coverage the higher the public emotions will run so each will inflame the other. This does give the nuclear flask an added attraction for terrorists, probably more as a subject of a threat than of a physical attack. This attraction should be reduced by the Nuclear Materials (Offences) Bill and by better public information to discount any exaggeration of the danger.

The emotive factor does, however, have a compensating effect. If the terrorists did carry out their threat and did put lives or public health at risk, the public revulsion against them would be so intense that the political cause they espoused would be discredited, probably irrevocably. They know that and the authorities know it too, so the prospects of successfully calling the terrorists' bluff over a nuclear threat would be better than for most other forms of threat. If the threat came in from an individual or group which did not care whether it was politically discredited or not, or from a psychopath or mentally deranged person, it is unlikely that they would have the sophistication or resources to overcome the immense physical difficulties of carrying out a threat involving a nuclear flask. Governments and other authorities, therefore, could and should be ready to stand firm against such threats, from whatever quarter. The fact that no such threats have been made either with nuclear material or with other emotive material such as LPG or HCN suggests that terrorists realise that these threats might not be credible and would be far less effective than simpler operations using guns or bombs against human beings.

CONCLUSIONS

Nuclear flasks are very unlikely to be selected as targets for attack, theft or hijack. If they were attacked the likely health hazard would be low and could be satisfactorily controlled by existing emergency procedures. However, there could be considerable disruption and public alarm, although this should be reduced by the enactment of the Nuclear Materials (Offences) Bill and by better public education about the nature and degree of the hazard.

The main form of security lies in the sheer bulk and robustness of the nuclear flask itself and hence in the extreme difficulty that any terrorists would have in getting sufficient explosive inside it to do any damage. Lifting the lid would require a crane and it is difficult to see how that could be done at sufficient distance for the terrorists themselves to avoid being exposed to radiation. Moreover, to the group concerned, the political price of carrying out such an attack would be so great and so disproportionate to any actual damage it could do that such a threat is far less credible than almost all other potential terrorist threats.

On security grounds there is no justification for redirecting irradiated nuclear fuel away from London and other urban areas. To deny British Rail the use of the many rail routes through London would in fact increase the overall danger: it would make it more difficult for British Rail to avoid sending other dangerous loads on the same routes as nuclear flasks. There would thus be a greater risk of complex accidents, fires, disruption and public alarm.

25 Implications at the Local Level

DAVID HALL

INTRODUCTION

The aim of the Town and Country Planning Association, which is an independent voluntary body concerned with planning of the environment, is to promote policies that will achieve a better environment both in the physical sense and in the economic sense, i.e. ensuring that scarce resources are spent to best effect. So far as energy policy in general is concerned and nuclear power policy in particular, we have an interest because certain aspects of these policies could conflict with our objectives. We are concerned not only with getting the policies right, but also with getting the decision-making processes right to ensure that there is a democratic vehicle for adequate public debate and for there to be sufficient information available to all and sundry, so that society at large has the right tool-kit to make these important collective decisions. The debate about nuclear power has increasingly become a debate about central versus local decision-making and the extent to which legitimate interests of people at the local level can be reconciled with those of the nation as a whole.

I propose to look briefly at the policy and political implications for local government as well as at the more detailed implications for local government of the management and transportation of radioactive waste. In looking at the policy and political implications, I shall be assuming that elected local authorities represent the people in their area every bit as much as elected MPs, or the elected national government and thus have a legitimate reason to concern themselves with how *national* policies do or will affect the area as well as with the local technical, economic, social and administrative consequences of those policies. The practical implications at the local government level therefore include such matters as how a local authority or group of local authorities should organise themselves so as to influence national policy on an issue as well as how to cope with its effects. As with so many major issues these days, that of nuclear power and the management and transportation of the wastes from it thrusts us into the 'central' versus 'local' debate.

PUBLIC DEBATE

A dominant theme in that debate is the extent to which, on any issue, the public in general, and the relevant authorities in particular, are adequately informed about all the relevant facts and there is a satisfactory forum for discussing them. In the case of radioactive waste management the Government itself said in its recent White Paper (Cmnd 8607, July 1982) that 'the public must be kept fully informed about what is being done, and there must be proper scope for public discussion'. Unfortunately, however, the White Paper did not say how, where or when that public discussion will take place.

Indeed, the Government seems more concerned to stifle than to stimulate public discussion. There has been scant debate of these issues in Parliament. The idea was rejected of publishing a Green Paper which would at least have made it possible to assess public opinion on the proposals before issuing them in a White Paper. The Royal Commission on Environmental Pollution which would have provided an ideal forum for an objective, independent examination of the Government's present policies, was not invited to consider them. Moreover, the Government has not only declined to provide any financial assistance for objectors wishing to question Government policy at the Sizewell B public inquiry but the Department of Environment witnesses on radioactive waste management issues will 'not be prepared to answer questions directed to the merits of Government policy' (Transcript, 1 June 1982, p. 14). This is contrary to the stance taken by the Department of Energy, whose witnesses have been prepared to respond to such questions. It is my contention therefore, quite simply, that as matters stand there will not be 'proper scope for public discussion' on this issue unless a series of quite specific steps are taken to bring it about.

Radioactive waste management is an issue of great importance which generates strong emotions. It must therefore be debated openly.

Nor is it enough simply to make information available to the public. The public must be assured, through open discussion and independent investigation, and through detailed consideration of the possible local effects by their local authorities, that the Government's policies are soundly based, and that public health and safety are adequately protected.

As a minimum therefore I suggest the Government should take the following steps to provide proper scope for public discussion; moreover, I urge local authorities also to press for these measures:

(1) arrange for a full debate on its White Paper in the House of Commons as soon as practically possible; and refer the matter to a House of Commons Select Committee;
(2) request the Royal Commission on Environmental Pollution to review the conclusions and recommendations of the September 1976 report (the Flowers Report) in the light of subsequent developments;
(3) instruct Department of Environment witnesses at the Sizewell B public inquiry to answer questions directed to the merits of Government policy.

THE POLICY FRAMEWORK

The underlying argument behind these recommendations does of course apply equally to the nuclear power issue as a whole, not only to the waste management question. Although the nuclear power programme has been the subject of Parliamentary debate and examination of a House of Commons Select Committee, the development of Government policy on nuclear power has been characterised at various times by secrecy, ad-hocery and domination by interests vested in the nuclear industry. This has led, *inter alia*, to misinvestment on a grand scale and a huge opportunity lost as the result of alternative investments have been foregone. So far as public debate is concerned, the chief characteristic seems to have been the mass production of glossy, biased literature by the nuclear industry which in the absence of adequate public or private funds to produce more balanced material can only be described as brainwashing.

There is an urgent need, therefore, to improve the decision-making process in relation to national energy policies and major energy projects as has been recommended on many occasions.[1] In particular, the Government should provide itself with some form of Standing Commission on Energy to provide independent policy advice on a systematic and continuous basis, instead of depending on advice from the industry itself and the spasmodic consent that is available from Parliament and independent experts. The public inquiry process of major projects also needs changing so that the merits of Government policy can be challenged at public inquiries; and, above all, so that the case *against* a major project can be as vigorously scrutinised and researched as the case in its favour, by means of financial assistance being made available for objectors.[2]

I turn now to deal in more detail with radioactive waste and its practical implications at the local government level.

LOW-LEVEL WASTE

Looking first at low-level waste, it is estimated that between now and the end of the century, Britain will produce an average of more than 20,000 cubic metres of low-level waste each year. Low-level wastes are mainly materials with short radiation lives. Some of the wastes are in liquid or gaseous form but, except at Sellafield (formerly known as Windscale), these are of relatively minor importance. Most of the wastes are in solid form. They come mostly from the nuclear power industry, but some also come from the health service, civil research laboratories and businesses using radioactive materials.

Where the solid wastes are of very low activity, they are usually buried on ordinary landfill sites, and under the Radioactive Substances Act 1960 there is a statutory duty on local authorities to accept such wastes on sites which they provide. There are also a number of disposals of such wastes on privately owned sites. At present, the licensing conditions for private sites cease when operations finish, but pollution may continue or arise after the cessation of operations when the operating company no longer exists or is not in a position to meet its liabilities.

In May 1982, these arrangements were criticised in a report by the Association of County Councils (ACC).[3] In that report, the ACC declared that

> the administrative regime is insufficiently sensitive to the fears of ordinary people; although the local authorities are informed of authorisations where no special precautions are required, there is no

consultation upon its effects, no information given to the public, and the authorisation over-rides both the site licensing procedures under the Control of Pollution Act and the wishes of local authority owners of waste disposal sites.

A point of particular concern was that the local authorities 'have no information on the level of monitoring of leachate from waste disposal sites taking low-level deposits; the Department of the Environment keeps its own records, and monitors where it feels there is a need or where concern is expressed locally'.

The Government's White Paper was not able to take account of the ACC report because of its timing. Nevertheless, some of its main recommendations clearly call for speedy adoption, in particular that:

(1) the Department of the Environment and the Scottish and Welsh offices should be required to consult the local and water authorities before granting any authorisations under the Radioactive Substances Act and to hold local hearings where there are objections;
(2) there should be regular monitoring of waste disposal sites by the local and water authorities or, by agreement, by the Radiochemical Inspectorate, and the results should be freely available;
(3) private site licence conditions should extend at least 10 years after the cessation of tipping operations where radioactive or other hazardous wastes are involved, with continuing liabilities covered by bonding arrangements.

Solid wastes which are more heavily contaminated and unsuitable for burial on landfill sites are normally stored for a few weeks until they have cooled and the level of radioactivity has fallen sufficiently for them to be moved. Nuclear power stations and the major nuclear research centres package these in special containers on the sites where they are gathered but wastes generated by research or medical activities using isotopes or by industrial operations which result in low-level contamination are sent to Harwell to be packed into special disposal drums there.

In all, about 250 tonnes of low-level waste requiring special treatment are collected in Britain each year; when it is packed into concrete and steel containers, the total weight of material for disposal rises tenfold to about 2500 tonnes. Some of the containers are taken to

Drigg, a site in Cumbria owned and operated by British Nuclear Fuels Limited (BNFL), where they are buried in a shallow concrete trench, but the majority are dumped at sea at a spot 500 miles south-west of Land's End and 340 miles north of Spain.

SEA DUMPING

The first discharge of low-level waste at sea was in 1949, since when both the quantity and the radioactivity of the dumped material has grown inexorably. Between 1950 and 1960 the number of curies increased from 3½ to 279. In 1967, 724 tonnes holding 1600 curies were dropped, in 1971, 2200 tonnes holding 9000 curies, and in 1982, 2696 tonnes holding 108,000 curies. The sharp increase in radioactivity levels has been partly due to the increased dumping of intermediate as well as low-level waste at sea; for instance, 400–500 tonnes of intermediate level waste were included in the load discharged from the MV *Gem* in August 1982.

The first international dumping on the present site was carried out in 1967 under the banner of the OECD's Nuclear Energy Agency (NEA). Existing arrangements allow the OECD's nuclear members one ocean drop a year under strict arrangements regarding the type of waste and its packaging. West Germany, Sweden, Italy and France all ceased dumping radioactive waste at sea in 1974, and Holland has decided to stop dumping nuclear waste at sea in late 1983. Only Britain, Belgium and Switzerland are continuing to do so. The use of the present site was last reviewed by the NEA in 1979, which concluded that it could safely continue to be used for this purpose until the next review in 1984.

As the recent meeting of the so-called London Dumping Convention demonstrated, there is now growing international opposition to the present arrangements for dumping nuclear waste at sea. The Scandinavian countries are opposed to them. Spain is unhappy about them, and last September the European Parliament voted to stop the dumping of nuclear waste at sea. Where Britain's waste will go if dumping at sea is either prohibited or severely restricted is not known, but the decision clearly has implications for local government. The White Paper merely stated that 'in view of the remote location of Drigg and the extent to which it will be required for BNFL's own wastes, additional sites suitable for burial of this type of waste should be identified,' and the recently established Nuclear Industry Radioactive

Waste Executive (NIREX) is said to be currently planning new facilities.

One problem with the sea-dumping programme is that there appears to have been no systematic monitoring of the effects of this maritime waste disposal. Although there have been periodic reviews by the OECD Nuclear Energy Agency and at the recent meeting of the London Dumping Convention a UK proposal for a review of the Scientific and Technical considerations was accepted, these measures scarcely seem to go far enough when one considers that hitherto such monitoring as has taken place has been reliant on mathematical models of ocean behaviour. The reliability of such models has been questioned by academic experts, and their findings seem to have been at variance with recent studies of American waste dumped in the sea up to 30 years ago.

In a policy statement published last December[4] commenting on the White Paper the Town and Country Planning Association said that having regard to the growing international opposition to this dumping at sea, the inadequate and apparently conflicting evidence concerning its possible effect, and the virtual impossibility of retrieving the waste once it has been dropped, the Government should:

(1) commission a detailed study of the possible environmental and ecological effects of the present programme;
(2) defer any plans to increase the quantity or radioactivity of waste dumped at sea (including the replacement of the MV *Gem*) until this study has been completed and publicly debated;
(3) identify new land disposal sites as a matter of urgency, so that there is adequate time for public discussion on the general issue and for public local inquiries on specific proposals prior to the next NEA review in 1984.

INTERMEDIATE-LEVEL WASTE

The term 'intermediate-level waste' is used to cover a wide variety of wastes produced in the form of sludges, cladding, resins and assorted debris, whose radioactivity levels are higher and longer-lasting than in low-level waste, but not so high that the waste needs to be stored in specially cooled tanks. Until now, such waste has simply been stored on the surface, either at Sellafield (Windscale) or at the nuclear power stations themselves. However, the quantity is already large and is increasing rapidly, and the present sites are running out of space. Some 35,000 cubic metres have accumulated since the nuclear programme began, and this will have doubled by the end of the century, not allowing for any wastes which may arise from the decommissioning of plant. In volume, it will then be sufficient to cover two football pitches to a depth of 7 metres.

The White Paper stated that

> the lack of suitable disposal facilities for intermediate-level wastes is the major current gap in waste management, and it is important that it should be remedied. . . . Work is now proceeding on the basis of bringing into operation by the end of the decade facilities which should be able to accept a high proportion of the wastes suitable for disposal. The probable forms are an engineered trench at a depth of about 20–30 metres and a modified mine or purpose-built cavity at greater depth.

The 'engineered trench' will be the first priority and will probably require a site of up to 100 acres. It will be required by 1986. A clay subsoil is likely to be preferred because of its impermeability. There have been rumours that it is likely to be located at Harwell, which is situated on a bed of clay, is an existing nuclear establishment, and is where many of the low-level wastes are now sorted and packaged. The 'purpose-built cavity at greater depth' is likely to be needed by 1990. There have been suggestions that it might also be located in the clay deep beneath Harwell, or alternatively in brine cavities in the former salt mines of Cheshire.[5]

In view of the widespread alarm which either proposal is likely to cause throughout the whole of the hydrological area within which it is to be situated, not to mention its public importance nationally, the ordinary public local inquiry process will not be adequate for evaluating it. Thus I suggest that a special planning inquiry commission should be set up to consider this proposal; that the appropriate county council should commission a detailed environmental impact assessment of it before the commission hearings begin, and that financial assistance should be provided to representative local groups within the hydrological area concerned so as to enable them to participate in the public inquiry process.

In its White Paper, the Government announced that it intended to set up a Nuclear Industry Radioactive Waste Executive (NIREX), consisting of BNFL, CEGB, SSEB and UKAEA, to coordinate the disposal activities (other than high-level waste) of these four waste-producing organisations. The activities of NIREX will be subject to the control and supervision of the regulatory bodies – the authorising departments under the Radioactive Substances Act 1960 for the disposal of waste, and the NII under the Nuclear Installations Act 1965 for the licensing of nuclear sites and the safety of on-site operations.

In my view it is unfortunate that the Government has decided to leave arrangements for nuclear waste disposal in the hands of the main waste producers, rather than make them the responsibility of an independent statutory body, as recommended by the Royal Commission on Environmental Pollution in 1976. Now that this proposal has been implemented, it is essential that the role of the regulatory bodies be strengthened in order to ensure the safety of the public and the protection of the environment. In particular, therefore, membership of the Radioactive Waste Management Advisory Committee should be expanded to include representatives of local and water authorities and national environmental organisations.

HIGH-LEVEL WASTE

High-level wastes are ones in which the temperature may rise significantly as a result of rapid radioactive decay. In the UK, these are the highly active liquid residues which arise from the first stage of the reprocessing of spent reactor fuel. Although relatively small in quantity, they present major problems of long-term control.

When fuel rods are changed at a nuclear power station, the spent fuel is kept for at least 90 days in cooling ponds at the power station before being transported by road or rail to Sellafield (Windscale) for eventual reprocessing. The waste contains the most powerful radiotoxic substances, the actinide series, only one of which (plutonium) is extracted for future use in fast breeder reactors, or for military purposes. The remaining wastes are kept underground in double-walled, stainless steel tanks, where they undergo constant stirring, cooling and monitoring. About 1000 cubic metres of such waste are now stored in tanks at Windscale, plus a small amount of lower concentrations at Dounreay in Scotland, and by the end of the century it will be increasing at a rate of about 200 cubic metres a year.

Since many of the fission products and by-products, which are the principal source of heat, are both long-lived and highly toxic, successful disposal requires the total isolation of the wastes for at least a thousand years, with no risk of their ever contaminating air, soil or water. To safeguard against this, it is proposed to convert the liquid wastes into solid boro-silicate glass blocks, which will then be enclosed within stainless-steel containers. Work is now going ahead on the design of a vitrification plant at Sellafield (Windscale), planning permission having been granted by Copeland District Council in 1982, and it is expected to come into operation in 1987. These glass blocks will then be stored at the surface for a period of at least 50 years, until the heat they emit (2½ KW per ingot initially) has been much reduced and the shorter-lived radionuclides have decayed.

By the end of the century Britain is likely to have accumulated about 10,000 steel bottles, like milk churns, each filled with a radioactive glass ingot, and the number will go on increasing year by year until some means of permanent disposal is found.[6] How and where this will take place is not known. The White Paper says that the Government is simply 'leaving the decision on disposal to a future generation'. In its view, 'we in the present generation have a clear moral duty to formulate the options (for disposal) as we see them at present, and to develop the supporting scientific and technical knowledge, so that they (i.e. future generations) will be better placed than we are to make the eventual choice.'

But surely we have a moral duty to do more than merely formulate options. Is not the morally correct course to avoid creating more waste until we are sure that an option exists which is practical, safe and publicly acceptable? No such option has yet been demonstrated. Until it has been so demonstrated, there should be no further expansion of nuclear energy in this country. As the Royal Commission on Environmental Pollution said in 1977, the issues posed by waste management should be 'fully considered at the outset of a nuclear programme, not dealt with many years after the decisions on developments that lead to the wastes have been made and when options may have been effectively foreclosed'.

NUCLEAR-WASTE TRANSPORT

Between 1962 and 1974, some 4500 consignments of irradiated fuel, weighing about 9000 tonnes, had been moved over 1½ million miles by rail to Windscale. There is now an average of 350 nuclear rail journeys per annum in Britain, about one every day of the year, and the number will increase in future years if more nuclear power stations are opened. Of late, there has been growing concern about the possible safety risks involved, including the possibility of terrorist attacks.

Arrangements for moving waste around the country by rail, road and sea have at no time been the subject of any public inquiry. They certainly will not be properly assessed at the Sizewell inquiry. The CEGB's Statement of Case says that 'at this stage it is premature to specify which design of flask will be used to transport irradiated fuel from Sizewell B,' and it has recently been confirmed that it may still be undecided when the inquiry has ended. In the March 1982 issue of *Local Government News*, it was revealed that neither local authorities, nor the police, who carry joint responsibility for emergency planning, are forewarned of when and where a shipment of nuclear fuel by rail is going to take place, and that the fire services have little knowledge at station level of appropriate emergency procedures. Clearly some tightening up is required, and I would suggest that the police, fire and highway authorities should be formally consulted on the routing and frequency of regular movements of intermediate and high-level waste, on the design and safety of the flasks, and on the procedures to be adopted in event of an emergency; moreover, provision should be made for the Department of Transport to hold a public local inquiry if, following such consultation, a local authority formally objects to the proposed arrangements.

CONCLUSION

In conclusion, I believe that the public and the local authorities should be involved in the formulation of nuclear waste management policies, and that those policies should have full regard to the twin needs of ensuring public safety and protecting the natural environment, both now and in the long-term future. The White Paper is deficient in both these respects. The management of nuclear waste is undoubtedly an emotive issue, and it is one which is causing widespread fear and concern. It is vitally important that it be discussed openly and fully so that people can understand the implications for themselves, for the places in which they live, and for future generations. I believe that local government has an essential part to play in this process.

NOTES

1. 'Energy policy and public enquiries', TCPA Policy Statement (February 1978); and D. Pearce *et al.*, Decision-making for Energy Futures (Macmillan/SSRC, 1979).
2. 'Financial assistance for objectors at major public inquiries', TCPA Policy Statement (February 1979).
3. 'Radioactive waste', Report by Association of County Councils (May 1982).
4. 'Radioactive waste management', TCPA Response to the White Paper (Cmnd 8607) (December 1982).
5. Reports by David Fishlock in *Financial Times*, 28 April 1982 and 10 August 1982.
6. *Hansard* (24 March 1980).

The author wishes to acknowledge the background research and writing by John Blake, Vice-Chairman of the TCPA Executive Committee, who wrote the original TCPA Policy statement on which this paper is based.

26 A Trades Union Viewpoint

PETER JACQUES

The Trades Union Congress attempts to bring together the views and expressions of just over a hundred unions which together organise over 11 million workers, some of whom are in the business of working in nuclear plants, some who are not; some who are involved in the transport of nuclear waste and some who are not; some who are involved in the emergency services and some who are not. Thus, many individual trade unions take very different views on this issue. But the trade union movement as a whole advocates the integrated expansion of all energy resources including coal, oil, gas, nuclear power and renewable resources, hydro, tidal power, wind and solar power. We stand quite clearly and unequivocally for energy expansion and against the contraction of the energy industries. Nevertheless, we do not accept that that means energy at any price.

All forms of energy production pose risks to workers, to the public and to the environment in different ways and at different points in the overall energy cycle. The health and safety hazards of mining, for example, are well recognised as are the risks of working in the North Sea to extract oil and gas. The construction of dams, power stations and energy installations generally put construction workers at risk. Energy distribution systems pose risks to urban populations. The waste from fossil or nuclear fuels pose long-term problems for the environment.

Some would agree that these penalties are the inevitable price which must be paid for the advantages which industrialisation based on energy brings. But the scale and consequences of these penalties are by no means fixed and are not to be regarded as acceptable. Trade unions have always believed that they can be reduced progressively as technology and expectations advance. This in fact is the key principle underlying the discussion on health and safety and environmental issues that is set out in Chapter 12 of the TUC's 'Review of energy policy'. Coal, on which our industrial revolution was based, provides perhaps the best example, or certainly the earliest example, of this trade union philosophy in practice. Although mining remains one of the most hazardous industries, the great advances which have been made in accident and disease prevention, particularly since the nationalisation of the industry in the late 1940s, can be attributed overwhelmingly to the increasing efforts of the trade union movement to secure improvements at both local and national levels. Further improvements in mining health and safety are still to be achieved. However, it is self-evident, given the numbers who continue to be killed and seriously injured in that industry every year, that we still have a long way to go. Blood cannot be accepted as the price of coal. The same conclusions can also be drawn about nuclear safety.

In the 22 years of Britain's nuclear experience the rigorous system of regulatory control that has been involved to control risks in this industry has been developed largely in response to public, including trade union, concern; but this does not mean to say that continuing efforts are not required to reduce risks to the lowest possible level. In fact, given the high level of public concern about nuclear safety, this in itself dictates the need for greater efforts than in the past. Concern about the transport of irradiated fuel between reprocessing plants and nuclear generating stations, while representing until recently a relatively small part of the nuclear controversy, illustrates very well the wide range of interests involved. Residents in urban areas through which working flasks are transported (the majority of people living along railway lines tend to be working people) have expressed real concern about the probability and the consequences of major accidental releases of radioactivity in transit. Workers in the transport industries have demanded adequate protection to ensure effective radiological control.

All these concerns are real and legitimate and find expression within the affiliated membership of the TUC. It is the TUC's job to ensure that these understandable concerns are accurately reflected at national level in the consultative machinery established in connection with nuclear safety. The TUC, for example, is in fact represented on the Health and Safety Commission, that Commission's Advisory Committee on Safety in Nuclear Installations, the Commission's Technical Working Party on Radiation and the various sub-groups and the

Radioactive Waste Management Advisory Committee. Through all these bodies the TUC actively pursues the attainment of the highest possible standards of nuclear risk control but we would be rightly labelled as complacent were we to suggest that all scope for improvement in nuclear safety had been exhausted.

Indeed against a background of Government cuts in health and safety, including the Health and Safety Executive's Nuclear Installations Inspectorate (which plays a central part in the nuclear field), the possibility of continued advance can be called into question. As a result of TUC and other representations there has been a marginal expansion of the NII's manpower resources. But ultimately the future of the Inspectorate is closely linked to that of the Health and Safety Executive as a whole. In the coming months the TUC will be mounting a major campaign against the cuts in the Health and Safety Executive and pressing for an expansion of resources including in the nuclear field. Anyone who attempts to follow the nuclear debate cannot escape the fact that it is conducted in very complex technological and economic terms and many of our members in the trade union movement sometimes wonder how to keep up with this debate.

But it would be naive in the extreme, I think, to imagine that the motivations of either side in this debate are always entirely rational. On both sides of the argument convictions about the advantages, or the penalties of nuclear power stem, it seems to many of us, from deep-rooted intuitions about the potential of nuclear fission itself. To the nuclear technologists it could be argued that controlled nuclear fission merely represents another way of boiling water to drive turbines. The problems involved for them are essentially technical and are capable of technical solutions. To the opponents, nuclear fission seems to represent an interference with vast natural forces, the harmful consequences of which are beyond man's innate capacity to control. This view has been greatly strengthened in recent years by the growing understanding throughout all sections of society of the destructive power of nuclear weapons and the linkage which existed historically between the development of nuclear generating and weapons technologies.

But a nuclear reactor, of course, is not a bomb. Neither is the control of nuclear operations purely a technical issue. But at both extremes of this debate each proposition is equally unacceptable. I think it is the failure to understand why this is the case that prevents real progress towards a mutual understanding being achieved between both sides. This is equally true when dealing with the specific issue of transport of irradiated fuel. Much effort is expended by the industry in explaining that the risks of transport of irradiated fuel through cities are qualitatively no different from the risk presented to urban populations by the transport of conventional energy materials, such as flammable liquids and gases or toxic materials. But such assertions tend to ignore the reasons underlying public concern.

The generation of heat from nuclear fission is outside practically everybody's experience and imagination. Man has clearly lived with thermal energy in the form of fire for millions of years but for less than half a century with energy harnessed from nuclear fission. In addition, the potential hazards of radioactivity are perceived as qualitatively different from other kinds of environmental contamination. After all, ionizing radiations are incapable of detection by the normal five senses. Radioactive materials with very long half-lives, which may remain in the environment and embody unknown long-term consequences, inspire real fear in ordinary people and are therefore not considered equivalent to other kinds of risk. Those who have had the opportunity or otherwise to be able to discuss the rather different risk between, for example, blindness and deafness or between mental handicap and epilepsy and so on will know that people have a different perception between those different forms of illness.

This cannot be changed simply by explaining that only a very small exposure to ionizing radiations arises from man-made sources. At the other extreme many of the opponents of nuclear energy steadfastly refuse to adopt the same methodological disciplines of the nuclear technologists when they themselves attempt to assess the levels of assurance of safety in nuclear operations such as fuel transportation. Understandably perhaps, comprehension and consequently critique of nuclear safety theory tend to be somewhat superficial. Suspicion often stems from the belief that the effort expended by the industry to explain its case is motivated by self-interest and bias. Yet it is no answer to dismiss the science of nuclear safety as so much pro-nuclear ideology. It is all too easy to be simply negative. But science and technology which made nuclear power a practical possibility can and must also be made to yield practical solutions to the problems which nuclear power in turn creates. And this is particularly true in the transport issue.

The conjectural risks associated with nuclear power, such as loss of containment of nuclear fuel in transit, are often described in terms of low-probability/high-consequence risks. They are presented as similar in nature to extremely rare catastrophic events, such as the collision of large meteorites with the earth or equally unlikely disasters from man-made sources. The attempt is then made to assess the acceptability of such risks to exposed populations by comparing them with the medium-probability, medium-consequence risks faced by people in the normal course of their day-to-day activities. Examples often quoted are risks of accidents or ill-health arising from transport or leisure activities. That this approach to risk perception and acceptance is woefully inadequate can be seen from the extremely marginal impact which it has in convincing sceptics that the risks presented by nuclear power should be lived with. High-consequence risks are qualitatively different from low-consequence risks regardless of their probability precisely because they are viewed differently by those exposed to them. The question, 'what if something goes wrong?' inevitably weighs much more heavily in the popular imagination than the question 'how often?'. Questions about the potential consequences of an accidental loss of containment of the contents of a railborne nuclear flask in an urban area are bound to attract more attention than the questions about the likelihood of such an event happening at all. If the consequences of such an event are thought to be so disastrous that no level of probability could be viewed as acceptable under any circumstances, such doubts are not removed simply by describing probabilities of such accidents occurring as the inverse of very large numbers – particularly when the numbers are derived from perhaps crude theoretical models which are beyond the ability of most people to evaluate. Neither is it a solution to point equally high-consequence risks to urban populations which, because of the unevenness of perceptions, do not attract the same level of concern as the nuclear fuel transport issue. The perception of risks among workers and the public in general about nuclear safety must be accepted as factors affecting the priority to be attached to reducing such risks to the lowest possible level. Ultimately it is ordinary people not probability calculations that decide whether or not risks can be lived with. Probability is an aid to judgement and no more.

The technical issues arising from public concern about transport of nuclear fuel through urban areas have been reasonably well identified for some time and have been set out in the GLC's Statement of Case to the Sizewell B inquiry. Basically they relate to assessing and minimising both the probability and the consequences of an accidental or malicious loss of containment of radioactivity from a flask during its journey by rail through a populated area. An additional issue of importance for railway staff has been the question of routine exposure to radiation from flasks during the transport operation. The electricity unions, those who are involved with the putting of the materials into the flasks and unloading at the other end, have been particularly involved in that area and considerable attention has also been focused on the issue by the railway unions, British Rail and the National Radiological Protection Board, and there is now a widely held view that levels of exposure of staff are of a relatively low order under normal operating conditions. The unions concerned, both of the electricity unions at the beginning of the journey and at the end, and those who have the responsibility for the transportation of the materials, have made considerable efforts to ensure that their memberships and their normal operational responsibilities are taken into account. One of the issues they have had to concern themselves with is the possibility of a major loss of containment incident. Reasonable doubts continue to be expressed about the integrity of flasks under abnormal and accident conditions. Three scenarios have attracted attention – the accidental high-speed impact with structures during transit, prolonged exposure to heat, and vulnerability to sabotage.

The fact that impact testing of flasks has been carried out on models suggests that insufficient resources may have been available for the kind of full-scale trials which have been carried out in other countries, although the argument that conclusions drawn from models can be reliably scaled up is very hard to disprove. Nevertheless there are still continuing concerns from work people that the full-scale trial should take place. Similarly, behaviour of the contents of the flask during a prolonged fire, perhaps accompanied by a loss of cooling water, must inevitably be based on theoretical models which, even incorporating assumptions, will rely on a large element of informed judgement about the manner in which components of the flask will behave under fault conditions. Similarly the extent and pattern of distribution of the contents of the fuel flask in the event of an accident must remain a matter of conjecture, although again some projections have and can be made.

Given these uncertainties which have been expressed by some trade unionists and the level of concern which exists, there would appear to be a very strong case for further work to demonstrate that the current understanding of these risks and their control is adequate. Derailment of conventional cargoes is not uncommon and although there have been no major catastrophes in Britain involving dangerous substances, events such as the fire at Mississuaga have shown what can happen when things go wrong with major quantities of hazardous material in transport. In these situations there can be no substitute, I think, for a full and open presentation of all the scenarios and all the calculations in a way that is meaningful to those who are at risk, whatever the scale of risk may be. It is they ultimately who have to be convinced about safety. From this point of view the TUC welcomes the further studies commissioned by the GLC with the National Radiological Protection Board and hopes that full engineering and operational information will also be made available by the CEGB to enable them to be completed.

Additional safety measures must also be included in the debate. The question on the use of preferred routing away from centres of population is fundamental. But it may not be as easy to implement as may be imagined. Routing outside London may well mean fuel shipments passing through other towns where the inhabitants are just as likely to be concerned about safety as are Londoners. Operational and emergency safeguards have been improved and no doubt can be reviewed. Unions representing the emergency services have very strong views on this. In the long-term technical changes to flask design and mounting or even the provision of a secondary containment may well also be appropriate but there is little point in carrying out measures if the public are not kept informed. Reassurance can only be achieved by frank and honest disclosure of information.

This brings me to perhaps the most important issue underlying this debate, which is the issue of accountability. It is axiomatic in the trade union approach to safety that risks to workers and the public from industrial activity must be controlled by those in charge of such activity. In the case of rail transport of nuclear fuel this means the CEGB and British Rail, but others, too, have a major role to play including the Department of Employment, the Department of Transport, the Nuclear Installations Inspectorate and the Health and Safety Commission. That system of control, however, and the legislation which underpins it, must be seen to be accountable to those who are at risk, whether they are at work or in the community at large.

The trade unions reject quite categorically any approach to safety based on technological elitism or straightforward old-fashioned paternalism. While it is the role of experts in the field, including the nuclear field, to identify risks, to assess control measures and to advise, it needs to be emphasised that it is the right of those who are at the sharp end of the risk to reach agreement with those in control of risk-creating activities, about whether or not the controls are adequate. I therefore return to my central point that efforts to reassure the public must be based on full information being made available to the public's representatives, be they trade unions or local authorities, or other people, who have to reach a judgement.

27 Open Forum

JOHN SURREY

All the preceding chapters are written by people who in some sense are 'experts'. As some of them have noted, there can be a wide gulf between 'expert' assessments and how the man in the street perceives the risks. Public attitudes will in the end determine whether the transport of irradiated fuel in particular and nuclear power in general are socially acceptable. This chapter presents the range of views expressed from the floor during the GLC Conference and questions and answers which supplement the specialist contributions in the previous chapters. As far as possible, I have tried to capture *all* the viewpoints that were expressed – by politicians, trade unionists, anti-nuclear campaigners and unaligned individuals. The only omissions I am aware of were questions which are answered in the preceding chapters. Although it cannot be assumed that every statement made from the floor necessarily represents the views of the speakers' organisations, the speakers' affiliations are given because they give important clues to opinions among the types of organisation which collectively are likely to reflect or shape public attitudes generally.

At all large gatherings to discuss a controversial topic almost inevitably the debate spills over to encompass numerous related concerns. The GLC Conference was no exception – after all, the transport of irradiated fuel is part of the nuclear power debate and nuclear power is part of the wider decision-making for energy and the environment. The fact that this chapter contains a wide spread of popular concerns indicates some of the problems of decision-making in this area. A further indication is given by two strands in the discussion which have much in common with nuclear power debates in other countries over the past decade. On the one hand there is the strand that runs 'Now that the "experts" have told us the facts, I'm more confused than ever, so let's ban it altogether,' and on the other hand, 'Not all my doubts and fears have been removed by the "experts"; therefore we need a public inquiry and/or a Freedom of Information Act and/or direct participatory democracy at the community level; only thus can the people decide.'

Complex technological questions, where a rather remote chance of something nasty happening must be balanced against the benefits of proceeding and where those who bear the risk are not necessarily those who receive the benefits, pose some of the most difficult problems for modern democratic societies. How should the decision-making be organised? How far should it be left to the 'experts' and to elected Parliamentary representatives? How far is it possible and desirable to involve other interested parties, notably trades unions and local communities, and what weight should be given to outspoken minority opinion compared with that of the unconsulted, 'silent majority'? Does consultation automatically confer the right of veto?

This problem in relation to complex technological decision-making has important similarities with other conflicts between representative democracy and direct or delegated democracy in other spheres, e.g. debates over capital punishment, over nuclear weapons policy, and over methods of electing leaders of political parties. As long as western societies manage to avoid the collectivist extremes at either end of the political spectrum, the problem of how to organise decision-making so as to reconcile the conflicting aims of democracy and efficiency will remain with us. Democracy requires continuing debate; but the need for debate must not stifle action, and action must be accompanied with appropriate accountability. It is a difficult tightrope to tread.

The debate starts with contributions from two leading GLC Members which illustrate the type of political context in which complex technological issues arise and have to be handled.

POINTS OF VIEW

Ken Livingstone (Labour Leader of the GLC) said that irradiated fuel transport through London was a matter of concern to all parties on the GLC. It had become an issue in October 1980, during the previous Conservative administration. Following consideration of the problem by the GLC's Public Services and Safety Committee, the GLC stated its concern to the Department of Transport, British Rail and the CEGB and suggested alternative routes avoiding London. During

1981 the GLC was asked by other Councils (including Lewisham, Wandsworth, Barking and South Yorkshire) to hold a public inquiry and the GLC repeated its concern in evidence to the House of Commons Select Committee on Transport. During the same year the GLC declared London a nuclear-free zone and it came to favour the idea of a special conference on irradiated fuel transport in order to give the opportunity for rational analysis of the fears that people have on the subject. All his experience showed that neither machines nor people were 100 per cent safe and accurate:

> there is great concern in London, and that concern is not diminished by a welter of statistics to show that such and such cannot ever happen. Those people who live near the routes through which spent fuel rods are taken through London can see for themselves the state of the rolling stock and just how vulnerable those particular lines could be to attack. . . . we are perhaps more vulnerable than anywhere else in the world to a terrorist group intervening in the transport of spent fuel rods through thc capital, and with devastating effect.

He concluded that if anything came out of the Conference that brought nearer the day when this threat to London is removed, all his GLC colleagues would say that the Conference had been worthwhile.

Neville Beale (GLC Opposition Spokesman on the Environment) drew attention to the political motivation behind the calling of the Conference; it was no coincidence, he said, that it was held just as the Sizewell inquiry was getting under way. Nevertheless, many speakers had achieved high standards of scientific objectivity – especially Lord Flowers, who had concluded that irradiated fuel transport was essential and a minimal risk. He wondered whether the concept of a nuclear-free zone, besides excluding nuclear weapons from London, was also meant to exclude both nuclear fuel from passing through London and radio-isotopes for medical work. If so it would preclude taking nuclear current from the grid and eventually it would necessitate switching off, since it would not be possible to reprocess spent fuel or to dispose of the waste products; and the banning of radio-isotopes would mean that radiation therapy in London hospitals would cease. He also pointed out that the Conference had cost London ratepayers £110,000 and that the GLC had budgeted nearly £1m of ratepayers' money for nuclear-related matters (including £135,000 on opposing the Sizewell PWR, £37,000 on publicity about nuclear weapons, £37,000 promoting 'Peace Year', £43,000 on grants to 'peace' groups, £214,000 for 'Peace Year' events, plus 60 per cent for loss of Government grant on the net cost). He added that: 'Some of you may recall a row started by the Labour Party in Parliament recently when it was suggested that the Government might be about to spend a similar sum promoting its policy for national defence. Obviously, there are double standards in these matters.'

John McDonald (Health and Safety Executive) said that the 1974 Health and Safety at Work Act had created the Health and Safety Executive (HSE), and the Health and Safety Commission to which the Executive reports, with wide responsibilities for protecting workers and the public from risks arising from any work activity, including irradiated fuel transport. The Commission contains representatives of employees, trades unions and local authorities and has a number of specialist advisory committees (including one on the Safety of Nuclear Installations and another on the Railways Industry); it prepares legislation for approval by the appropriate Ministers – the Secretary of State for Transport in the case of irradiated fuel transport. The HSE is responsible for enforcing the 1974 Act in a wide range of industries (except the railways, which come under the Department of Transport's Railways Inspectorate). In the field of radiological protection the HSE has a close relationship with both the NRPB and the Safety and Reliability Directorate of the Atomic Energy Authority, both of which provide risk analyses when requested by the HSE.

Mr McDonald stated that since joining the European Community in 1973 the UK has been bound by the European Atomic Energy (EURATOM) Treaty which requires Member States to establish safety standards against risks arising from any kind of work with ionising radiation, including irradiated fuel. The HSE is enforcing those standards and in November 1982 it published detailed proposals on radiological protection which were prepared by a working party that included representatives of the CEGB, British Rail and the trades unions. The proposals relating to irradiated fuel would require British Rail to conduct a survey of the hazards arising from this traffic, to prepare contingency plans for accident situations and to submit both for approval to the HSE; the same requirement will apply to all nuclear installations throughout the country. However, this did not mean that the HSE lacks confidence in existing practices. The new proposals would supplement and reinforce, rather than replace, present arrangements. The HSE proposals were publicised through the press and a consultative document had been sent to the Local Authority Associations (the Association of County Councils, the Association of Metropolitan Authorities, the Association of District Councils in England and the Convention of Scottish Local Authorities).

Terry Segars (Fire Brigades Union) said that since 1981 his Union had been opposed to the transport of irradiated fuel through urban areas; however, if any accidents occurred the firemen would continue to do their duty as in the past.

Celia Pillay (Electrical Power Engineers' Association) said that the CEGB had provided all the information requested by her Union and that the Union was satisfied about the safety of its members who load and despatch the flasks at nuclear power stations. However, she regretted that the CEGB were not persuaded of the need to conduct tests similar to those by Sandia in the US, for she felt that such tests would do far more than any number of scientific treatises to reassure the public on the question of safety.

Hazel MacPherson (Paddington NUR) said that two important changes had recently occurred on the railways. Firstly, track maintenance standards had fallen; whereas previously, the permanent way staff examined the track daily, since 1981 it was done only weekly. On the London Midland Region, which includes Willesden (the junction through which passes the irradiated fuel en route to Sellafield), accidents had increased by 50 per cent over the past year. Secondly, British Rail's intention to remove guards from freight trains would increase transport risks generally, including those associated with irradiated fuel.

Ron Blanchard (London Borough of Hackney) thought that the cost of the GLC Conference was a pittance compared with the amount the CEGB is spending to promote nuclear power. People should beware of believing that, because scientists say there is only a very remote chance of something happening, it is never going to happen; after all, Alexandra Palace, which was not supposed to burn down once, has burned down three times this century, and numerous dams and bridges throughout the world have collapsed: 'the more you allow things to happen, the more likely it is that there will be a final catastrophe. Realising how little we know, I would think every Council Member should be pressing for a Freedom of Information Act.'

Chris White (Environmental Health Division, London Borough of Hackney), referring to Lord Flowers' analogy of the Rolls-Royce and the starting handle, said that whilst the starter may never fail, the battery may – many people had mourned the passing of the starting handle on that score. Lord Flowers and Dr Cox had done nothing to allay people's suspicions that the 'experts' might omit something vital from their assessments. He agreed with the Friends of the Earth view put by Des Wilson on the need to consider *all* energy options, not just those coloured by biased and limited views, and the need for an integrated approach to energy use and resource conservation. He added that 'the concept of twenty people pulling the plough in order to conserve the tractor's fuel may be part of a totally new way in which we should be regarding the use of our available resources'. He was particularly concerned about the long-term effects of energy policy decisions and identified the key question as 'When will the build-up of CO2 (due to fossil fuel combustion) become unacceptable and when will nuclear waste accumulations reach dangerous proportions?'

Tony Desoatoy (Association of County Councils) urged that local authorities, being responsible to their local electorates and responsible for the highway authorities and emergency services, should be consulted about the routing of irradiated fuel in their areas. He also proposed that decision-making responsibilities for evacuation should be clarified, that British Rail should be subject to statutory controls (in particular mixed loads involving the irradiated fuel flasks and inflammable or explosive materials in other wagons should be banned), and that the carriage by air of any radioactive material should also be banned.

Dr J. Croll (University College, London, and Consultant to the GLC on the mechanical strength of nuclear fuel flasks) expressed strong reservations about the integrity of the 'Excellox' cylindrical flasks which are used in the US to transport PWR spent fuel and will be used for the same purpose in the UK if the Sizewell PWR is built. He said that this design had not been tested in the UK and the US test information is not available to the CEGB. The 'competent authority' (the Department of Transport) were saying that these flasks were safe without having any relevant test details. Further, because the US tests relate to road transport, they have concentrated on head-on impacts. However, the most plausible rail accident was derailment leading to a side-on impact with a bridge or tunnel abutment. He argued that a cylinder is less likely to withstand a side impact than a head-on impact of similar force. In the event of fire, the balsa-wood shock absorber acts as a heat shield, so that if the flask has been heated to a very high temperature, the end-plate will be less hot; as the massive cylinder expands, it will be resisted by the end-plate. He estimated that a temperature differential of only 100–150°C would be sufficient to rupture the end-bolts on the flange.

Note: The Department of Transport has since reminded the organisers that its paper states that:

> where national certification is necessary for designs originating abroad, the Department (as competent authority) requires the submission of a safety case accompanied by full engineering drawings just as for our own national designs. Such cases are subjected to full assessment by the Department's technical staff.

Tony Kennedy (Birmingham CND) revealed that the Department of Transport had privately admitted that there is no monitoring of other hazardous materials to ensure that they would not be on the same track as trains carrying nuclear fuel flasks. He also thought it strange that the Department of Transport were evidently unaware of the HSE's new requirement that British Rail should submit a hazard survey and contingency plans to deal with accidents to nuclear flasks. He could not accept that, merely because the 'experts' say that the flasks cannot break or be broken into, they should go through urban areas. Especially as any extra cost would be minimal, there was no reason why the flasks should not be restricted to designated routes. He noted that uranium ore from Namibia goes by road from Dover via London to Preston, apparently without any monitoring.

Margaret Savage-Jones (London Region Waste Transport Campaign) felt that transportation of spent fuel should not be considered in isolation from the back-end of the fuel cycle as Lord Flowers had urged. The US speakers had identified dry storage as an alternative to reprocessing and it was a pity that no information had been given on the CEGB's dry storage experience at Wylfa. The US contributions were particularly useful because of the speakers' knowledge of and involvement with the issues. She regretted that the British speakers had relied on CEGB handouts for their information and that the CEGB, British Nuclear Fuels, British Rail and the Ministry of Defence had not been officially represented at the Conference. She added that prior consultation with campaign groups could have avoided the bias towards the 'official view' which had emerged at the Conference and would have resulted in greater participation by the unions and lay people. Her organisation supported the GLC in including spent fuel transport under its 'Nuclear Free Zone' policy and she urged the GLC to support such groups as her own which need funding.

QUESTIONS AND ANSWERS

The second part of the Open Forum consists of questions from the floor and answers given mainly, but not exclusively, by a panel drawn from the speakers; it also includes some written answers provided after the Conference. In the main the topics complement rather than duplicate those covered by the speakers in the preceding chapters.

(1) **Should an accident occur, do those who bear responsibility fully understand the emergency arrangements?**

R. O'Sullivan: Primary responsibility rests in the UK with the consignor – the CEGB for spent fuel despatched from CEGB nuclear power stations. The agreed emergency arrangements would be carried out by the consignor in cooperation with other organisations which are also interested in the transport of the irradiated fuel. As the 'competent authority' under the IAEA regulations, the Department of Transport must ensure that adequate emergency arrangements exist; this is done through the Department's assessment and certification procedure (see Chapter 8). Irradiated fuel has separate emergency response arrangements – 'The irradiated fuel flask emergency plan'; this has recently been the subject of a Home Office circular to chiefs of police and chief fire officers throughout Britain.

Mr Reynolds (Superintendent, London Ambulance Service) confirmed Commander Cree's statement that, in the event of accident involving irradiated fuel, anyone exposed to an injurious level of radiation would be sent to one of a number of London hospitals specially staffed and equipped. The designated hospitals are St Mary's (Paddington), the Middlesex (Mortimer Street), Mount Vernon (Rickmansworth), the Hammersmith, Westminster and Charing Cross Hospitals, the North Middlesex (Edmonton), St Bartholomew's, University College, Oldchurch (Romford), the Royal Free, the Royal Marsden (Fulham and Kingston) and St George's.

(2) **Do the police and other emergency services have the protective clothing needed to deal with an accident involving radiation release?**

Commander Cree: It is not intended to expose the police to radioactivity and they have no protective clothing. The fire brigade would be on the scene very quickly and they do have protective clothing. Shortly afterwards the GLC Scientific Officer would arrive and give expert advice to guide the fire brigade.

(3) **Could the fire brigade cope with the types of fire that might result from an accident involving a nuclear fuel flask?**

Harbans Malhotra: Those responsible for setting safety standards must pay full attention to the interactions between the fire and the system which is exposed to it and the wide variety of situations which can develop for a given type of accident and a given type of structure. The Fire Brigades Union must seek the maximum information and those responsible must be in a position to provide it.

Bryan Bennett (Hereford and Worcester County Council): As a result of public expenditure cuts, fire services in areas such as Hereford and Worcester have been greatly reduced. Many towns, e.g. Evesham and Malvern, are left with only part-time cover and the training of firemen to deal with special emergencies has been cut to the absolute minimum.

(4) **Do US studies confirm that certain types of transport accident, e.g. involving a large release of caesium, would necessitate long-term evacuation and do plans providing for long-term evacuation exist?**

Marvin Resnikoff: Criteria for evacuation and clean-up are given in US Nuclear Regulatory Commission reports. Whether evacuation takes place is essentially a *political* decision to be taken by elected officials in the US. If everyone were to be evacuated for a long period from a contaminated area, large economic costs would arise – whole sections of a city might be closed, businesses might go bankrupt.

Leonard Solon: It is primarily a public health question and I say that on public health grounds it is necessary to keep irradiated fuel out of cities.

Commander Cree: As in the US there has been no accident in the UK requiring such a decision. In the UK a decision on the extent and duration of any evacuation would be taken on the advice of experts, in the first instance CEGB health physicists. I am not aware of any plans, however, for long-term evacuation.

Simon Turney (GLC Labour Member): I would first point out that the Home Office has secret contingency plans in the event of a nuclear weapons attack – one involving the evacuation of 12 million people, the other involving the evacuation of 35 million people; however, I do not think the details are available to the Metropolitan Police or any other police force in the UK.

Ken Shaw (NRPB, Chilton): The NAIR Scheme (National Arrangements for Incidents Involving Radioactivity) is available to support the police in any incidents involving radioactivity. The first stage of the scheme is based on local physicists – in London, mainly hospital physicists. The second stage comes into operation if the problem goes beyond the competence of a single expert; at this stage a team is available to give support. The NRPB has in fact produced radiological protection emergency criteria. Criteria do therefore exist for radiological protection purposes during incidents, although they do not necessarily constitute 'plans'. Clearly, if individual doses represent no hazard, no action is required. Above the 'no action' dose level one has to take into account that evacuation itself involves risks – you might put people at greater risk by evacuating them. Depending on circumstances, you would have to consider whether the hazard of radiation exposure exceeded that of evacuating the people. Evacuation is only one possibility and in particular circumstances it might be better to keep the people sheltered in their own homes.

(5) **Why is there such a difference of facilities among ports handling foreign shipments of irradiated fuel, especially between Barrow which has a special fire-fighting tower and an enclosed docking facility and Immingham and Harwich which have neither?**

Department of Transport (written answer): As explained in the Department of Transport paper the required safety standards are essentially concerned with ensuring that the appropriate safeguards are already 'built-in' to the design of the package (flask). This applies irrespectively and independently of the mode of transport and of the particular facilities there may be at any ports used.

(6) **BNFL maintains a specially equipped Land-Rover to patrol the West Cumbrian rail link between Barrow and Windscale (Sellafield) to**

be used in the event of accident on that line. Why is it that spent fuel travelling to other parts of the UK is not afforded this extra back-up system?

Department of Transport (written answer): If by back-up system is meant the arrangements for responding to an accident, these are nationally organised and Cumbria is treated on precisely the same basis as other areas. A specially equipped vehicle is based in each part of the country at appropriate establishments participating in these arrangements.

(7) **How many types of flask are in use in the UK, how many of them have been subject to full-scale impact and fire tests, and how many of them have been subject to bans in New York and elsewhere in the US?**

CEGB (written answer): The CEGB uses two generic types of flask, both cuboid, to carry irradiated fuel from Magnox and AGR power stations. Other types of flask moved in the UK carry PWR, BWR and SGHWR spent fuel; all these are of cylindrical construction.

Simon Turney (GLC Labour Member): According to a Home Office Circular, three types of flask are being used in the UK. The CEGB and South of Scotland Electricity Board use an 8 ft × 7 ft × 7 ft flask. Nuclear Transport Ltd use flasks which are cylindrical (6 ft diameter, 18 ft long). The third type is the UKAEA 1120 design which is also cylindrical.

Robert Jefferson: None of the flasks in use in the UK have been subject to full-scale tests of the type conducted by the Sandia National Laboratories in the US.

(8) **Given that the cost of a flask is a trivial part of total nuclear fuel cycle costs, would it not be far more satisfactory to conduct full-scale rather than reduced-scale model tests?**

Robert Jefferson: The question is whether it is worth spending a great deal of money for something which provides very little additional benefit. Whilst it might be comforting to test a full-scale flask, you would need to define the one test which would make that flask acceptable to you. Once it had been rigorously tested, the flask could not be used for a different test.

(9) **How long does a nuclear fuel flask last?**

CEGB (written answer): The flasks are essentially simple containers and are subjected to a detailed examination every two or three years as well as checks before every journey. The flasks do not wear out in the ways that limit the life of some other engineering structures and they could have a virtually indefinite engineering life. However, the design of a flask can become obsolete with, for example, changes in the design of power station handling arrangements, with the advances in the design of flasks or with changes in regulations. The decision on the retirement of flasks thus depends on such considerations.

(10) **When would transport of irradiated fuel for Sizewell B be expected to begin?**

CEGB (written answer): Assuming that the CEGB obtains the necessary consents to build Sizewell B, a start on site for the main foundations might be possible in April 1985. On this basis, the power station would be commissioned in autumn of 1992 following a 90-month construction programme as used by the CEGB for appraisal purposes. The initial core would operate for about 2 years without replacement, and then about one-third of the assemblies would be replaced at approximately annual intervals. With the provision of high-density racking in the storage pond, irradiated fuel assemblies from eighteen such discharge cycles could be accommodated on site. Since reprocessing cannot take place until at least 5 years after discharge, irradiated fuel would be kept on site for at least 5 years, and possibly much longer. The CEGB therefore does not expect routine movements of spent fuel from Sizewell B until after the turn of the century. A possible exception to this is that it may be considered desirable to move single assemblies from the site for post-irradiation examination. Such fuel would be cooled for at least 6 months in the storage pond; only small quantities, if any, of irradiated fuel from Magnox reactors would be transported to Sellafield by the end of the century, given the expected lifetime of Magnox reactors.

(11) **Are flasks likely to be contaminated on the outside; if so, are railwaymen likely to be affected and how do workers obtain information of this sort?**

John McDonald (HSE): Last year the HSE commissioned the NRPB to conduct a survey of exposures arising from the transport of radioactive materials. The results will be given at the PATRAM (Packaging and Transport of Radioactive Materials) Conference in New Orleans next month; later they will be published in the UK. The results show that the doses normally incurred by workers and the general public in the transport of irradiated nuclear fuel are far below the relevant dose limits recommended by the International Commission on Radiological Protection and endorsed by the NRPB in the UK. Workers seeking safety information should ask their local safety representative who would then raise it locally or refer it to his union headquarters if it were a national question. There is a long-established consultative machinery between unions and management in the railway industry. Workers should be able to get satisfactory answers to their questions by using the available channels.

(12) **Are other hazardous cargoes allowed to travel on the same rail routes as nuclear fuel flasks and have petrol and chlorine tankers etc. been tested as rigorously as nuclear fuel flasks?**

CEGB (written answer): British Rail has produced a chart showing permitted and prohibited mixes of dangerous goods, and incorporated this into the operational computer control system to minimise the chance of multiple hazard accidents. Minimum separation distances between the locomotive and rear of the train are specified, and in some cases more than one hazardous substance may not be carried on the same train. In the case of loaded flasks, explosives, flammable gases, highly flammable liquids, spontaneously combustible substances and organic peroxides may not be carried on the same train.

Fred Millar: There have been numerous accidents involving dangerous materials in the US in recent years. In Waverley, Tennessee, a railcar was derailed and the resulting explosion killed seventeen fire-fighters as well as the Fire Chief and Chief of Police and it levelled the town. After three further disastrous fires the US railroad industry put heat shields round the huge tank cars and used special coupling devices to prevent the coupler from one tank car from puncturing the one ahead or behind. The history in the US has been that stringent precautions have been introduced only after serious accidents have occurred. However, the fact that we should be worried about other hazardous materials is no reason for ignoring the safety of nuclear fuel flasks.

(13) **Do the current regulations address the problem of deliberate rather than accidental damage?**

Robert Barker: In the design of nuclear fuel flasks the requirements for meeting the shielding and the package test standards result in a container which will coincidentally resist attacks of a malevolent nature. The design requirements, however, are not specifically intended for that purpose; they had in mind transport safety requirements only.

(14) **Given that the long-term radiation effects of the Christmas Island bomb tests 30 years ago are only now coming to light, how can we be sure that the accidents which have occurred in transporting radioactive materials, which are said not to have caused any injuries, will not produce long-term effects?**

Morris Rosen: The old question with radiation is: 'What is the threshold for radiation damage?' I think most scientists, if they were to look at any exposure that has occurred from transport, would conclude that the likelihood of injury is very small. There is plenty of evidence that indicates that small exposures of radiation are really not harmful.

(15) **Is it true that a paper presented at the PATRAM Conference in 1978 showed that after impact and fire tests on a nuclear fuel flask, there was a fracture which led to radiation release?**

Robert Jefferson: Actually there were two fractures on the outer skin, each about four-thousandths of an inch wide and 6 inches long. After 100 minutes into the test, molten lead from the interior was extruded into the cracks. The test was concluded after 123 minutes, three minutes after the pylons holding the system at the optimum height above the flames softened and the whole system rolled over. A minor portion of the lead shielding was lost as a result of the whole system turning over. We had drilled holes along the spine of the flask to put thermo-couples into the lead to measure the temperature at various points and when it rolled over lead poured through those holes. Despite both sources of lead loss, there was no serious loss of shielding from the flask.

(16) **Is it true that in an accident to a flask containing Magnox fuel, in which the fuel burns uncontrollably, all the caesium would be released, leading eventually to several thousand fatal cancers?**

Roger Clarke: You cannot separate the probability of the accident from the caesium release and both numbers must be specified to make an estimate of radiological risks. I would guess, however, that for a serious accident there would be a lower probability of releasing 100 per cent of the caesium than of releasing only one-thousandth part of it, which was the assumption in my paper. In addition, the consequences would not necessarily increase linearly with the amount of caesium released because release in, rather than after, an intense fire may disperse the caesium.

(17) **Why are nuclear risks specifically excluded from insurance policies?**

CEGB (written answer): At the outset of the nuclear power programme in the UK it was decided by Parliament that the overall responsibility for nuclear liability and insurance for all nuclear installations in the country would rest with Parliament. Accordingly, provision is made in the Nuclear Installations Acts, 1965 and 1969, for compensation for personal injury or damage to property caused by contamination by radioactive material or by toxic explosive or other hazardous properties of nuclear material. The Licensees of a nuclear installation (for most UK power stations, the CEGB) have absolute liability for injury or damage under the Acts and must provide the first £5 million of cover for any accident at a nuclear installation. There is, therefore, no need for insurance companies to provide such cover. A breach of statutory duty under the Acts makes the Licensee of a nuclear installation absolutely liable for injury or damage from nuclear incidents (even if they were unavoidable). The Licensee must make provision either by insurance or otherwise (subject to Treasury approval) against their liability. If the total amount of money exceeds £5 million, then the Acts make Parliament responsible for providing moneys to meet liabilities in excess of £5 million up to an aggregate of £50 million for any one occurrence. Any claims in excess of £50 million would be met out of funds provided by such means as Parliament may determine.

(18) **What are the costs and benefits of the present system in the UK of transporting irradiated fuel for eventual reprocessing compared with interim dry storage at nuclear power stations and eventually disposing of it without reprocessing?**

Gordon MacKerron: The answer for the UK depends largely on whether there is to be a serious attempt to introduce fast breeder reactors. The reprocessing option makes sense only if it is decided that fast breeders are a good investment. The CEGB seem reluctant to commit themselves to a date within the next 25 years when they might want a fast breeder reactor. I think that is one reason why the CEGB Statement of Case for the Sizewell inquiry suggests that they will not do any reprocessing of Sizewell PWR fuel until 2030, after the reactor shuts down. I know of no reliable British cost estimates of the different fuel cycle options, including dry storage; but French estimates suggest that storage and disposal of spent fuel (the 'once-through' cycle) could well be cheaper than the reprocessing option provided that one gets rid of the idea that fast breeder reactors are somehow 'essential'. The CEGB is considering whether to build a large dry store for AGR spent fuel, although this may simply be part of the business of bargaining between the CEGB and British Nuclear Fuels Ltd over the costs of reprocessing. Compared with on-site storage a central dry storage facility would probably be considerably cheaper; therefore the use of dry storage would not necessarily eliminate the need to transport irradiated fuel.

David Hall: Because we do not have clear answers to such questions, my preferred solution would be to have a moratorium on nuclear power, then we would simply be discussing how to deal with the waste already generated for a limited period. I don't know whether dry storage would be acceptable to the relevant local authorities. An irony of the Windscale Public Inquiry was that the inspector dismissed dry storage as being technically infeasible; only a few years later it is seen as a viable option.

(19) **Given that irradiated fuel transport is so small a proportion of nuclear fuel cycle costs, why not re-route this traffic away from urban areas?**

Gordon MacKerron: My calculations suggest that the cost of irradiated fuel transport from all CEGB nuclear power stations probably amounts to less than one-half of 1 per cent of the CEGB's total nuclear fuel expenditure. If it were decided to re-route trains through rural areas it would hardly affect the economics of nuclear power – the sums are negligible in terms of the dimensions within which the CEGB operates.

(20) **Given that there seems to be a gulf between 'experts' assessments'**

of the risks and the public's perceptions of the same risks, has anyone studied the scientists' own perceptions of risks?

Terence Lee: Over the past three years the Royal Society has had a study group containing some distinguished scientists and a psychologist (myself). The scientists have been made increasingly aware that the side of the coin they were concerned with was only one side. They still think it is the head side, but changes are taking place. The role of the scientist was epitomised by Lord Flowers (see Chapter 2) – I found it quite stimulating because it was such an excellent representation of the attitudes of the contemporary distinguished scientist.

(21) **Was the practice of allowing trucks carrying irradiated fuel on the streets of New York within the IAEA safety guidelines or was it the result of 'cowboy' practice (as was apparently the case with the accident at the Three Mile Island reactor)?**

Leonard Solon: It was ill-advised counsel combined with bureaucratic inertia resulting in the continuation of bad practice. I do not think that it contravened the letter of any specific regulation, but it certainly departed from Congressional legislative intent with regard to the protection of public health.

(22) **Doesn't the need for certification and police escort of irradiated fuel shipments in New York actually increase the danger of sabotage?**

Leonard Solon: Very few shipments are currently taking place anywhere in the US. There is no public announcement – disclosure is made only to the public health, police and environmental protection authorities. Whereas previously vehicles containing radioactive materials would enter New York City and sometimes get lost, the police escort ensures that they duly arrive at the correct destination. Safety has therefore improved.

(23) **If irradiated fuel is banned from New York, why are nuclear materials for military application allowed to move through New York?**

Leonard Solon: On this point the Health Code regulation says that

> This section shall not apply to radiation sources shipped by or for the United States Government for military or national security purposes, or which are related to National Defense. Nothing herein shall be construed as requiring the disclosure of defence information

> or restricted data defined in the Atomic Energy Act of 1954 and the Energy Reorganisation Act of 1974.

There is no question that, in terms of the juridical processes, our regulation would be found insupportable or unconstitutional if we endeavoured to tell the Department of Defense where and when they can move nuclear weapons. However the caveat with respect to the issue of certificates is germane to this point:

> It is intended that such certificates – the certificate of emergency transport – will be issued for the most compelling reasons involving urgent public policy or national security interests transcending public health and safety concerns, and that economic consideration alone will not be acceptable as justification for the issuance of such certificates.

Such certificates are intended to be issued for Hectacurie and Kilocurie Cobalt-60 and Caesium-137 tele-therapy sources employed in therapeutic radiology and biomedical research or educational purposes and for medical devices designed for individual human application, for cardiac pacemakers containing plutonium 238, prometheum 147 or other radioactive material. 'This subsection is not intended to apply to small quantities of specified radioactive materials intended for therapeutic radiology and biomedical research or educational purposes.' Therefore there are situations that transcend or which dominate public health and safety, but they're so rare and infrequent that we have not had to exercise that caveat yet.

(24) **Were the more stringent regulations which were introduced in France in 1980 occasioned by a serious accident at the La Hague reprocessing plant?**

Colonel Berthier: There is no problem of civil security in France and no event at La Hague or elsewhere occasioned the 1980 regulations. The incidents referred to were some very minor leaks at La Hague which affected no one inside or outside the plant. They had nothing to do with the regulations. The fact that since 1980 we have had new regulations against sabotage is probably related to the international context, perhaps to comply with IAEA or EURATOM recommendations; the Americans in particular are very conscious of the risks of sabotage and the French authorities felt they should follow other countries in this respect.

(25) **Doesn't the fact that world uranium resources are very small and world coal resources very large indicate that we should place less emphasis on nuclear power and more on coal?**

Gordon MacKerron: The main reason that people think there isn't much uranium is because no one looked for it after nuclear weapons stockpiles had been built up (around 1960) until the mid-1970s when it was necessary to find more for civil nuclear power. When people began looking in places like Australia and western Canada, they found a great deal. The notion that we are running out of uranium is, I think, a false one – I would be very surprised if we run out of uranium for a very long time indeed. That is not to say that we should be complacent about exhaustible resources, whether uranium or coal; but if you don't like nuclear power you have to remember than one of the arguments designed to overcome fears about potential uranium shortage is that fast breeder reactors will use uranium around fifty times more efficiently than thermal reactors. Unless uranium became extremely scarce and it had to be extracted from seawater, almost inevitably the cost of uranium per unit of power generated will be far cheaper than coal. The real question is whether the technology needed to harness the uranium to generate electricity will counteract the cheapness of the uranium fuel. Although nuclear capital costs have been rising very sharply, the argument that uranium will remain considerably cheaper than fossil fuel per kilowatt hour of electricity generated is a strong one.

(26) **Would not the banning of nuclear power aggravate the problem of acid rain due to coal-fired power stations, a problem which is spoiling large areas of northern Europe?**

Hugh Mild (FOE): Acid rain is a serious problem and it partly results from British coal-fired power stations without scrubbers to remove the sulphur dioxide. Scrubbers should be put on and a CHP (combined heat and power) programme with fluidised bed firing would substantially reduce the total amount of coal burned. Therefore opposing nuclear power doesn't imply a worsening of the acid rain problem.

Des Wilson: It will be far cheaper to eliminate the acid rain problem – the solutions exist, it's just a question of investment – than to introduce nuclear power.

(27) **What is the scope for local authorities to promote energy conservation and renewable energy schemes?**

Simon Turney (GLC Labour Member): The scope is determined by the progress which is made with local authority housing programmes and that is constrained by the financial policy of the central Government.

(28) **How can the average citizen, who is not well up on technical matters and is perhaps a little intimidated by the air of secrecy surrounding nuclear power matters in particular, feel more in control of events?**

David Hall: Discussion of national issues which are complex and have far-reaching consequences needs to be carefully structured to ensure that everyone with an interest can participate. A beginning was made with the Energy Commission, but that was disbanded. Parliamentary Select Committees are also useful. However, MPs are often badly served – the Commons debate on the Windscale inquiry report was one of my most depressing experiences. At the local level there needs to be a much more balanced introduction of information, if only to counterbalance the glossy, one-sided information which is currently being poured into the schools. I feel that the GLC Conference has revealed some basic inadequacies about the arrangements for transporting irradiated fuel. Some questions have not been answered or have been shunted from one Department or organisation to another. It is unsatisfactory that workers do not have readily available to them the information necessary to allay their anxieties. All this confirms the need for a public inquiry on the issue. However, should an inquiry take place, it would put in people's minds the idea that an accident might really happen and that would lead to a questioning of all nuclear power policy. The industry does not wish the idea to be placed in people's minds and I think that is one reason why the CEGB refused to attend the Conference.

Index

Withdrawn
3
571

Peril at Price Manor

 For information address HarperCollins Children's Books, a division of HarperCollins Publishers, 195 Broadway, New York, NY 10007.
www.harpercollinschildrens.com
Library of Congress Control Number: 2022049162
ISBN 978-0-06-324953-0
Typography by Catherine Lee
23 24 25 26 27 LBC 5 4 3 2 1

First Edition

PERIL AT PRICE MANOR

LAURA PARNUM

HARPER
An Imprint of HarperCollins*Publishers*

In memory of my dad and all his gory tricks

PERIL AT PRICE MANOR

1

THE FLOWER DELIVERY GIRL

A shrill scream pierced through the otherwise quiet morning in the sleepy town of Bedlamton. It was the kind of scream usually associated with axe murderers, vengeful spirits, or toxic-sludge monsters. The only monsters in sight, however, were the ones in the movie posters plastered all over the walls of the tiny bedroom where Halle Thompson stood alone, facing down her own reflection.

Her scream was followed by an equally alarming yell from the kitchen down the hall.

"Halle, cut that out! You'll give the neighbors a heart attack!"

She sighed. If the neighbors were disturbed by her scream, her mother's yelling was definitely not helping.

A moment later, the upstairs neighbor pounded on the floor in protest, adding to the ever-increasing racket of the morning.

Halle chose to ignore both her mother and the upstairs neighbor. She had just produced what she considered to be the most perfect of all her perfect horror movie screams. It had the pitch, it had the duration, and it even had a very subtle but very convincing little quiver at the end that had come from somewhere in the back of her throat.

She took a deep breath and let out another bloodcurdling scream.

Not quite as good but still pretty perfect as far as she was concerned.

Her mother's groan echoed down the hall, along with a chuckle from her aunt Rhona.

"She's going to scream herself into the movies one of these days. You'll see," Aunt Rhona insisted.

"I'll give her something to scream about if she doesn't hurry up and get dressed," her mother muttered. Then, louder, she added, "We've got twenty minutes to get downstairs so we can open the shop on time."

Halle smirked at her reflection. She was already dressed, wearing her comfiest pair of pants and a too-big T-shirt with a cartoon depiction of the Soggy Bog Beast on it. She was just taking some extra time to practice. How would she ever get into the movies if she didn't practice? Georgia and Dakota practiced movie scenes all the time during recess, in front of a captive audience, no less. They took the same acting class as Halle on Sundays, but Georgia and Dakota were the elegant kind of actresses—the kind who could smile and laugh and sometimes even cry. Halle wasn't interested in being that kind of actress. She was going to be a horror movie Scream Queen. Scream Queens had to do all that and more. They had to tremble and gasp and run like crazy! And they always had to trip and fall while running, which Halle was especially good at (although not necessarily on purpose). But above all, they had to be able to produce those ear-piercing horror movie screams that were actually super hard to do in real life.

Now it was time to give her vocal cords a rest and work on facial expressions. She studied the Scream Queens in the movie posters all around her. There was the blond schoolteacher sticking a rusty key into the attic door in *The Children of Deadwood Orphanage*. Halle knew the expression well: Curiosity. She faced the mirror, tilted her head to one side, furrowed her eyebrows, and frowned slightly.

When she was satisfied that she had pulled it off, she moved on to the next poster, *The Manic Marmoset Invasion*. It showed a young lab assistant watching helplessly as the handsome scientist battled a ferocious monkey with wild red eyes and sharp fangs. This one was a bit trickier: Worry. Halle turned back to the mirror and opened her eyes wide. She slanted her eyebrows toward the center of her forehead and experimented, raising them a little higher and then a little lower again until, finally, they were at just the right spot.

Next was one of her favorites: Damsel in Distress. It wasn't just a facial expression; it was an all-out faint. In the poster for *Swamp Devil of the Deep*, a bathing suit–clad beachgoer lay passed out in the arms of a slimy green monster. Halle placed one hand over her heart, fanned herself with the other, and swooned. She closed her eyes, buckled her knees, and collapsed in a lifeless heap onto the rug. After a dramatic pause of at least seven seconds, she opened her eyes and sat up again. Her performance had been perfect. She'd even remembered to keep her eyeballs completely still under her eyelids.

As she gazed around the room for another Scream Queen to imitate, her eyes landed on the vintage poster hanging directly over her bed. It was Raven Rose, the mighty demon huntress. She stood tall and straight with one foot resting atop an enormous slain beast. A gleaming ruby-encrusted dagger hung from her belt, and her smooth, jet-black hair blew dramatically to the side. Cuts and dirt covered her skin, but she was proud and defiant. Halle studied her face. Raven Rose's chin jutted forward, and the hint of a smile spread across her lips. One sculpted eyebrow arched slightly higher than the other. This was no Damsel in Distress. Raven Rose had confidence. She had bravery. She had pluck. This was an honest-to-goodness, save-the-world *Heroine*.

Halle got up, placed her hands on her hips, and rested one foot on a large stuffed panda bear that was lying on the floor. Her hair was dark like Raven Rose's, though definitely not as sleek and shiny as the demon huntress's. Halle was taller than most twelve-year-olds, though, which was at least something. She straightened her back and squared her shoulders, emulating the Heroine pose. But as she thrust her chin forward the way Raven Rose did, she lost her balance and teetered backward, crashing into the dresser.

"Ow," she said, rubbing her bony backside where a bruise would surely appear soon. It looked like she'd be sticking with Damsel in Distress for now.

"Halleeeeee!" her mother bellowed from the kitchen.

Halle sighed and shuffled out of the bedroom. "Coming, Mom."

Halle's mother and Aunt Rhona were clearing their plates from the table when Halle entered the kitchen. Halle dropped a slice of bread into the toaster and heated a pan for scrambled eggs.

Aunt Rhona handed her a drink. "Warm honey water. If you keep screaming like that, you're going to hurt your throat."

"Thanks." She took three big gulps and then set the glass on the table.

Aunt Rhona grinned and raised her eyebrows. "I heard Maximus Price is back in town this week. Better make sure those flowers look extra nice today."

"The flowers always look nice, whether he's home or not," Halle said. "You *know* I take my work very seriously." She poured the eggs into the pan and stirred them around with a spatula. At the mention of Maximus Price, however, a funny feeling squirmed in her stomach, like wriggling worms were having a party inside her.

Her mom filled the sink with soapy water. "Well, if you don't hurry up with your breakfast, Lenny will have to deliver one of my arrangements to Price Manor."

Aunt Rhona chuckled and shook her head. "No offense, but your flowers are much too cheerful for Maximus Price. They're fine for weddings and Easter Sundays but not for the most famous horror movie maker of all time—the creator of the Mutant Mole Man, the Fire Kraken, and Raven Rose. That weekly flower arrangement is Halle's job. She's the only one who can do it."

Halle smirked and the worms in her stomach settled down.

Aunt Rhona held up her phone. "Speaking of Raven Rose, guess what movie's playing tonight."

Halle checked the image on the screen. "*Raven Rose and the Isle of Empty Souls*—good one!"

The old-timey movie theater in Bedlamton specialized

in classics and never failed to show one of Maximus Price's older films every weekend. People who came to town for the quaint shops and rustic charm would then stay for dinner and one of the local celebrity filmmaker's movies.

Her mother rolled her eyes. "Haven't you guys seen that one about a dozen times already? Why don't you see something different? A nice romance with singing and dancing."

Halle pressed her lips into a thin line. "Maximus Price movies do have romance in them. Raven Rose is always having to save her true love, Geraldo, from all kinds of fiends and monsters."

"But do they have singing and dancing? No."

Halle sighed. Georgia and Dakota preferred the singing-and-dancing movies too. "Mom, this one is sooo good. All these people fall under a trance. They can't talk or communicate at all, and they walk around like zombies. Everyone thinks the people have lost their minds, but really they're under the spell of a horrible mind-control demon."

"That's right," Aunt Rhona jumped in, "and they get shipped off to this deserted island because everyone thinks whatever happened to them could be contagious—"

"Until Raven Rose's true love, Geraldo, gets put under the spell and sent off to"—Halle paused and wiggled her fingers at her mother—"the Isle of Empty Souls," she said in a deep, spooky voice.

Her mother rolled her eyes again. "All right, all right, I get the picture. Now let's get moving before I send you off to the Isle of Empty Souls!"

The phone on the counter blared, and Halle startled, knocking over her glass of honey water.

Her mother shook her head as the liquid spread across the table. "Better clean it up fast."

Halle grabbed some paper towels and sopped up the mess while her mother checked her phone.

"Oh no. It's a text from Lenny. He's got the flu and can't do deliveries today."

"Poor guy. He was looking a little pale yesterday," Aunt Rhona said. "Maybe Elsa can fill in."

"Not this weekend. She's away."

Aunt Rhona frowned. "Well, I can walk a few of the deliveries around town this morning, but you know I can't drive that big van." She looked down at her own tiny frame and shrugged. "My feet don't even reach the pedals."

"I know. And thank you. Luckily most of today's orders are local. You could probably get them done before the tourists show up for the afternoon candy rush. We may have to cancel the other deliveries, though."

Halle's stomach worms started up again. "But what about Maximus Price's order?"

"I don't know . . . ," her mother said. "As much as I hate to, I suppose I'll have to call his staff and give Mr. Price a refund for this week."

Halle's eyes bulged. Those flowers were her one connection to him. If they cancelled this week's order, he might get upset and decide to use some other florist. She shook her head. "Uh-uh. I'll make the delivery."

"Halle, that's six miles outside of town! It would take you ages."

"I'll ride my bike. We can strap the flowers inside the basket."

Her mother crossed her arms. "I don't think so. That bike is kind of old, and your flower arrangement is usually pretty big. You'll be wobbling all over the road. Not to mention getting there is all uphill."

"Please, Mom. I can do it. I ride with my heavy books to school all the time."

"That's nowhere near six miles."

"But . . ." Her mind raced. She had to convince her mother. Especially since Maximus Price was back in town—she might even get to meet him! "But, Mom, I already dyed the roses. If you cancel this week's delivery, they would just go to waste. I mean, who else would want black roses?"

Her mom sighed. "You've certainly got a point there." She hesitated, then checked the kitchen clock. "All right, you can do it. But let's get moving. With you two out doing deliveries, I'm going to have a busy morning by myself in the shop."

2
THE WALL CLIMBER

Paisley Price's eyes shot open as a shrill scream echoed through the winding halls of Price Manor. *Dad must be back*, she thought. She was not the least bit alarmed by the scream. Poor Mrs. Pennywhisker was always wailing and hollering, thanks to her father's habit of placing gory movie props around the house: severed body parts, puddles of fake blood, mechanical beasts operated by remote control. He had even tricked Mrs. Pennywhisker last month by moving a full-length mirror from the drawing room to the wall just outside the study. When they came out from their calligraphy lesson, Mrs. Pennywhisker had taken one look at her unexpected reflection in the mirror and run shrieking down the corridor.

Yes, when Dad was home, Mrs. Pennywhisker's screams were as regular as sousaphone practice in the morning, tapestry weaving before lunch, and roller-skating in the afternoon.

Paisley sat up in her loft bed and stretched—first her arms, then her legs. She pointed her toes straight in front of her, grabbed her feet, and lowered her forehead to her knees for a count of twenty before shaking her legs out again.

Her father had returned home just in time. In less than twenty-four hours, it would be her and Argyle's birthday, and no matter what his filming schedule was like, he'd never miss that.

A mix of nervous and excited thoughts swam through her head of what their father might have in store for them tomorrow. She tucked her feet under the railing of the loft and began counting sit-ups. When she got to sit-up number twenty-seven, she checked the clock to see exactly how many hours it would be until they officially turned thirteen. It was 8:06 a.m.

But that couldn't be right! Paisley was accustomed to

a strict daily schedule that began precisely at seven o'clock with a sharp knock on her door from Mrs. Danworth, the housekeeper. Not only had Mrs. Danworth failed at her morning wake-up call, but Mrs. Pennywhisker had also not shown up to pick out a fresh set of clothes suitable to the day's weather and activities, and Mrs. Paprikash, the cook, had neglected to ring the breakfast bell at seven thirty!

Having finished her sit-ups, she dangled her legs over the side of the loft. On one end was a ladder, and on the other, a spiral slide. However, Paisley opted to take the most challenging way down: the rock-climbing wall. She reached for the colorful plastic grips fastened to the wall and stretched her legs down, wedging her toes between two smooth crevices. Once in place, she maneuvered her way down the plastic grips effortlessly until she reached the bottom. She could practically do it with her eyes closed. The loft wasn't very high, and her entire bedroom floor was covered in springy tumbling mats—her father had insisted on it.

Without thinking, Paisley headed for the armchair where Mrs. Pennywhisker always laid out a set of clothes for the day, but the chair was empty. Of course. Mrs.

Pennywhisker was probably still cowering in a corner somewhere downstairs.

Paisley crossed the room and poked her head out into the hallway. “Hello?” she called.

The corridor was empty and still. The only sound was the ticking of the large grandfather clock in the entrance hall downstairs.

Now what was she going to do? She was already late for breakfast, but going down in pajamas and bare feet would definitely be frowned upon. Should she pick out her own clothes for the day?

She walked over to her dressing room door and tentatively opened it. Inside were racks and racks of shirts, pants, dresses, and skirts organized in sections for formal, semiformal, casual, sporty, and rugged, all sorted by color and season. In the back, there was even a whole rack of fluffy terry cloth bathrobes in every shade of pastel imaginable. Paisley sighed. In order to choose her clothes for the day, she would need to know what was on the daily schedule, what the weather forecast was, and probably what was on the menu as well. After all, one couldn't very well eat pea soup while wearing a white cashmere cardigan.

In the corner, she spotted a pair of fuzzy pink monster feet slippers. At least she could put those on. Mrs. Danworth always insisted on keeping one's feet covered, and since Paisley had no idea where any actual shoes were stored, these would have to do.

After pulling on the slippers, she sat down in front of her vanity mirror and drummed her fingers. Her black hair was a tangled mess, but Mrs. Pennywhisker was not there to comb it out and braid it. She picked up the comb and attempted to run it through her hair, but it got stuck in the knots. The harder she pulled, the more it hurt. She dropped the offending comb back on the vanity and left the dressing room in defeat.

With no idea what to do next, she paced around her bedroom, checking the clock every few seconds. With each minute that ticked by, her stomach clenched further. If breakfast was late, then sousaphone practice would be late. And that would make calligraphy lessons late too. Or would something have to be skipped altogether?

She wrung her hands. It was now 8:28. She would give them until 8:30, then she'd go find out where everyone was. Two minutes. One hundred twenty jumping jacks . . .

3

THE BOY AND THE FISH FOOD

In the bedroom at the other end of the corridor, Argyle Price poked his head out from his nest of goose-down comforters. It wasn't Mrs. Pennywhisker's scream that had awoken him but his grumbling stomach. Had he missed the breakfast bell? Wrapping one of the comforters around his body, he lugged himself out of bed and made his way across the dark room, where he plunked down into his swivel chair. He extracted one hand from his blanket cocoon and felt around in front of him until he located a panel of switches and knobs, then pressed the main power button.

Click.

Bright green and orange lights flashed in the darkness on the console in front of him. He flipped a lever.

Whirr.

A rush of warmth blew across his bare feet as a space heater oscillated hot air around the room. He emerged from his comforter and turned another knob.

Swish.

The curtains glided open. Daylight seeped in through the brightening windows, first illuminating the endless array of buttons and switches on the console, then the wall-to-wall shelves crammed with science books and, finally, the enormous aquarium at the back of his room. He flipped a few more switches and the aquarium lit up with bright colorful lights. At least thirty species of neon and silvery fish flitted through coral arches and vibrant aquatic plants.

"Hello, friends," he said as a few of the fish darted in and out of the hull of a miniature sunken pirate ship with a tiny skeletal captain at the wheel.

The fish weren't quite what he'd had in mind when he asked his father for a pet, but after a bit of research, the aquarium had turned out to be fascinating. If he didn't have to spend his days learning useless things like "Textiles of Twelfth-Century Romania" or "Identification of Native Rocks of East Anglia" with Mrs. Pennywhisker, he'd probably spend all his time in his room, relaxing in his swivel chair with the aquarium water bubbling in the background and his nose buried in a good book on tropical marine life.

He swiveled back to face his console. It had once been the command center of the mother ship in the movie *Killer Snot Monsters from Outer Space*. His father's special effects team had repurposed it so each button and flashing light corresponded with a different gadget in his room. He turned to the monitor and clicked over to the temperature readout for the aquarium: seventy-two degrees. A little on the low side. He adjusted the dial that activated the aquarium heater, then swiveled his chair to the far end of the console and turned a manual crank. A mechanical arm gripping a container of fish food popped out of a little door in the wall. As he continued turning the crank, the mechanical arm lifted the fish food container up over the aquarium, tilted, and shook some flakes directly into the water.

The fish darted to the surface to snatch up their

breakfast. They devoured the food in a matter of seconds and circled about, looking for more. Argyle turned the crank again and shook more flakes into the water. The fish zipped around, gobbling up the bounty.

His stomach growled. Now that the fish were fed, it was his turn. He rolled his swivel chair around the room, searching drawers and shelves for snacks—crackers, cookies, pieces of toast—anything he'd squirreled away over the past few days. Unfortunately, Mrs. Danworth, the housekeeper, usually made a daily sweep of his room, searching out every last morsel of food stashed there.

"You're going to attract rats," she would say, "and rats spread diseases like the plague."

It had actually been fleas that spread the bubonic plague—Argyle had read about it recently—but he wouldn't bother correcting her. She'd still insist on ridding his room of crumbs.

She didn't know all of his hiding spots, though. Argyle rolled his swivel chair over to a floor lamp in the corner and reached up under the lampshade, where he had concealed a granola bar yesterday, but it was no longer there. And there was nothing tucked under his mattress or inside his pillowcase or behind his collection of *Oceans Weekly* magazines either.

His stomach growled again, and he could feel his legs weakening. He rolled around the room, his search turning up nothing, until he came to a stop in front of the mechanical arm with the fish food.

He pulled the arm toward him and sniffed at the container. It smelled a little like tuna. The tiny flakes probably tasted something like salty, fishy potato chips. He released the fish food from the mechanical arm and held the container up over his open mouth. He was about give it a quick shake when Paisley walked in.

"Argyle! What are you doing?"

He started back in surprise, his cheeks growing hot. "Err . . . nothing. I'm just . . ."

Paisley marched over and snatched the fish food away. "Hungry? You need people food not fish food. I'm sure breakfast is ready by now. It's after eight thirty."

Argyle frowned and checked his clock. "I didn't hear the breakfast bell."

"We must have slept through it. Mrs. Danworth never came to wake us up. But I heard Mrs. Pennywhisker screaming, so I'm pretty sure Dad's home."

Argyle's eyes lit up. "Just in time for our birthday tomorrow."

His sister smirked. "Are you ready for it?"

He rubbed his hands together. "Always."

"Come on, let's go downstairs." Paisley cartwheeled toward the door. "Oh, and don't bother getting dressed. There's no time. We're already way behind schedule."

He rolled his eyes. Getting dressed hadn't even occurred to him, and if Mrs. Pennywhisker wasn't here to tell him to do it, he definitely wasn't going to. But he did need food. He powered down all the lights and the space heater. "Okay, but if breakfast isn't ready yet, I'm coming back for the fish food," he said, only half joking.

With a salute to his aquarium friends, he rolled his swivel chair all the way to the bedroom door before finally forcing himself to his feet and following Paisley into the corridor.

4

RED AND BLACK ROSES

Halle stood at the worktable in the small back room of Main Street Flowers and Candy surrounded by shelves full of bright, cheerful flower arrangements in purples, yellows, oranges, reds, and blues. But Halle was concentrating on the vase in front of her, carefully inspecting each velvety petal of a dozen black roses. Yesterday, while Georgia and Dakota had probably been hanging out at the mall after school, Halle had picked out the most flawless white roses from the refrigerated case against the back wall, trimmed the stems, and set them in a bucket of water mixed with dye and plant food. Overnight, the flowers had soaked up the color and were now a deep inky black.

Satisfied with the dye job, Halle selected and trimmed a dozen bloodred roses as well and some wispy tendrils of young, fresh ivy to complete the arrangement. After she placed the last flower in the vase, she stood back and tilted her head to one side, then the other, making sure the red and black roses were evenly spaced and that the arrangement was symmetrical. Her hand trembled as she reached out to fix one of the ivy vines. For the first time, *she* would be handing over her creation. Maybe into the hands of Maximus Price himself. She took a big breath in and out and gave both arms a good shake. That steadied her hand, and she made the adjustment.

There. It was the most perfect of all her perfect flower arrangements. She carefully draped protective plastic film around the whole thing and stapled it together at the bottom of the vase. Normally, she'd set the arrangement on the shelf by the back door for Lenny to load into his van with the other deliveries, but not today.

Halle's stomach worms started dancing again. If she didn't pull herself together soon, she imagined the worms would explode right out of her guts—just like in the maggot scene in *The Masked Mummy*. She took some more deep breaths while focusing on her flower arrangement. Cut roses stayed fresh only for a brief amount of time before they began to wilt. The longer she waited, the less perfect they would be.

Leaving the flowers undisturbed on the worktable, she hurried to the front of the shop to let her mother know she was ready. She raced through the door and skidded around the counter, only to trip over one of the many buckets of flowers on the crowded shop floor. Water spilled across the linoleum tiles, stopping just inches from the stylish boots of the two girls standing at the candy counter, Georgia and Dakota.

Halle's mother, busy helping another customer at the cash register, glanced at Halle impatiently. "Better get the mop."

Georgia and Dakota smirked at each other, and Halle's face grew hot.

"And make it quick," her mother added, handing the customer his change. "You've got a long bike ride ahead of you." She turned to the two girls and smiled. "Halle's going to make a delivery at Price Manor today."

Dakota cringed. "Eww! That place is super creepy."

"Yeah," Georgia said. "I wouldn't be caught dead anywhere near there."

Halle was very aware of how Georgia and Dakota felt about anything related to Maximus Price and his horror movies—even if her mother was oblivious. But that wasn't going to stop her. She grabbed the mop from the supply closet and started soaking up the spilled water. "Well, I think Price Manor is amazing. Maybe I'll even meet Maximus Price himself! I could show him my scream." She paused her mopping and inhaled deeply, preparing to let out one of her signature screams.

Georgia clapped her hands over her ears, and Dakota winced, blurting, "No, don't!"

Halle caught her mother's scornful eye and released her breath. Georgia whispered something to Dakota, who covered her mouth with her hand. Halle pressed her lips into a thin, straight line and squeezed the mop into the bucket.

The store phone rang, breaking into the heavy silence that had quickly sprung up between the girls. Halle's mother answered the call, and Halle resumed her mopping. "So, I guess you guys are going roller-skating tonight, right?"

"Of course," Dakota said. "Where else would we be on a Saturday night?"

Georgia picked a rainbow lollipop from a stand on

the candy counter. "We haven't seen you there in a while, Halle." She snickered and glanced over at Dakota.

Halle narrowed her eyes, remembering what had happened the last time she went roller-skating. "Saturday nights are movie night for me and my aunt Rhona. We always see whatever horror movie is playing. It's for my research."

Georgia raised her eyebrows. "Research?"

Halle leaned the mop on the counter and straightened to her full height. "Yep. And in addition to our acting class, I've been running at the track on Tuesdays and Thursdays. Scream Queens have to be able to run fast."

"Yeah, okay," said Georgia. She looked sideways at Dakota.

Dakota forced a smile. "We take tap."

"I'm not sure how useful tap-dancing would be if an axe murderer is coming after you."

The two girls exchanged glances. Georgia put the lollipop back in the stand. "We should get going. My mom's waiting for us in the car."

"Yeah, we gotta go." Dakota eyed Halle's outfit. "Nice shirt."

The two girls burst into giggles and hurried out the door.

Halle slumped her shoulders as she looked down at her Soggy Bog Beast T-shirt. There was a time when Georgia and Dakota would have thought the caricature was cute. "At least Maximus Price will appreciate it."

Halle's mother ended her call and came out from around the counter. She took the mop and finished wiping the floor. "Don't get your hopes up, Halle. Lenny has never even seen Maximus Price. The butler always takes the deliveries."

Halle shrugged. "Yeah, he told me that." She headed into the back room and stopped in front of her flower creation. It was dark and haunting and beautiful, like a cemetery in the moonlight. She picked up the vase and hugged it to her stomach, keeping the worms at bay as she carried it back through the front of the shop.

"But Mom . . . you never know."

Her mother opened the door for her, and Halle glanced up at the wooded mountainside where the very top windows of Price Manor beckoned to her from above the tree line.

5

MRS. DANWORTH

Paisley was having second thoughts about her enormous fuzzy monster slippers. They were cumbersome to move around in, and her cartwheels were all off-balance. She didn't dare walk around in just her socks, though. Mrs. Danworth would be sure to scold her with such horrors as, "You might slip and fall and split your head open!" or "You might step on a rusty nail and end up with lockjaw!" Not that there were ever any rusty nails lying around Mrs. Danworth's immaculate floors. And what in the world was lockjaw anyway?

Even with the impractical slippers markedly slowing

her progress, she was still way ahead of her slowpoke brother as they descended the grand marble staircase. Midway down, she turned back to check on Argyle's progress and noticed he had put neither shoes nor socks on and was walking around barefoot, without a care whatsoever about what Mrs. Danworth might say!

She shook her head and continued down the stairs, nearly taking a tumble when the unwieldy slipper clipped the very edge of the next step. Grabbing the banister, she made a mental note to change into more sensible shoes as soon as she could locate some. For now, she'd have to concentrate very hard on her feet. It wasn't so much the splitting her head open that worried her but rather the admonishment that would surely come from Mrs. Danworth.

The rest of the way down, Paisley was so focused on her feet that she didn't even notice Mrs. Danworth lying facedown on the marble floor of the entrance hall until she had nearly stepped on her.

"Oh, look!" she said when she finally saw the housekeeper. "No wonder Mrs. Pennywhisker was screaming. Dad's convinced Mrs. Danworth to play dead."

A thick puddle of translucent pinkish slime oozed

out from under the housekeeper's head.

"Good morning, Mrs. Danworth," Argyle muttered, shuffling across the entrance hall in the direction of the breakfast nook.

"Hold on, Argyle. It looks like Dad went to a lot of trouble, setting her up like this. Even though he already scared Mrs. Pennywhisker, he probably wanted Mrs. Danworth to stay put until we were able to check out his handiwork too. Let's have a good look so she can get on with her day."

Argyle grumbled, clearly not wanting to delay his breakfast for another minute.

Paisley ignored him. Their father had been away for weeks scouting locations for his next movie, and she had missed his antics. She bent down and dipped her finger in the viscous slime. "Ooooh, new special effects. Help me turn her over."

They grasped the housekeeper's skirt and rolled her onto her back.

"Ew, yuck!" Paisley said, jumping back.

Argyle's eyes lit up. "Whoa! What is that?"

Covering most of Mrs. Danworth's face was a lumpy grayish-pink blob about the size of a softball with slimy tentacles that wrapped around her head and coiled into her hair. Paisley leaned in for a better look.

The gelatinous creature pulsated gently as slime oozed from tiny pores in its skin. She poked it and then quickly pulled her hand back when the thing quivered. "It must be a prop from Dad's next movie. What's it called? *Planet of the Prawn People*?"

"It's not very prawn-like," Argyle said. "Prawns are oblong and have legs and exoskeletons. This is more like a cephalopod. It has a bulbous body and tentacles. Kind of like a squid. Mmm. I wonder if we're having calamari omelets for breakfast."

Paisley stood up and wiped the slime from her finger onto her pajama pants. "I don't know, but I guess we should find out. We're way behind schedule. Great job, Mrs. Danworth. See you later." As she stepped over the housekeeper, her enormous slipper grazed the woman's leg and she nearly tripped. "Oh, and Mrs. Danworth, when you get a chance, could you help me find some better shoes? These things are nearly impossible to walk in."

The housekeeper remained motionless on the floor. She did not bother to remark on Paisley's poor choice of footwear or to scold Argyle for his lack thereof.

Paisley shrugged and the two made their way to the breakfast nook.

But when they got there, the breakfast table was bare.

Argyle groaned. "I'm going back to bed."

"Don't be silly. We just need to find Mrs. Pennywhisker. She'll know what to do."

Argyle plopped himself down in the nearest chair and laid his head on the empty table. "I'm too hungry to move."

Paisley sighed. As much as it bothered her to be behind schedule, she knew there was no way Argyle would function until he had a proper breakfast. He was what their father called "delicate."

"Well, then," she said, putting her hands on her hips, "I guess we're just going to have to find the kitchen."

6

BANNED BAGELS

Argyle mustered up the energy to follow his sister through the meandering hallways of Price Manor. They had never been to the kitchen before. All of their meals were brought to them in appropriate places like the dining room, the breakfast nook, or the tearoom, and there were always snacks available in the study. The kitchen was somewhere off in the east wing, where he and his sister never really had any reason to go. He pictured an enormous room with platters of crudités, multitiered plates of cookies and pastries, cornucopias overflowing with fresh fruit, and the cook, Mrs. Paprikash, standing in the center of the room with a silver tray of sandwiches in her outstretched arms. He quickened his pace.

They wound through the corridors, past antique suits of armor standing guard in the hallways, demonic marble busts lurking in alcoves, and heavy wooden furniture clawing at the carpets with their beastly carved feet. Argyle wasn't paying attention to any of that, though. His last meal had been more than twelve hours ago (not including the roll he'd smuggled into his pocket at dinner last night, which he had devoured in his bed at around 11:45). Since it was breakfast time, there would probably be a steaming platter of scrambled eggs waiting in the kitchen. Maybe with bacon. And hash browns on the side. And orange juice and sweet summer sausages and warm buttered toast. No—toast was kind of boring. English muffins would be better. Or bagels. But bagels were strictly forbidden at Price Manor.

He'd had a bagel once, many years ago. Their old nanny had given him one. He fondly remembered the initial crunch of the toasted outside and the soft chewiness of the warm inside. It was topped with cream cheese and fresh chives from the herb garden. But then Mrs. Danworth had walked in and discovered the contraband.

"Where in the world did that come from?" she'd shrieked, and quickly disposed of the bagel. Not long after that, the nanny was gone too. His father insisted they were too old to have a nanny at that point. What they needed was a governess—someone to oversee their education—and that's when Mrs. Pennywhisker came along. Argyle couldn't help but wonder, though, if the bagel had been the real reason the nanny was let go.

Argyle thought about that bagel often. If it hadn't been for the nanny, would he even know bagels existed? And if bagels existed in the world outside of Price Manor, what else was out there that he didn't know about? Price Manor had its own library, of course, and he and Paisley did plenty of reading, so they knew about worldly things like ski resorts and sushi restaurants, but when Argyle had once asked if they could go out for sushi sometime, his dad had simply arranged for sushi to be delivered to the house instead. And when Paisley asked if they could go skiing last winter, he'd simply bought them some skis and had his team of set designers install an actual ski lift along the sloping west lawn. He gave them everything they ever needed or wanted. His one requirement, though, was that they always stay within the stone walls of Price Manor. It was, in his words, "The Safest Place on Earth." Paisley was always delighted when their dad accommodated their every wish. Argyle, however, couldn't help but compare their dad's substitutions to the pictures he saw in his books and magazines—pictures of smiling groups of friends on a ski trip together or seated around a big table at an authentic Japanese restaurant.

Argyle understood why his father insisted on this. He and his sister had only been in diapers when the tragedy had happened. He didn't remember his mother at all, but he knew the story well. He and Paisley had been home with the nanny when their mother set out for a drive in the convertible that Saturday afternoon long ago. She'd been driving along a coastal road with the top down when she pulled over to a lookout point on a high cliff to snack on a bagel with smoked salmon and capers. A group of bird-watchers armed with fancy binoculars and expensive telescopes were there getting a rare look at a non-native great speckled waffle-billed gull when the bird swooped into their mother's uncovered car to snatch a bite of the aromatic snack beside her. The gull had been so pleased with its loot that it let out a triumphant squawk, which alerted its bird buddies nearby about the tasty treat. Suddenly, she was bombarded by a flapping, squawking storm of feathers, beaks, and claws, and the entire bagel was devoured within seconds.

The bird-watchers had been so delighted to get a glimpse of not one but an entire flock of great speckled waffle-billed gulls in action that they did nothing but ooh and aah as they pointed and ogled through their binoculars and telescopes. Then some of the gulls decided Argyle's mother, who wore a bright pinkish-orange dress with dark green buttons, sort of resembled a giant bagel with salmon and capers herself. They began to claw and peck at her with their strong talons and waffled bills. She swatted and kicked at the attacking beasts as the bird-watchers stared in horror. When they finally ran to her aid, it was too late. His mother, in her distress, had accidentally knocked against the emergency brake, and the car rolled off the cliff into the craggy ocean waters below.

His dad was devastated. After the loss of his wife, Maximus Price immediately feared for the fragile lives of his toddlers. But what should he protect them from? The ocean? Convertibles? Birds? Bagels? Everything became a threat. Danger was all around. And that was why he had moved his family to the enormous manor behind high stone walls, far, far from the seaside, with an army of scarecrows guarding the grounds. And no bagels.

There, Argyle and his sister were completely safe from the dangers of the outside world and had everything they ever wanted at their fingertips. The devoted staff looked after them and ensured that their father's safety precautions were met. Paisley, who had climbed and scampered about like a spider monkey from the moment she could walk, needed lots of high-energy activities, while Argyle, who had a "delicate condition," needed a steady supply of healthy snacks and meals to keep his energy up along with plenty of engaging low-energy activities. They were put on a strict schedule to accommodate both children's needs equally and fairly. It was a demanding task, but their father and the staff at Price Manor banded together to ensure the twins' optimal mental and physical growth while at the same time maintaining their absolute safety.

Now Argyle followed Paisley into a part of the house they had never been to before. It wasn't that they weren't allowed in the east wing; it was just that Mrs. Pennywhisker filled their hours with every kind of activity ranging from calligraphy and soapmaking to synchronized swimming and roller-skating. Here in the east wing, the rooms and hallways were much smaller and narrower, but they were modern and functional compared to the sprawling gothic grandeur of the rest of the house. He couldn't help stopping to open every closet door, marveling at shelves and bins labeled with things like "Furniture Waxes" and "Summer Table Linens." He even discovered an entire room just for doing laundry. Wicker baskets were lined up neatly against one wall, and a load of wash churned away in an industrial-sized washing machine against the other.

But laundry and furniture waxes were not their current quest, as Argyle's sudden lightheadedness reminded him, so they continued searching until they finally pushed through a swinging door and discovered a large room with a gleaming porcelain sink, spotless marble countertops, and an abundance of shiny stainless-steel appliances. Argyle's heart sank, however, when he rounded the island counter and spotted the cook lying still as a stone on the floor, covered in slime, with another one of those gooey gray blob things across her face. She held a wire whisk in one hand, as if she had been just about to scramble some eggs for them.

With his hopes for breakfast suddenly dashed, he turned to Paisley with a frown. "It looks like Dad's convinced Mrs. Paprikash to scare Mrs. Pennywhisker too." He bent down and lifted one slimy tentacle away from the cook's eye. "Helloooo," he sang. "Mrs. Paprikash? It's just us. Is breakfast ready?"

The cook did not answer.

The wet tentacle wrapped itself around Argyle's finger and squeezed. "Whoa, it moves!" he said, shaking his hand free and stepping away. The tentacle flopped back into place over Mrs. Paprikash's eyelid. "Cool special effects! I wonder if it's thermal sensitive or if it's operated by remote control." He looked around for the source.

"Mrs. Paprikash?" said Paisley. "You're doing a wonderful job. I'm sure Dad would be very proud of you. But Argyle is super hungry, and we're way behind schedule. Maybe you could just fix up some toast for us and then go back to lying on the floor again? We won't tell."

Mrs. Paprikash still made no reply.

Argyle's eyes roamed over the assortment of high-tech gadgets and shiny appliances lining the countertops, and he soon forgot all about Mrs. Paprikash and the gooey, moving blob. Fancy juicers. Mixers. Fryers. All with varying settings and temperature controls. He ran his fingers lightly across the buttons of the automatic veggie chopper and the high-performance eighteen-speed blender. His hunger got the best of him, however, when he spotted a bowl of fresh fruit on the countertop. Unfortunately, no one had bothered to slice the apples or peel the oranges or do whatever needed to be done to a pineapple to make it edible. He finally settled on grapes, which he discovered could simply be plucked off the stem.

7

THE BUTLER

Halle wobbled and weaved as she pedaled her bicycle up the steep, winding road toward Price Manor. Her mother was right. This was not like riding her bike to school. Her legs burned from the uphill effort, and her breath was quick and labored. Not only that, but she could barely see over the humongous flower arrangement strapped into her basket. She had secured the vase with as many bungee cords as she could find, but every bump in the road made the basket rattle and the rose petals quiver.

The woods grew denser the farther up the mountainside she went. She steered away from the side of the road to avoid jostling the basket over the bits of bark and gravel that lined the pavement. After all, this flower arrangement could possibly be the key to her future profession. She couldn't risk having it damaged.

A car sped past her from behind, horn blaring, and Halle's heart leaped into her throat. She swerved to the right, nearly veering off the road. As she attempted to straighten out again, she hit a large pothole, which shook the bike so much that several rose petals fell from their blooms. Halle gasped and squeezed the hand brakes. She hopped off the bike to inspect her arrangement, her chest heaving.

It looked okay, but that swerve had been a close call. She took a big breath in and out and looked ahead of her. Price Manor couldn't be that much farther up the road. It might be better to walk the bike from here.

She pushed the bike forward, steering away from cracks in the asphalt and constantly checking behind her for cars. After several quiet minutes, her heart rate finally slowed again. Sunlight dappled through the dense overhead canopy, and birds sang from hidden perches in the trees.

Lenny had often told her about visiting Price Manor,

and she frequently rode past it in the car with her mom, but she had never been able to see much beyond the high stone wall surrounding the estate. According to Lenny, the mansion was set back in the woods up a long gravel driveway lined with scarecrows. The butler, who had crazy white hair and wore a tuxedo with a purple bow tie, always answered the door. Still, if Lenny was sick, then butlers could get sick too, which meant there was definitely a chance she might meet Maximus Price. She could picture the scene perfectly: the massive doors of Price Manor opening up to reveal Maximus Price himself.

Halle brought her bike to a full stop and practiced what she would say. She placed one hand over her heart and imitated the surprised look of one of the Scream Queens from the posters on her wall back home (eyebrows raised, jaw slightly agape). "Oh! Hello, Mr. Price!"

"Who are you?" he would say.

Halle attempted Bashful, biting her lower lip and pointing her eyes toward the ground. "Oh, I'm just the flower delivery girl." She raised her eyes again and smiled, then stuck her arm forward, pretending to shake hands with her idol. The bike teetered, and she quickly grabbed the handlebar again. She set the bike on its kickstand and resumed her rehearsal. "It's such an honor to meet you, sir. I'm a big fan of your movies. Here are your flowers." She mimed handing him the flower arrangement and did a little curtsy.

Then she turned, pretending to leave, but whirled around and faced the imaginary Mr. Price once more. She switched her facial expression to Coy (a sweet smile and a flutter of the eyelids). "You know, I can scream pretty loud. Loud enough to awaken the Bog Beast."

Halle was counting on the fact that not too many other twelve-year-olds knew about the Bog Beast. *Raven Rose and the Curse of the Soggy Bog Beast* was Maximus Price's very first film. It had come out before Halle was even born and long before his later films became blockbuster hits. But it was Halle's favorite, and it had all started with a scream.

Still, she imagined that Maximus Price might wave her away and start to shut the door. He probably got this kind of thing all the time. And that's when Halle would show him.

She cleared her throat daintily, then opened her mouth and let out another one of her bloodcurdling screams. Throughout the forest, birds rose out of the trees and flapped away, pulling Halle from her reverie. With a smile reminiscent of her Raven Rose poster, she took hold of the bike's handlebars once more and kicked up her kickstand. Her heart was pounding again, but this time from excitement.

She forged ahead with renewed energy, and in a few minutes came to the colossal stone wall enclosing the estate of Price Manor. The massive wrought iron gate was flanked on either side by two ugly winged and horned gargoyles. Until now she had only ever seen them whizzing by through a car window. But here she was, standing face to face with the monstrous beasts, who greeted her with their blank gray eyes. Halle leaned her bike against the wall and peered through the iron bars at the property within. Beyond the gate, a heavy mist rolled across the ground. There hadn't been any mist at all on her way up the road. It was super creepy.

And she couldn't wait to see more!

She hoisted the heavy latch of the gate that barred the entrance. The hinges groaned as she leaned against the bars to push it open. Surely there must be someone in Price Manor who could oil the gate to make it easier to open. The creakiness definitely added to the haunting ambiance, though.

She pushed her bicycle inside, marveling at the scenery. It was even more incredible than she had imagined: the fog, the damp gnarled trees, the army of faceless scarecrows lining the long gravel driveway, and finally, the vast and magnificent stone manor resting at the very top of the hill at the end of the drive. Clouds cast a dim shadow over the woods, and a steady, cold breeze blew through the tree branches, causing them to creak and moan.

She steered her bike slowly forward, not wanting to jostle the flowers too much on the rough gravel. Each scarecrow loomed over her as she continued up the driveway. Whether they were truly there to frighten away birds or were strictly ornamental, she couldn't say, but even though they had no eyes, she got the eerie feeling they were watching her.

Now Halle couldn't help but imagine herself in one of Maximus Price's movies—the lone girl walking toward the creepy old mansion. She could almost hear the suspenseful soundtrack. Violins mixed with the slow, very low notes of a piano. She tried out another movie poster expression, the one from *Lair of the Demon King* in which the innocent young waitress walks down a dark alley while an ominous figure lurks behind her in the shadows: Wary (neck craned forward, wide eyes scanning her surroundings to the left and right). Maybe Maximus Price was watching her from a window at this very moment!

When she finally came to the end of the driveway, she parked her bike in front of the wide stone steps of the main entrance. Her fingers trembled as she disentangled the bungee cords that secured the flowers inside the basket. She pulled off the protective plastic film and set the arrangement carefully on the ground, making a few quick adjustments, first to the flowers and then to her hair, smoothing down the flyaways. Then she lifted the arrangement into her arms and, with a deep breath, walked straight up the steps toward the massive oak door.

A brass face—a devil with an enormous protruding tongue—was mounted in the center of the door. A door knocker. Lenny hadn't even mentioned that! She reached for the heavy tongue, the knocker part, but just before she touched it, a rustling sound caught her attention. Halle turned toward the noise. There was a large bush on the other side of the steps. Probably a squirrel or a bird was moving around in the branches. Nothing to worry about.

She went for the door knocker once more, but as she lifted the brass tongue, the rustling came again. She froze. This time, the whole bush shook. There was definitely something there, and it was a lot bigger than a squirrel or a bird.

Without making a sound, Halle set the door knocker back in place. Her eyes were glued to the bush. The shaking grew more and more intense, and she thought she could detect some other sound too. A gasp or a wheeze? Definitely not the kind of sound a bird or a squirrel would make. Goose bumps prickled her flesh as she backed down the steps, clutching the flowers to her chest. She imagined some large animal—a huge mangy dog with vicious yellow teeth or a rabid raccoon drooling and frothing at the mouth.

As she reached the bottom of the stairs, the bush fell still, though the flower arrangement continued to quiver uncontrollably in her trembling arms. With as little noise as possible, she tiptoed over to peek into the space between the bush and the steps.

Halle recoiled. It was neither a dog nor a raccoon. It

was a man! His body lay tangled in the underbrush with only his legs sticking out. She backed away, her heartbeat drumming in her ears. As she took another step backward, her foot caught on an uneven paving stone. Down she went, sending her most perfect of all perfect flower arrangements hurtling through the air. The vase hit the ground and broke into a hundred pieces, leaving an unrecognizable mess of stems, petals, glass, and water on the stone pathway. But before Halle could react to this tragedy, the bush shook again, and the man began to stir beneath it. A cold chill spread through her whole body.

The man rolled over slowly and raised himself onto his hands and knees. Halle craned her neck for a better look, but he was still half hidden by the stairs. He raised one hand up onto the balustrade and, with what seemed like a great effort, pulled himself to his feet, his back toward her. His dark clothes were streaked with dirt and ripped in several places. His thick white hair was matted and littered with twigs and leaves. He seemed old. Very old. Maybe he was hurt.

"H-h-h-hello?" she stammered. "Are you okay?"

As he turned toward her, Halle's jaw dropped open. Something was on his face. A grayish-pink blob that covered his eyes and nose. It clung to his head with slimy tentacles threading through his hair and coiling around his neck just above a lopsided purple bow tie.

This was the butler! But what was that thing on his face?

With his eyes completely covered by . . . whatever it was, there was no way he could see, and yet his face was aimed directly at Halle as if staring straight at her with a predatory interest. She inched backward, not daring to breathe. He took a few stumbling steps toward her, somehow sensing her presence, then stopped. Halle's eyes widened. The gelatinous body of the creature began to throb and swell. It was growing, doubling in size and dripping slime all over the butler's once-fancy suit.

The creature's body expanded while its tentacles still hugged his head. It writhed and twisted until a split appeared down its back, and in a gush of thick slime, a whole new creature emerged from inside the first one. The new creature glistened and pulsated gently, as if taking its first breaths. Then, with unexpected energy and surprising speed, it leaped from the butler and landed inches away from Halle's feet.

Halle opened her mouth to scream, but no sound would come out.

8

BACKWARD BICYCLES AND BIRTHDAY CAKE

Even if Halle had been able to scream, Argyle would have never heard her over the roaring motor of the kitchen blender. He was, at that very moment, in the middle of concocting a strawberry-banana-yogurt–maple syrup breakfast smoothie. Unfortunately, with their current state of unsupervision and his lack of experience with kitchen appliances, he hadn't realized he needed to put the lid on the blender. This resulted in a giant splatter of thick pink smoothie all over the walls, ceiling, and even on poor old Mrs. Paprikash on the floor. The cook was doing a terrific job not reacting to the horrible state of the kitchen. And it was lucky that Paisley had insisted on aprons and hairnets, which she'd found hanging on the wall just moments before he'd started blending.

Argyle shut off the blender. He licked some of the splattered smoothie off his pajama sleeve and grimaced. Something wasn't right. Maybe he should have removed the banana peel before dropping the whole thing into the blender.

"What a mess," Paisley said, scraping gooey blobs of yogurt off her monster slippers with a spatula. "Mrs. Danworth is not going to like this."

Argyle ran his tongue across the rim of the blender. The smoothie was a bit of a disappointment, but he could at least salvage what was left to fill his tummy. He scooped a blob off the counter and held his finger out to Paisley. "Want some?"

"I think I'll stick with grapes, thank you very much."

She gave up on her slippers and began smearing sticky, yogurty clumps of banana across the floor with her apron. "Ugh. This isn't working. I'm just making things worse!"

Argyle turned to Mrs. Paprikash on the floor. "Uh . . . we could use a little help here," he said.

Mrs. Paprikash hadn't moved an inch the whole time they'd been in the kitchen. The peaceful rise and fall of her chest made it look like she was simply napping, covered in slime. But there was no way she could have slept through the noise of the blender.

"Hey, Paisley, don't you think this is a little weird? I mean, I know Dad's been away for a few weeks, but this is going a bit too far just to give old Mrs. Pennywhisker a good scare." He inspected the creature on the cook's face more closely. "Besides, there are much easier ways to scare her."

Abandoning her attempt to clean the floor, Paisley knelt beside him. "You think there's something else going on?"

"I'm not sure. I mean, Mrs. Danworth and Mrs. Paprikash have more important things to do than helping Dad scare Mrs. Pennywhisker. Like cleaning floors and cooking breakfast."

Paisley met his eyes. "Do you think . . . could it possibly be . . . ?"

He nodded. "Yep. It's got to be our birthday prank."

"But our birthday isn't until tomorrow."

"Right, but Dad knows we'd be expecting something tomorrow. I bet he's trying to throw us off by doing it a day early."

Paisley furrowed her brow. "I guess that's possible."

Argyle leaned back against the kitchen counter. There wasn't a single birthday he could remember that hadn't involved some kind of practical joke. Instead of presents and treats there was nothing but tricks and pranks. How could their dad resist when their birthday fell on April Fools' Day?

The birthday pranks had started out pretty small. The earliest one he could remember was the time his father brought in an elaborate, multitiered birthday cake, and after sitting through a rousing chorus of "Happy Birthday" in four-part harmony from the household staff, the twins cut into the cake—only to find it was full of horribly smelly cheese from the cheese cellar. He and Paisley couldn't help but gag as they watched their father devour an enormous slice. Their father loved cheese—the stinkier the better. Argyle never ventured to try birthday cake again after that.

Another year, their father gave them shiny new bicycles, but when they tried them out, they discovered they only rode backward instead of forward. They were unsteerable and completely useless.

After that, Argyle and Paisley became suspicious of anything having to do with birthdays. Every April 1, they were on guard, which meant that each year the birthday prank had to be more and more elaborate than the year before. For instance, last year, on their twelfth birthday, their dad had convinced the whole household staff to speak to them only in gibberish for the entire day. No matter how much the twins pleaded and griped, not one person in the house spoke a word of sense until the day was over, this time replacing their customarily vivacious "Happy Birthday" song with an impressively well-rehearsed rendition of "Hobbity Brathity" in four-part harmony.

Argyle folded his arms across his chest. "That's why no one called us down to breakfast this morning."

"Or woke us up or picked out our clothes," Paisley added.

"They're probably all lying around somewhere covered in these slimy cephalopod things."

"What about Mrs. Pennywhisker, though? There's no way she'd be lying around in a puddle of slime with one of Dad's weird props stuck to her face."

Argyle frowned. His sister was right. Mrs. Pennywhisker wouldn't want to be anywhere near their dad's gore fest. Even if she knew it was a prank. "Maybe she's waiting for us in the study, too afraid to go anywhere else."

"If so, she's probably been waiting for ages." Paisley

glanced at the clock. "My goodness, it's nearly ten o'clock! We'd better get moving."

Argyle wiped up a glob of yogurt that was oozing down one of the cabinet doors and popped it into his mouth. Now that his stomach wasn't so grumbly, he was ready to face the day—especially if the day was filled with further discoveries of the household staff covered in slime and icky invertebrates.

9

THE NOTEPAD

Paisley stashed her ruined slippers in a corner of the kitchen along with the soiled aprons and hairnets. Even her socks were covered with pink yogurty smears, so she discarded those as well.

As they made their way back toward the west wing, Argyle was once again slowing them down. He kept poking his head in different rooms and closets as if they had all the time in the world. He did discover a mudroom with assorted jackets, boots, and gardening tools that must have belonged to Eddie, the groundskeeper. Still worried about a lecture on cracked skulls and lockjaw,

Paisley slipped into a pair of his large rubber rain boots. They were a slight improvement over the slippers, though now instead of too wide, her footwear was too long. She still couldn't do proper cartwheels, and her feet made a ca-chunk ca-chunk sound as they meandered back to the larger, more familiar part of the house.

When they finally arrived in the study, Mrs. Pennywhisker was not there. There were no activities laid out on the table, and the chalkboard where they usually consulted their Schedule of the Day was bare.

Paisley tapped her fingers on the table. "Now what do we do?"

Argyle opened the snack drawer and found one measly granola bar. He flopped down in a cushy velveteen armchair and tore open the wrapper. "Whatever we want, I guess. No one's here to tell us what to do."

"We can't just sit here," she said, ca-chunk ca-chunk ca-chunking around the room.

"Why not?" Argyle said with his mouth full of crunchy oats.

"Because! Even if Mrs. Pennywhisker is hiding somewhere, she would want us to keep busy."

"Doing what?" Argyle gestured at the chalkboard.

"There's nothing on the schedule."

"Sousaphone lessons. We always have sousaphone lessons on Saturday mornings."

"But Mrs. Pennywhisker's not around to teach us."

"Well, we can at least practice until she shows up," Paisley said.

"Who knows when that will be. If this is our birthday prank, the household staff may keep this up all day and even into tomorrow. I wouldn't be surprised if she decided to just leave for the whole weekend."

"Then we'll practice for an hour and move on to something else. Like calligraphy. Hang on—I should write this down." Paisley grabbed a notepad and pencil from the desk. She wrote in fancy, loopy letters:

Schedule of the Day

Sousaphone practice

Calligraphy

"What else can we do today?" She tapped the pencil eraser on her chin.

"We could go back to the kitchen and make sandwiches," Argyle said, throwing his granola bar wrapper in the trash can.

"Yes, that's good. It should be about lunchtime when

we finish those activities. What else? Oh, I know! We could work on our weaving. We're up to letter R in our alphabet tapestries. Mrs. Pennywhisker would be so impressed if we finished it up all on our own."

"And snacks. We always have a snack break between activities."

"You mean *you* always have a snack break between activities. It's not actually part of the schedule. It's more like a given."

"I'm just trying to make sure you don't forget, since Mrs. Pennywhisker isn't around to get the snacks for us."

"I'm sure you'll have no trouble with that. Hey, I know! If Mrs. Pennywhisker is still missing after that, we can relax for a while with a little roller-skating and rock climbing."

She added the items to the list.

Make sandwiches (12:00 p.m.)

Tapestries

Roller-skating

Rock climbing

Argyle's head dropped back on the cushion. "There's nothing relaxing about roller-skating and rock climbing. What about some reading time?"

"It's Saturday, silly. We don't need to do schoolwork on the weekend."

"I don't mean reading school stuff. I mean reading for fun."

Paisley ignored her brother. She checked and rechecked her list but couldn't think of any other activities to add that didn't require Mrs. Pennywhisker's supervision. The governess usually took care of every imaginable detail of their lives, from sharpening their pencils to tying their shoes. Shrugging, she tucked the notepad and pencil into her pajama pocket. "Well, it's a start. We can always add more later. Come on. Let's go to the music room."

She pulled her groaning brother up from the armchair and marched him out of the study. On their way to the music room, they wandered back into the entrance hall, where Mrs. Danworth was exactly where they had seen her last. The puddle of slick, pink slime had spread even farther across the floor.

"My goodness," said Paisley. "I don't know how she can stand this. She usually won't even leave so much as an eyelash on her immaculate floors."

"Dad must have given her strict orders to ignore the

mess and remain absolutely still. You know how serious he gets about our birthday prank."

Paisley walked up to the slime puddle. "Mrs. Danworth, if you think this is bad, you should see the kitchen. Then again, maybe you shouldn't." She was goading the housekeeper, hoping to detect a twitch or a wince—any kind of response—but Mrs. Danworth's only movement was the gentle rise and fall of her chest.

"That creature thing must have a built-in slime generator," Argyle said. "Maybe we could turn it off." He walked to the very edge of the puddle, his bare toes stopping just shy of the goo. "Never mind. It's too far to reach."

"That's all right. I can jump over to the other side." Paisley took a running leap across the puddle, but the enormous boots threw off her balance and she ended up landing in the puddle, splattering more slime across the marble floors and wood-paneled walls.

"Oh no!" she said, fearing a harsh scolding. "Sorry, Mrs. Danworth."

However, no scolding came. She'd probably never hear the end of it when the prank was over, though. She took a few squelching steps through the muck.

"Ugh. At least these boots are slime-proof."

"See if you can find a switch on that thing."

Paisley bent down next to the housekeeper to examine the creature. "It's kind of like a little octopus."

"No. Octopuses have arms. This thing's got tentacles."

"What's the difference?"

"On an octopus arm, there are two rows of suction cups. If you look closely, these appendages are suckerless, more like those of a nautilus, but a nautilus has a shell. This is a"—he counted—"a tredecapus!"

"A tre what?"

"A tre-DEC-a-pus. It has thirteen tentacles. Remember when we studied Latin prefixes last year? 'Tredeca' means thirteen."

Paisley reached down and tugged on the body of the creature. It was warm and slippery. Too slippery to get a grip.

"Aha!" Argyle said, making Paisley jump.

"Aha, what?"

"Thirteen tentacles because of our thirteenth birthday!"

She smiled and nodded. "Very clever. But I can't seem

to turn this thing off. It's too slimy, and there's no switch of any kind."

"Never mind that. It's probably powered by remote control, which means Dad's got to be somewhere close by. Let's look for him."

"But what about our sousaphone practice?"

"I have a better idea." He walked the long way around to the other side of the puddle and plucked the notepad out of Paisley's pocket. In sloppy, uneven letters at the bottom of the list, he wrote:

FIND DAD

10
DIRE CIRCUMSTANCES

Halle crouched behind a moldy rain barrel in the vegetable garden, heart thumping, legs trembling. After she'd finally gotten to her feet and sprinted away from the butler and that horrifying creature, she'd scrambled around the side of the house and dived for cover in the first place she could find. She peeked out to see if either the butler or the creature had followed her, but there was no sign of them.

Along this side of the manor, the mist had lifted, the cold breeze had died away, and she was out of view of those creepy scarecrows. Her breathing steadied, and she tried to make sense of what she had seen. It was just like a horror movie. There was a terrifying creature, and she was the unsuspecting victim. The butler had been attacked by that thing—whatever it was—and it was either feeding off him or using him as a host to multiply. Or maybe both, like in *Attack of the Pod People*! Now there was another one out there somewhere, probably in search of a host too. And since she was the only one around, that meant it was after her.

But this was not a horror movie. This was real life. There was no scary background music, no atmospheric lighting, and worst of all, no Aunt Rhona sitting in a comfy movie theater seat next to her. These were Dire Circumstances.

If Aunt Rhona were here, she'd know what to do. Every Saturday after going to the movies, they always stopped at the ice cream parlor for dessert. They'd squeal and shriek as they relived the scariest scenes, causing the other patrons to shoot them annoyed looks. Aunt Rhona loved to analyze the movie characters and talk about how they would most certainly have survived if they'd only done things her way. She believed she could survive almost any horror movie situation. She wasn't an idiot like those movie characters. There were rules. They should have not gone into the woods or not checked out that noise in the basement or definitely not walked backward.

Halle smacked herself in the forehead. She had been an idiot. She had walked backward! And that had resulted in the destruction of the most perfect of all her perfect flower arrangements. She sank back against the stone wall of the house. Now this was a whole other problem. Not only was there a weird, possibly deadly creature on the loose, but those flowers were a standing weekly order at her mother's shop. What if Maximus Price decided to use a different supplier because she messed up the delivery? Not only would it sever her one connection to him, but it would mean a big loss for her mom's business.

It wasn't her fault, though. It was that creature. She just needed to tell the truth. She needed to find Maximus Price and explain what had happened. Halle's stomach began worming around again. This was her chance! She couldn't talk to the butler, so she had to find Mr. Price.

She took a step out from behind the rain barrel, but then an icy chill ran through her. This was not what Aunt Rhona would do. She quickly crouched back down in her hiding spot. Aunt Rhona's number one rule was Avoid Peril at All Costs. And the creepy butler and that nasty creature definitely counted as "peril." If she were to go back and knock on the front door, that slimy little thing would be sure to pop out at her from somewhere.

But maybe there was another way into the house.

Halle cautiously peeked out from behind the rain barrel again. Nothing was stirring. She slunk out and surveyed her surroundings. A little way beyond the vegetable garden was a clothesline stretched between two posts. To her left was the path that led to the front of Price Manor—to "peril"—and to her right, the path continued beyond the manor to what looked like a landscaped garden complete with marble statues and fantastical topiary. The creature could be hiding anywhere in there. She turned and gazed up at the house. Thick ivy covered the stone facade, weaving and clawing its way toward the eaves. A few yards away, however, was a small door.

Halle crept down the path toward the door. It was quiet. Too quiet. There was no movement. No birds flying overhead, no chipmunks darting through the grass, not even a single leaf blowing in a breeze. That meant the creature would surely pop out at any moment—just like in the movies. She quickened her pace to the door, but as she reached for the handle, it sprang open. Halle jumped back. A young woman balancing a large, full laundry basket on her hip emerged from the house and shut the door behind her.

Relief washed over Halle. "Thank goodness! Can you help me? I need to speak to Maximus Price. It's urgent!"

The woman turned to Halle in surprise. "Who are you?"

"Oh, I'm just the flower delivery girl. I brought Mr. Price's arrangement but—"

Something flashed across the path. Halle gasped and leaped back. The woman, who was adjusting the heavy basket of laundry in her arms, looked up. "Are you all right?" she asked.

Halle's eyes darted around the garden. "There's something. . . ."

The laundry woman set the basket down and turned to open the door again. "Why don't you come inside with me? We can—"

A sudden blur of grayish-pink flew out from behind a shrub, straight at the woman's face. She yelped as the thing wrapped its tentacles around her head. Halle froze in terror.

The woman fell to the ground, clawing at the creature

as it enveloped her head and muffled her cries. Its slimy tentacles and slippery, gelatinous body slid right through her frantic fingers. She squirmed and struggled, her movements gradually slowing until, a moment later, she lay completely still.

Halle's legs buckled. She fell to the ground, staring at the woman. If there was a time to scream, it was now, but her voice, her breath, and all the moisture in her mouth had disappeared.

The woman's chest rose and fell. She was breathing. The thing had shifted away from her mouth, though it completely covered her eyes, nose, and forehead. Halle had to get somewhere safe. She knew what would happen next. She had seen it once already. She scrambled up from the ground and dashed through the door into the house.

11

SILVER BOXES

Argyle now led the way with a spring in his step. For the first time in his life, he was unsupervised and schedule-less. There was no one telling him where to go, what to do, and when to do it, and now that they had figured out their birthday prank, it was like being on an adventure. He didn't know what ghastly scene they would stumble across next. In fact, when they rounded the corner to the main corridor, they found a thick trail of slime smeared down the long hallway.

“Aha,” he said. “Someone must be making this mess. If we follow the trail, it could lead us to Dad.”

"Makes sense to me," Paisley said.

The grisly trail meandered all over the first floor of the house. As they crisscrossed through the rooms and corridors, Argyle kept his eyes peeled for further evidence of their father's trickery.

"I spy with my little eye something beginning with the letter E," he said, peering through the sunroom windows out to the back patio.

"Eddie!" Paisley said as she joined him at the window.

The groundskeeper was sprawled across the bocce court amid the blue and red balls. His straw hat lay upside down in the dirt, and one of their father's creatures was wrapped around his head.

Next, Argyle spotted Brewster, the chauffeur, through another window. He was flopped over the steering wheel of the Rolls-Royce out by the garage. The car door hung wide open, and though Argyle couldn't see his face, he had no doubt that one of their father's nasty tredecapuses was in there with him, oozing slime all over the leather seats.

"This is better than a scavenger hunt!" he said.

As they continued to search for their father, Argyle was most interested in exploring the rooms at the east end of the house where they rarely had any opportunity to visit. They found a room full of props from their father's movies: rubber body parts, fake weapons, and some highly realistic robotic animals that could be operated by remote control. He had a terrific time maneuvering the animatronic animals around. There were rats and snakes and a remote-control bat that could even fly. He spent several delightful minutes pitting two vicious robotic dogs against one another while Paisley bounced rubber eyeballs against the wall in a disturbing game of wall ball.

After that, they wandered into a dimly lit room with a round table and high-backed leather armchairs. The dark green walls were decorated with a series of paintings of kings, queens, and knaves with patterns of hearts, spades, diamonds, and clubs.

"This must be where Dad and his buddies play pinochle," Argyle said. He made his way to the back of the room, where there was an odd stainless-steel door with a small window in the center. He peered through the window and could just make out a little room lined with shelves. He opened the door for a better look, and a foul smell wafted out.

"Blech! Cigars," he said.

"Dad has a room just for his cigars?"

"It must be the walk-in humidor. He told me about it once. His cigars have to be kept at a certain temperature and humidity or else they'll get even grosser than they already are."

Paisley waved her hand in front of her nose. "Ugh, they stink. Let's get out of here."

They turned to leave, but Paisley grabbed his arm. "Look," she said, pointing to a side table. On it were two small silver gift boxes tied with red satin ribbons. Each had a tag with one of their names on it. "Presents!" she said, picking up the one bearing her name. She gave it a little shake. "What do you think it could be?"

Argyle picked up his as well. "I don't know. Probably something Dad brought back from his trip." He turned the tag over. "On second thought, maybe not."

He showed the message to his sister:

Happy thirteenth birthday!
Love always,
Dad

He put the box down on the table and stepped back.

"It can't be that bad," said Paisley, turning hers over in her hands. "I mean, it's pretty small."

"Or it could be really bad. We could open them up and trigger some kind of reaction. Like a whipped cream explosion!"

Paisley laughed. "I don't think so. I think we should open them."

Argyle shrugged. "I suppose now's as good a time as any."

"At the same time, okay?"

They each held their box at arm's length and slowly untied the ribbon. Once the ribbons were removed, they counted to three and carefully lifted off the tops, wincing. Nothing happened.

"Wow!" said Paisley, peering into the box. She pulled out a sleek black-and-silver watch with a handsome leather wristband. "I've been wanting a watch for a while. Now I can keep track of our schedule."

"And now I can see exactly how long it is until snack time." Argyle fastened the watch around his wrist and admired it.

"These are really nice birthday presents," Paisley said.

He bit his lower lip. "Maybe a little too nice."

Just then they heard a crash from another part of the house.

"What in the world was that?" Paisley asked.

"It could be Dad. Let's go!"

They hurried out of the card-playing room and took off in the direction of the sound. At the end of the hallway, the corridor divided.

"Which way?" Paisley asked.

Another loud bang answered her question, and they turned right. A trail of slime led from the kitchen door to the far end of the corridor.

"This way," Argyle said.

They followed the slime trail straight into the laundry room, but it wasn't their father that they found.

"Mrs. Paprikash!" Argyle said.

The cook stood with her back to the twins, still holding her wire whisk. She did not respond to her name at all. Instead, she seemed focused on something in the corner. Argyle followed her gaze, past the overturned ironing board and the container of spilled laundry detergent. There, trembling in the corner, was a girl he had never seen before.

12

THE RUNAWAY LAUNDRY BASKET

Paisley stepped around Mrs. Paprikash to get a better look at the girl cowering in the corner. She had no idea when she had last seen someone her own age besides her brother. The girl was tall and thin with dark eyes and wavy hair and an absolutely terrified expression on her face. But she wore very sensible shoes. Running shoes, in fact.

The girl immediately started blathering. "Watch out! It can multiply! It got the woman outside—and the man in the bush too! I can bring more flowers tomorrow on my bicycle—have to find Maximus Price!" Her eyes widened as Mrs. Paprikash stepped closer to them. "But first, run—or hide!" She grabbed one of the wicker laundry baskets and pulled it over her head.

Paisley turned to Argyle with a puzzled look. No doubt their father had hired this girl as part of their birthday prank. But she wasn't making any sense.

Paisley approached the laundry basket and peeked underneath. "I think you've made a mistake. Gibberish was last year. This year it's tredecapuses. You should be covered in slime or something."

The girl scrunched up her nose in confusion. "Huh?"

"What?" Paisley said.

"Uhhhh . . . ," Argyle added.

A squelching, slurping noise came from Mrs. Paprikash, and Paisley turned to see the tredecapus on her face pulsating rapidly.

"We need to get out of here!" the girl whispered.

"Ohhh . . . ," Paisley said, nodding. The girl was trying to get them to believe the tredecapus was real. "Don't worry. We know she's just pretending. Right, Mrs. Paprikash? Aren't you supposed to be lying in the kitchen?" As Paisley addressed the cook, she noticed the creature had doubled in size since the last time they saw it.

"You've got to listen to me," said the girl. "That thing will attack. I saw it happen outside!"

"Outside? Where?" Paisley walked over to the garden door and peeked through the windowpane. A figure was sprawled across the path next to a basket piled high with sheets. "Oh, that's just Ms. Tuttle. She comes on Saturdays to do the laundry. She likes to hang the sheets outside to dry on the clothesline for 'extra freshness.'"

"Let me see," Argyle said. "I forgot she'd be here today."

Paisley reached for the doorknob.

"Don't open it," the girl pleaded. "Number one rule: Avoid Peril at All Costs!"

A smile spread across Paisley's face. So this was how it was going to be. There were rules. The birthday prank was far more complicated than they'd thought. She faced the girl again. "Okay. I like rules. Is this a game?" She was about to take out her notepad to jot things down when Argyle pulled the door wide open.

"Are you crazy?" the girl shrieked. With the laundry basket still over her head, she jumped up and kicked the door closed. "We gotta get outta here!"

Paisley's jaw dropped as the girl took off, speeding

past Mrs. Paprikash and the throbbing creature on her head. As the girl ran down the hall, she bumped blindly into walls like a billiard ball ricocheting from one rail to another.

This was fascinating. Props and slime were one thing. A real-live actual kid was another. Excitement bubbled up from somewhere in Paisley's belly all the way out to her fingertips and toes. She didn't know whether to jump, skip, or spin in a circle. She turned to face Argyle. His bulging eyes and open mouth told her he was just as intrigued as she was. Without wasting another moment, she grabbed his hand, and they dashed out of the laundry room after her.

13

INSIDE PRICE MANOR

Halle tore down the hall, straining to see through the thin cracks in the woven wicker laundry basket. She slipped and slid across puddles of slime and turned blindly around corners. She had no idea where she was headed—she just needed to get as far away from the laundry room as possible. Footsteps slapped the floor behind her. Was it those kids, or was it that lady with the creature on her?

She fought the urge to run faster. Without being able to see properly, she could end up slamming into something—or someone! But there was no way she was going to take the laundry basket off her head. She wasn't about to let that thing come anywhere near her face.

There was light up ahead—a doorway? Halle propelled herself toward it, hoping at least for better visibility there than in these dark, winding hallways.

But as she passed through the doorway, her foot caught on the edge of a thick carpet. She flew forward and crashed to the floor. The laundry basket bounced off her head and rolled to the other side of the room. Halle shielded her face with her hands as footsteps, heavy breathing, and another strange sound drew closer: laughter.

"Are you all right?" the boy asked.

Halle peeked through her fingers. The two kids hovered over her, grinning widely, their eyes round and bright. The woman with the whisk was nowhere to be seen, and the hallway behind them was empty—no creature in sight.

"Are we safe?"

"Of course we are," said the girl.

Halle breathed a sigh of relief and lowered her hands. "We have got to get help. That creature is attacking people. I saw it happen. It got the butler too. The man with the purple bow tie."

"Mr. Belloft?" the boy asked. "I can't wait to see him."

"Uh-uh. You don't want to see him. He was coming after me. With that thing on his face!"

The girl gave her a friendly smile and offered her a hand up. "It's okay. We know all about what's going on."

"You do?" Halle took the girl's hand, and they moved over to a cushiony sofa. She darted her eyes around the room, checking all the doorways.

"You're really good," the girl said. "You seemed genuinely scared. Even more so than Mrs. Pennywhisker was at the sight of a rubber hand in the soup tureen last month. Dad sure knows how to pick 'em."

Halle had no idea what this girl was talking about. Rubber hands? And what in the world was a soup tureen?

The girl kept going. "I'm Paisley, and this is my brother, Argyle. But I'm sure you already know that. You've probably been rehearsing for weeks if not months."

The boy flopped down on the other side of Halle and propped his feet up on an ottoman. "Don't worry. We figured it all out. You don't have to pretend anymore."

Halle shook her head. "Pretend what? I was nearly attacked!"

"Wait . . . aren't you part of the prank?" he asked.

"What prank?"

"Our birthday prank. That's why you're here, right?"

"What are you talking about?"

The boy eyed her with suspicion. "It happens every year on our birthday. We're twins, see? This year everyone's got these slimy cephalopods stuck to their heads. But no one's telling us what to do or anything. We're just wandering around the house on our own. It's pretty fun, actually."

Halle bit her lip. Could the creatures really be a prank? She thought back to when she had first arrived. The mist, the scarecrows, the shaking bush. It did seem just like a movie set. She began to relax a little, and for the first time, she took a good look at these twins, Paisley and Argyle. Not only were they still wearing their pajamas, but they were also kind of a mess. Their black hair was tangled and sticking out in places. The girl wore rain boots that were definitely too big for her, and the boy had what looked like a chunk of banana on his earlobe. He was slightly smaller than his sister, or maybe he was just slouching, but they both had the same slate-gray eyes. The same slate-gray eyes as Maximus Price. These were his kids!

Halle smoothed her hair and clothes and took in her

surroundings. Light shone in through a large picture window, reflecting merrily on colorful painted tiles of mythical creatures interspersed around the room: griffins, centaurs, three-headed dogs, and more. Through the window, there was a weathered stone patio and a bocce court, and beyond that, a landscape of spring blossoms complete with statues, marble benches, and flowering trellises. This was a much more charming view than the gloomy grounds at the front of the estate. Halle ogled the extravagant beauty all around her. She was inside Price Manor!

"Anyway," Paisley said, "if you're not part of our birthday prank, what are you doing here?"

"Oh, right. I'm Halle Thompson. I'm just the flower delivery girl."

"You mean the black and red roses?" Paisley asked. "Dad puts those under our mother's portrait in the hallway each week. She died a long time ago."

Halle raised her eyebrows. "I'm so sorry! I didn't know that."

"It's okay. We were really young when she died. We don't even remember her."

"Uh, well, unfortunately I dropped the flowers."

Halle looked down at her feet. "I'm really sorry about that. I can bring another arrangement tomorrow if that's okay."

"Of course!" said Paisley. "But how did you get in the house?"

Halle explained how she ran from the crazed butler and how the laundry woman was attacked right in front of her eyes just as she was about to let her inside.

"I don't blame you for being scared," Paisley said. "But don't worry. It's all our dad's fault. He can't really help it. He makes a living scaring people."

"Oh, I know who your dad is, all right," Halle said, growing excited. "I've seen every single one of his movies. I can't believe you guys are his kids. You're so lucky. You must get to watch his movies all the time!"

"Er . . . ," Argyle said, "we actually haven't seen any of them."

"You're kidding me!"

"They're all rated PG-13, and we're not thirteen yet. Dad takes movie ratings very seriously."

"But we'll be thirteen tomorrow!" Paisley said.

"Tomorrow? I thought you said today is your birthday."

"It's actually tomorrow. April first."

"Oh," Halle said, nodding. "April Fools' Day. So that's why your dad plays pranks on you. But then, why is he doing it today?"

"Just to be extra tricky, we think. He knows we'll be expecting it tomorrow." Argyle slapped his hands on his thighs. "You should meet him! He'll explain everything."

"Yes!" said Paisley, jumping off the sofa. "He's behind this whole thing."

At the mention of meeting their dad, Halle forgot all about birthday pranks, ruined flower arrangements, and slimy creatures. Maximus Price was here in this house, and she was going to meet him! The worm party in her stomach was back, and it seemed like they'd invited friends! She took a big breath and smoothed her hair and clothes. Then she frowned. "But wait . . . if the birthday prank is for you guys, then why was that . . . that . . ."

"Tredecapus?" Argyle said.

"Tredecapus . . . coming after me?"

The twins looked at one another, but neither had an answer.

14

A DEER IN HEADLIGHTS

The birthday prank was no longer at the forefront of Argyle's mind. An actual kid was in their house! This was way better than some slimy remote-control movie prop. He was dying to know more about her. Where did she live? What was her family like? Did she have any pets? Was she allowed to eat bagels? He'd read plenty of books featuring kids—both fiction and nonfiction—but this was the first time he could ever remember meeting one. He had so many questions! But the girl still seemed kind of nervous about the whole tredecapus thing. He had to prove to her that it was all just a crazy practical joke.

"Don't worry. Our dad will explain everything. The only problem is, we don't know where he is. We're pretty sure he's sneaking around here somewhere."

"Yes, and finding him is on our Schedule of the Day," Paisley said, pulling out her notepad.

"Want to help us?" Argyle asked.

Argyle detected the briefest hesitation from the girl, but then it quickly disappeared behind a wide smile.

"Sure," she said.

He led the way out of the sunroom and into the corridor. As soon as the girl—Halle, she'd said her name was—stepped into the hall, she took one look at the thick trail of slime smeared across the floors and stopped.

"I don't like the looks of this."

"Don't worry. It's all just part of the prank." He stepped over the sticky mess. "Check it out."

She took a few tentative steps forward. "I guess it does kind of look like the alien gunk that was streaked all over the spaceship in *Killer Snot Monsters from Outer Space*."

"Exactly," Argyle said, recognizing one of his father's movie titles. "Special effects."

"Your dad does this every year?" she asked, pressing her back against the wall to skirt the mucus-like puddle.

"Well, it's different every time," he explained. "We

never know what to expect. This year he's really outdone himself. You should see our housekeeper, Mrs. Danworth. She's lying in a huge puddle of this stuff in the entrance hall. She usually can't stand messes, but she didn't even move. She just lay there with the tredecapus clinging to her head like she was taking a nap or something."

"Hold on. . . ." Halle stopped walking again. "How long has she been lying there?"

"A couple hours, I guess."

"Uhh . . . so how many of those . . . things are inside the house?"

"Let's see, there's the one on Mrs. Danworth and the one on Mrs. Paprikash. But with Dad sneaking around, who knows how many there could be."

"And we saw a few more outside," Paisley added as she hopped over the trail of slime. "Brewster—that's Dad's chauffeur—he was in the car. And Eddie—he's the groundskeeper—he was out on the patio, just lying there in the bocce court."

"You mean the bocce court right outside those windows?" Halle asked, gesturing back toward the sunroom. "'Cause there wasn't anyone out there just now."

Argyle rubbed his hands together. "Ooh, this is

getting good. Maybe they're all 'waking up,'" he said, using finger quotes.

"Let's see if Mrs. Danworth is still in the entrance hall," Paisley said. "Come on. You'll love this!"

But Halle folded her arms across her chest and planted her feet firmly in place. "Uh-uh. There is no way I'm going to go see some lady lying on the floor, covered in slime, with a dangerous predator attached to her face. It could be sucking her brains out. And then it will replicate and come after us! Number one rule: Avoid Peril at All Costs. I have seen more than enough horror movies to know that."

Argyle shook his head. "Listen, you don't know our dad. He goes way over the top when it comes to practical jokes. He scares our governess just about every day when he's not traveling. And our birthday prank? It gets crazier and crazier every year. The whole household gets in on it."

"The laundry woman wasn't in on it. She was about to help me. She was going to bring me inside—until one of those things attacked her."

"She was just playing along," Argyle explained. "They do whatever Dad tells them to do. He was probably feeding her instructions through a wireless earpiece and planned the attack to scare you off."

"But even if he did, there's no way that thing is a prop. Props can't double in size or split in two or ooze liquid like that. It's got to be alive."

Argyle shrugged. "Well, I don't know of anything alive that can do those things either. Our dad's special effects team does amazing things. And it can all be done by remote control."

Halle remained rooted in place, chewing her lower lip. Argyle was used to dealing with his sister, who was a constant whirl of motion, *especially* when she was nervous. This girl was more like a deer in headlights—frozen in place. She did seem to perk up whenever they mentioned their dad, though.

"Once we find our dad, he'll explain everything. How about this? We'll skip the entrance hall, where Mrs. Danworth is. We've already been through there twice anyway. I know—let's try the medieval room! It's one of Dad's favorite spots in the whole house!"

"Yes," Paisley said. "That would be a perfect place for him to hide. Come on!" She twirled down the hall.

When Halle still hesitated, Argyle gave her an encouraging smile.

She glanced at the sunroom once more, then finally said, “Okay. I don’t want to be left here all by myself anyway.”

Argyle tipped his head in the direction Paisley went and swept his hand forward. “After you.”

Halle stepped cautiously down the hall.

15

THE MEDIEVAL ROOM

The medieval room was at the very center of Price Manor. Woven tapestries featuring knights and dragons hung from the walls, and shields, swords, and battle-axes were displayed in glass cases like in a museum. Several full suits of armor flanked the room, and in the middle was a long polished-oak tavern table with two intricately carved wooden chairs, one at each end. The table was set with hammered-metal plates, roughly forged cutlery, and bronze goblets. The circular chandelier that hung over the table held thick white candles instead of electric light bulbs.

Paisley marched straight up to the first suit of armor. "Nice try, Dad." She flicked up the visor of the helmet. It was empty. She checked the other suits of armor as well but with no luck. And he was not hiding behind the tapestries or under the table or behind any of the display cases.

"Maybe there's a secret passageway or something." She peeked behind several gilt-framed paintings of regal lords and ladies, only to find nothing but blank, solid walls. She sighed. "I was sure he would be here."

"Never mind," Argyle said. "We'll try somewhere else. Come on, Halle."

But Halle was busy staring up at one of the suits of armor. "This is *perfect*!" She knocked on the top of the helmet. "Is that real iron?"

Paisley walked over and examined the armor. "I'm not sure. Almost all the things in this room are reproductions made to look real for movies—except for those." She pointed to the two carved wooden chairs at the ends of the table. "Those are authentic medieval chairs from the thirteenth century. They were imported from Europe as a wedding gift for our parents. We're not even allowed to touch them, let alone sit in them.

They're very old and fragile."

Halle was still staring at the helmet. "Well, whether these are real or not, they'll protect our faces. We can wear them around while we look for your dad." She gave the helmet a tug, but it didn't budge. "I think it's stuck." She pulled harder.

Paisley peered around the back of the armor. "Hold on, there's a—"

As Halle gave one more hearty yank, the helmet came loose and the rest of the armor crashed to the marble floor with a great clanging sound.

"—latch," Paisley finished.

"Oops! I'm so sorry! But check it out." Halle placed the helmet over her head and looked around. She flipped the visor up. "It fits! It's a little stuffy, but who cares. In fact, this should be another rule: Protect Yourself." She walked over to the next suit of armor and reached for the helmet.

Paisley ran over before Halle could make even more of a mess. "That's okay! I'll get it!" She unhooked the latch at the back, and the helmet came off easily.

"Try it on," Halle said.

Paisley held the helmet up in front of her and stared

into its eye slits. She cocked her head to the side. "What are these rules you keep mentioning? Are you sure you're not part of our birthday prank?"

Halle removed the helmet from a third suit of armor without incident. She handed it to Argyle. "They're rules for surviving Dire Circumstances. Even if this is a prank, I'm not taking any chances." She flipped the visor of her helmet back down and paced around the room. "We've got to be smart. We've got to assume we're in peril. We've got to— Ow!"

She had tripped over one of the pieces of armor that had fallen on the floor and banged her knee against a display case. A silver chalice inside the case wobbled and fell over.

"Are you sure you can see with that thing on?" Paisley asked.

Halle rubbed her knee. "It's better than nothing."

Paisley eyed the helmet in her hands again. She glanced over at Argyle, who had put his helmet on and was now walking around with his hands outstretched so as not to bump into anything. She could practically hear Mrs. Danworth scolding them about taking a nasty spill and splitting their heads open. Although their father made the rules, the housekeeper was usually the one who enforced them—from always wearing shoes to not sliding down the banister or running in the halls. She never actually yelled or punished them, but her stern voice and glowering expression always left Paisley's nerves trembling.

Suddenly, Paisley's stomach did a flip-flop. They *had* broken a rule by running after Halle from the laundry room. It was a good thing Mrs. Danworth hadn't been there!

Paisley set her helmet on the table. "You guys can wear them if you want. I'll pass. But maybe you should at least think about keeping those visors up. Mrs. Danworth will give us an earful if you keep bumping into things like that. And no more running. We've got rules here too, you know. Mrs. Danworth always says—"

"Yowzers!" Argyle exclaimed, peeking out from under his raised visor.

"Yowzers? I've never heard Mrs. Danworth say 'yowzers' in her life."

"No, *yowzers* as in look behind you!"

Paisley turned around to see none other than Mrs. Danworth standing in the doorway in her slimy housedress with that icky tredecapus still attached to her face. Now they were in for it. There were bits of armor scattered all over the floor, one of the display cases had been disturbed, and Paisley was pretty sure they weren't supposed to be wearing the home décor on their heads.

Mrs. Danworth's eyes were completely covered by the slimy, pulsing body of the tredecapus. Maybe she couldn't see the mess they'd made. Yet she advanced toward them as if she knew exactly where they were.

"Mrs. Danworth, I'm so sorry. But don't worry. We were just about to clean all this up."

A low wheeze escaped the housekeeper's lips. She stopped and swayed slightly as the tredecapus undulated and oozed.

"Wow," Paisley said. "That thing is really gross." She looked over at Argyle and Halle for agreement.

"I think it's kind of neat," Argyle said. "A little cephalopod like that would be perfect for my aquarium."

Halle didn't say anything. Her eyes were wide, and her whole body seemed frozen in place.

Paisley turned back to Mrs. Danworth. Slime dripped from the creature's throbbing body. After a few seconds, its movements slowed, the oozing stopped, and the creature began to enlarge, filling itself with its own liquid. Paisley shivered with disgust.

The tredecapus quivered and grew until it blew up like a volleyball. It became so engorged that a split appeared across its back. Then, with a gush of pinkish liquid, the creature's back burst apart and a second creature wriggled its way out from the first. Paisley's mouth dropped open. The new tredecapus coiled its tentacles around Mrs. Danworth's collar, clinging to her like a sleepy toddler. The slit across the first one's back closed and sealed itself completely, as if it had never even been there.

"Let's get out of here!" Halle cried. She spun around and took off toward the doorway at the far end of the room.

"Wait—no running!" Paisley yelled after her.

But Halle paid no attention. She sprinted around the long oak table.

Paisley started to follow her at a quick walk, expecting Mrs. Danworth to scold them at any moment. "Come back, Halle! We need to clean up!"

A loud crash behind her caused Paisley to whirl around. Mrs. Danworth had lunged after them and knocked over one of the thirteenth-century chairs. It broke into pieces.

Paisley gasped as she took in the splintered bits of eight-hundred-year-old wood. The wedding gift for her parents. She slowly raised her eyes from the floor to Mrs. Danworth. Mrs. Danworth lurched forward with no regard for the irreplaceable—not to mention extremely sentimental—medieval chair.

"What do we do now?" Argyle asked.

Paisley turned to her brother. "We run."

16

THE RULES

Halle heard the slapping of Argyle's bare feet and the ca-chunk ca-chunk of Paisley's oversized rain boots behind her. She waited for them to catch up.

"Which way should we go?" she asked.

"Anywhere where there's no slime!" Paisley said, panic rising in her voice as she sped past Halle down the hall.

Halle lifted the visor of her helmet. A trail of slime zigzagged across the checkered floor tiles. Up ahead Paisley dodged the sticky goo with athletic speed and grace while Argyle swerved and stumbled behind her in his helmet until he finally tossed it to the floor. Behind Halle came the slow scraping sound of the housekeeper's footsteps. Halle raced after the twins, not even bothering to avoid the slime—she ran straight through it. A wider puddle lay ahead, and Paisley took a flying leap over it. Her enormous boot came loose in midair and twisted as she landed, sending her tumbling to the floor.

Halle slowed to a stop in front of the enormous puddle. "Are you okay? It never fails. Someone always trips and falls when they're being chased in the movies."

Paisley quickly kicked off both boots and jumped to her feet again. "I'm okay. Let's keep moving!" She ducked through a doorway.

Halle struggled to keep up as Paisley tore through the house, circling aimlessly through rooms and hallways that twisted and turned like a labyrinth. Argyle, who was already breathing heavily, began to lose speed. Halle surged past him, trying not to lose sight of Paisley.

"Wait! Slow down!" Halle called after her, gasping for breath.

Paisley slowed to a stop as Argyle caught up, and the three of them stood still, straining their ears. Not a sound could be heard. Satisfied, they all released their breaths.

"I think we're safe for now," Halle said, panting as she recovered from the run. She scanned the room for any movement. They appeared to be in a huge empty ballroom with crystal chandeliers, painted frescoes, and a bank of arched French doors that led out to a stone terrace.

Paisley wrung her hands and began circling the room. "Safe? We're not safe! This is definitely not a prank! Did you see how Mrs. Danworth broke that chair? She would never do that! And even if it was an accident, she didn't even seem to care! Or even notice what she'd done!"

"And did you see how that thing replicated?" Argyle added. "Even Dad's special effects team wouldn't be able to do that without some major CGI. But what *is* it? It's nothing like any animal I've ever read about!"

Halle had no answers, but she did have her wits about her. She swept across the room, closing the two sets of giant double doors leading out to the hallway and making sure the terrace doors were locked tight.

"What are we going to do?" Paisley cried. "We can't just sit here!" She continued to pace around the room, getting faster and faster.

"Well, walking in circles isn't going to help!" Argyle said.

Halle finished securing the room and then took a deep breath in and out. She approached the distraught twins and in a calm voice said, "You're right. It's not. These are Dire Circumstances. But we can't panic. That only leads to trouble. Now pay attention. It's time to go over the rules." She sat down cross-legged on the floor in the middle of the empty ballroom. The twins sat down in front of her and gave her their undivided attention. Halle adjusted the visor of her helmet so she could see them more clearly. She'd be sure to slam the visor back down again at the slightest sound or movement from any of the doorways.

"Okay, listen up. Rule number one: Avoid Peril at All Costs. Obviously that means the creature, but it also includes puddles of slime and anyone staggering around like a zombie."

"Even if it's our dad?" Argyle asked.

"Even if it's your dad. Rule number two: Protect Yourself." Halle banged on the top of her helmet for emphasis.

At this point Paisley pulled out her notepad and pencil and started jotting things down. Halle sat up a little straighter. The twins were actually listening to what she had to say. Georgia and Dakota usually rolled their eyes and wandered off anytime she started talking about monsters and murderers. She continued.

"From what we've seen of these creatures, they specifically attach themselves to people's faces. Protect your face and you might just save your brain from being sucked out. But until we can get those other helmets back, you'll have to improvise. A laundry basket. A towel. Even your hands might be enough to keep that creature from getting to your eyes, nose, mouth, or whatever it's after."

Paisley slowly raised her hand.

"You have a question?"

"Yes. What about shoes?"

"Shoes?"

"Yes. If rule number two is Protect Yourself, shouldn't we wear shoes? So we don't step on rusty nails or something?" She glanced at her now-bare feet. "I'm going to write it down."

"Sure, that's a good idea. In *Lair of the Demon King* the archaeologist's shoes washed away in a flooded tomb, and then later her bare feet got attacked by bloodsucking leeches." Halle shuddered. "That was awful. Shoes are definitely an asset. But maybe make sure they fit right," she added, remembering Paisley's mishap. Then she remembered her own mishap outside, when she'd tumbled backward on the stone pathway and ruined her flower arrangement.

"That brings us to rule number three: Never Walk Backward. In every horror movie I've ever seen, someone always walks backward—straight into a monster or a murderer." She glanced sideways and mumbled, "Or else they fall." While Paisley scribbled in her notebook, a new rule popped into Halle's head. "Rule number four: Don't Investigate. That includes strange noises, broken windows, and fiery meteors that fall from space like in *Danger in Devil's Wood.* All of that is just bad news." Halle was on a roll now. "Okay, rule number five—"

Paisley raised her hand again.

"Yes, Paisley?"

"Well, earlier today, we did hear a crash, and we did investigate. That's when we found you in the laundry room. But if we hadn't investigated, the creature could have attacked you. So, in theory, by investigating, we saved you."

Halle considered this. Paisley did have a point. Halle had been frozen with fear when the cook cornered her in the laundry room. If they hadn't shown up, she might be wearing a tentacle helmet right now instead of an iron one. "Okay, well, you guys didn't realize we were in peril at that point. Now you know. So let's just say, from now on, Don't Investigate. Because now you know we are definitely in peril."

Paisley furrowed her brow as she scratched out words in her notepad and furiously rewrote them.

"Okay, where was I? Rule number five: Never Go into Spooky Places. In the movies, that includes dark alleys, toolsheds, graveyards, and most certainly the woods. Those are the kinds of places where murderers and monsters lurk. And if you do find yourself in a spooky place, Get Out."

Paisley's hand shot up again. "Wait. Is that one rule or two rules? Or is 'Get Out' an addendum to rule number five?"

"I think it's more of a caveat," Argyle said.

Halle shook her head. "It doesn't matter. Just stay away from spooky places. Period."

This time, Argyle raised his hand.

"Yes, Argyle?"

"Okay. So, doesn't that mean we should leave here? Because, when you think about it, this whole place is spooky. We've got spooky furniture and spooky stone walls outside covered with spooky moss and even a bunch of fog machines to make it look like there's always spooky mist in front of the house. Our dad's made it a point to have it be as spooky as possible."

Paisley shook her head. "None of that spookiness is real. It's just our house. Besides, we can't leave. As Dad always says, this is 'The Safest Place on Earth.' That rule beats all the rest."

"Hmm," said Halle. "I don't know about The Safest Place on Earth, but I think Paisley's right. We shouldn't leave here because it could be worse out there. Who knows how many of those things are crawling around outside? And remember, this place is completely surrounded by the woods. Never ever, ever go into the woods. When I say spooky places, I'm talking about dark, creepy, abandoned places in the middle of nowhere. For example, what would you do if you're walking around at night with your friends and you come across a building with a chain-link fence and barbed wire all around it and lots of signs that say, 'DANGER, DO NOT ENTER' and 'KEEP OUT,' but your friends see a hole in the fence and want to check it out?"

"We don't have any friends," Argyle said.

Halle raised her eyebrows. "No friends? What about at school?"

"We don't go to school. I guess you could say we're homeschooled."

Halle shook her head. "Okay, we're getting off topic here. The answer is, you don't go into the creepy building because it's actually an experimental lab full of giant genetically mutated, killer rats."

"We don't have a building like that here!" said Argyle.

"I know, I know. Just . . . let's not go into the basement, okay?"

Argyle turned to his sister. "Do we even have a basement?"

"We do if you count Dad's stinky cheese cellar."

"Ugh. We're definitely not going down there!"

"All right. Let's move on," Halle continued. "If you do find yourself in peril and you can't leave immediately, rule number six: Run and Hide and Don't Come Out Until It's Safe. From what I can tell, this creature is making people slow and senseless. Almost like they're hypnotized or something. It's pretty easy to get away from them. It's the creature itself we need to worry about. When it's not attached to someone, it can move pretty fast, and it can even jump. We know there's at least one of those things inside the house that's looking for its next victim. We saw it burst out a few minutes ago. But there could be more."

Paisley looked up from her notepad. "Then what are we going to do? We can't investigate anything, and we can't leave the house."

Halle thought hard. They seemed okay in this room with all the doors closed. If Aunt Rhona were here, she'd probably tell them to stay put.

"We'll just wait right here where it's safe until someone comes to rescue us." She got up and walked over to one of the French doors. Through the glass she peered out over the misty grounds toward the front gate. Someone would come. Every movie had its hero. Heroes had to exist in real life too.

17

THE GRANDFATHER CLOCK

Argyle lay on his stomach, his chin propped on his hands, staring through the glass doors at the misty grounds beyond the terrace. To his left, Halle stood in front of another set of doors, also peering through the glass. Her arms were crossed, and she was drumming her fingers impatiently. Behind him, Paisley was counting jumping jacks. She was up to 413.

Everything outside looked surprisingly normal. Although Argyle couldn't remember ever having been beyond the stone walls of the estate, he knew that the manor rested at the top of a wooded mountainside. In the valley below was the town of Bedlamton. In the distance, another mountain rose up to form the horizon, but beyond that lay the unknown. Were the tredecapuses everywhere?

According to Argyle's textbooks, the world had seven continents. There were nearly two hundred countries and about 7.9 billion people. He knew that approximately 71 percent of the earth was covered in water, but he'd never seen an ocean, lake, river, stream, or even so much as a pond in person. The closest he'd come was the swimming pool at the center of the hedge maze in the back garden. He'd never been to a museum, a zoo, or a public library. And he'd never made a friend.

He glanced over at Halle. She probably had lots of friends. After all, she went to school, where, according to books he'd read, there were things like recess and group projects and extracurricular clubs and field trips. Maybe she even rode a school bus. An image of a school bus popped into his mind, packed with children—but when they turned their heads to look out the windows, they all had tredecapuses clinging to their faces. Argyle shook his head to get the image out of his brain.

He'd seen a real school bus in person once. When

he was younger, he used to stand at the iron gate, watching cars pass by on the road while Paisley ran laps up and down the gravel driveway. The bright yellow bus had rumbled by with the joyful sounds of singing, talking, and laughing floating out of the windows. Where had those kids been coming from? Where were they going? He'd looked for that bus again, day after day, when Mrs. Pennywhisker took them outside for fresh air, but he'd seen it only that one time.

Mrs. Pennywhisker never let him linger at the gate for too long. She would take his hand and lead him back closer to the house as Paisley weaved in and out between the scarecrows that lined the driveway. Eventually he stopped going to the gate. There wasn't any point.

Paisley finished her jumping jacks and started in on push-ups.

Halle moved to a different set of doors and resumed drumming her fingers.

Argyle had so many questions for Halle, but he didn't know where to begin. Not just about surviving Dire Circumstances but about everything. What were supermarkets like? How did parking meters work? Had she ever ridden an escalator?

But he didn't want to seem like an idiot. Then she'd

never want to be his friend.

He rolled onto his back and stared up at one of the chandeliers. Its arms stretched out like long wiry tentacles reaching toward him.

Suddenly the room filled with the sound of heavy bonging. Argyle bolted up.

"What's that?" Halle cried in alarm.

Paisley held her hand against her heart, recovering from the startling noise. "It's just the grandfather clock in the entrance hall. It strikes every day at noon and at midnight."

Now a new worry cropped up in Argyle's mind. "Uh . . . I hate to bring this up but . . . doesn't that mean it's lunchtime?"

"We can't think about that right now," Halle said. "We need to stay here where it's safe."

"But . . . we're going to have to eat at some point. I mean, how long do you think we'll be waiting here?" He glanced over at Paisley for support.

She smiled sympathetically and tapped her notepad. "Rule number one: Avoid Peril at All Costs."

Argyle sighed. He didn't want to complain. Halle would definitely not be impressed with that. He'd have to wait.

But now that he'd heard the clock strike twelve, his thoughts turned solely to food. They were in the ballroom, where his dad sometimes had fancy parties after his movie premieres. There was always an enormous buffet table loaded with all kinds of crudités and oysters and deviled eggs and filet mignon. And then there was the dessert table! His stomach gave a long growl as his eyes roamed the now-empty room.

He lay back down and stared up at the chandelier and its tentacle-like arms again. For a moment, his mind turned away from food to visions of Mrs. Paprikash and Mrs. Danworth with those tredecapuses on their faces. The image of the tredecapus replicating was horrifying and fascinating at the same time. When the creature's back split apart, a thin membrane had spread across the opening, like a layer of hot cheese stretching between two slices of pizza.

And now Argyle was thinking of pizza. His stomach growled again, and he buried his face in his arms.

18

BROKEN DISHES AND DROPPED KNITTING

Paisley couldn't keep still. As her mind whirled with worry, her body demanded to release all her pent-up energy. Where was their father? Who would come rescue them? Would the household staff be okay? Would *they* be okay?

Her arms were like jelly from the push-ups, but she still had energy to expend. Sit-ups were next. She scanned the room for something to hold her feet steady. She finally settled on the crack under one of the doors.

She wedged her toes under the door, lay back, and placed each hand on the opposite shoulder. Squeezing her abdominal muscles, she raised herself to a sitting position.

"One."

She lay back and lifted herself again. "Two."

Down and up. Down and up. Tightening and releasing. Counting steadily.

Her mind began to calm as she focused on the numbers. "Fifteen . . . sixteen . . ."

She closed her eyes as the burn spread across her stomach. "Twenty-eight . . . twenty-nine . . ."

She paid no mind to the sweat sliding down her neck. "Thirty-four . . . thirty-five . . ."

But that wetness on her left foot was odd.

At sit-up number thirty-six, Paisley opened her eyes. *"Yikes!"*

A tentacle slid under the door and wrapped around her ankle. She yanked her foot from the doorway. Three more tentacles slithered under the crack.

Halle and Argyle rushed over.

"It can't actually get in, can it?" Paisley cried.

The tentacles grasped the bottom of the door and tightened. The gray blob of the creature's body began to ooze through the crack.

"If it's boneless like an octopus, it can squeeze through impossibly small spaces," Argyle said.

"Then we'd better move!" Paisley scanned the ballroom. There was a service door at the back where the staff would go in and out during their father's movie premiere parties. "This way!" she cried, dashing for the door.

They scurried into the servery, a small room stocked with shelves full of china, crystal, table linens, and silverware. Argyle slammed the door shut after them. Paisley grabbed a tablecloth and pressed it into the crack under the door to block it. The three of them backed away and stared at the cloth. Nothing moved.

In the silence, Paisley became aware of a soft clicking sound. She turned. At the other end of the servery, a flash of movement flickered behind the frosted-glass window of a door.

Paisley raised a finger to her lips and pointed with her other hand. "Something's in there."

A shadow moved behind the door. A human-sized shadow making slow, undulating movements.

Halle tugged on Paisley's arm. "This way," she whispered, motioning toward the hallway.

"Wait," said Paisley. "We need to see who's in there."

"But what if it's somebody with one of those creatures on them? Remember the rules? Don't Investigate!"

Paisley stared at the figure on the other side of the door. Halle's rules were making everything so complicated. She took out her notepad and examined them. Clearly they needed to get out of there. She flipped to the previous page, and there was "FIND DAD" in big bold letters. Their dad could be the one behind that door. They *had* to find out.

She was just about to bring this up when Argyle marched over to the door and flung it open.

"Mrs. Pennywhisker!" he cried.

Relief washed over Paisley. Sure enough, there was their governess—perfectly creatureless—sitting on a low stool, clicking away on her knitting needles inside a small pantry.

Mrs. Pennywhisker let out a yelp when she saw them and then placed her wrinkled hand over her heart. "Oh my! Hello, dears. You startled me." A bundle of yarn lay on her lap, and judging from the length of the scarf trailing down to her feet, she'd been sitting there knitting for quite some time.

"Thank goodness we've found you! What are you doing in here?" Paisley asked.

"What does it look like I'm doing? I'm hiding. And I'm staying right here until your father gets rid of that horrible thing."

"Our father?" Paisley said. "So this *is* our birthday prank?"

"Your birthday prank? Oh, heavens no. It's not your birthday yet!"

"But then, what in the world is going on?" Paisley asked.

"Beats me. I've been in here all morning. I told your father I would stay right here until that terrible thing was out of this house."

"You've seen Dad?" Argyle asked.

"Of course I have. He arrived home this morning. And after he scared the daylights out of me, I gave him a piece of my mind."

"What did he say?" Paisley asked.

"He didn't say anything! How could he with that horrible thing covering his whole face like that?"

Paisley's heart sank. This couldn't be happening.

Halle came forward. "So, you saw Maximus Price,

and his face was covered with the creature?"

For the first time, Mrs. Pennywhisker noticed Halle in her medieval helmet. "Oh my! Who are you?"

Halle lifted her visor. "Sorry . . . I'm Halle Thompson. I'm just the flower delivery girl."

"Oh, for Mrs. Price's flowers. How wonderful! Mr. Price adores them. He says they're just perfect. It's so very nice to meet you."

Halle's cheeks reddened. "Thank you," she said with a shy smile. But her expression quickly turned serious again. "Where exactly did you see Mr. Price?"

Mrs. Pennywhisker tapped her chin with her knitting needle. "Now, let me think. It's hard for me to remember since I was so frightened. Let's see, first I screamed my lungs out and went running down the hall. But once I realized it was just Mr. Price up to his usual tricks, I yelled at him and pushed him into the—"

Mrs. Pennywhisker's eyes widened as she stared directly behind them. Then she let out a terrified yelp.

Paisley swung around. The tablecloth under the door was pushed aside, and several slimy tentacles were creeping their way through the crack.

"Mrs. Pennywhisker, you've got to come with us. It's

not safe here." Paisley grasped her governess's arm and guided her up from the stool.

In their haste, the yarn, knitting needles, and scarf dropped to the floor.

"Oh dear!" Mrs. Pennywhisker stooped down to quickly retrieve the items.

"Leave it!" Paisley cried. She whipped her head around to check the servery door. The tredecapus's slippery body was now oozing through the crack. At any moment it would be inside the room with them.

She pulled Mrs. Pennywhisker toward the pantry door while Argyle and Hallo urged them to move faster. But Mrs. Pennywhisker's legs were not quick or agile. Her feet got caught in the dropped knitting, and she stumbled. Argyle reached out to steady her while Paisley dived down to disentangle her legs from the long scarf. As Paisley tugged and pulled, the yarn caught on the buckle of one of Mrs. Pennywhisker's shoes. The situation was growing worse by the second! Paisley checked the door just as the tredecapus slid all the way through and scuttled out of sight behind the counter. There was no time to lose. She gave a mighty yank, and Mrs. Pennywhisker's shoe was freed. Together, she and Argyle led her quickly out of the pantry. But before they could make it to the hallway, a flurry of tentacles flew through the air. The tredecapus landed directly on Mrs. Pennywhisker's face with an audible splat. Her scream was muffled by the creature's body as she fell backward, out of Paisley's and Argyle's grasp.

"Mrs. Pennywhisker!" Paisley cried, kneeling at her governess's side.

Mrs. Pennywhisker flailed and squirmed as Paisley tried to help her up and Argyle helplessly grappled with the slimy creature. In moments it became clear their efforts were hopeless. Their governess stopped struggling and lay motionless on the floor except for the steady rise and fall of her chest. The tredecapus clung tightly to her head, its slippery tentacles laced together like the knitted yarn of her scarf.

Then came footsteps. Paisley swiveled her head from the fallen governess to the corridor doorway. Someone was coming.

"We've got to move," Halle said.

Paisley shook Mrs. Pennywhisker, attempting one last time to rouse her, but Mrs. Pennywhisker did not respond.

"Come on," Halle urged. "Let's go!"

Before they could go anywhere, though, Mrs. Danworth staggered into the room with a very engorged tredecapus on her face that looked just about ready to burst. Paisley scrambled to her feet and backed up against the shelves next to Halle and Argyle. She grabbed Halle's hand, and they inched toward the ballroom door. Paisley kept her eyes glued to Mrs. Danworth as she felt for the doorknob, expecting the creature to split and replicate at any moment.

Locating the doorknob, she yanked the door open, only to turn and see Mrs. Paprikash standing on the other side of it, wheezing and swaying. Paisley slammed the door shut again and pushed Halle and Argyle back against the shelves. Her eyes darted from one side of the narrow room to the other.

"We're trapped!"

Mrs. Danworth lurched forward, knocking over a huge stack of dishes from the counter. The plates shattered into thousands of pieces. Paisley gasped as shards of broken china blanketed the floor around her and Argyle's bare feet.

The housekeeper shuffled closer and then stopped as

the creature bubbled up to a massive size. Halle snapped her visor down and sank to the floor, while Argyle and Paisley stood frozen in place, surrounded by sharp bits of china in every direction.

Suddenly, instinct kicked in. Paisley turned and grabbed hold of the shelves behind her. She hoisted herself off the floor and climbed, scaling the shelves like a ladder. "Quick, up here," she told the others.

Halle still cowered on the floor, clinging tightly to the base of her helmet as if it might fly off on its own. Argyle surveyed the shelves, but instead of climbing up, his eyes settled on a chrome cake carrier on the counter. He reached for it.

"Argyle," Paisley shrieked, "this is no time for cake!"

The tredecapus on Mrs. Danworth split with a gush, and a new one emerged, shiny and wet. It flexed its tentacles and then drew them together, ready to pounce. Paisley cringed and braced herself for the attack. With her hands gripping the shelves, there was no way to protect her face.

The tredecapus sprang forward. In one swift motion, Argyle lifted the lid off the cake carrier and lunged, swinging the base up toward the creature. The creature slapped against the surface, and he slammed the lid down on top of it, trapping it inside.

"Gotcha!" he said.

Paisley let out a sigh of relief. But they weren't in the clear yet. Mrs. Danworth still blocked the doorway, and the shattered dishes strewn across the floor made any attempt at escape extremely dangerous. She climbed higher, knowing it wouldn't help, knowing the tredecapus could leap through the air with ease, but not knowing what else to do.

As she reached for the next shelf, a stack of place mats tumbled to the floor.

"That's it!" she said.

With one hand she began pulling down more and more place mats along with napkins and tablecloths and every kind of linen she could find. She tossed them across the floor, scattering them all over the jagged shards of china. When she'd thrown everything within reach, she scrambled back down and eased one foot carefully on top of the blanket of linens. The china crunched under her weight, but her foot was perfectly protected. She brought her other foot down and faced Mrs. Danworth. The tredecapus on her head was back to its normal size, and the split on its back had already healed. How long would it be before the tredecapus replicated again? The housekeeper herself seemed immobile, as if all her energy was being channeled to the recovering tredacapus. This was their chance.

"Come on—we've got to get out of here," Paisley said. Without a moment to lose, she hopped across the linen floor coverings, slipping past the dazed Mrs. Danworth and out into the corridor.

19

THE MOVIE PROP ROOM

It was a few seconds before Halle realized that Paisley and Argyle had made a run for it. With the visor down on her helmet, she could barely see anything that wasn't directly in front of her.

"Hey, new rule! Never Split Up!" she shouted, getting to her feet. She barreled hard into the counter. Ignoring the pain in her side, she adjusted her direction and tried again, this time knocking right into Mrs. Danworth. Halle reeled back with a shudder and then scampered out into the hallway. She followed the sounds of the twins' running feet, not daring to take her eyes off the narrow bit of floor she could see in front of her. When she turned the corner, she found them waiting for her.

"Poor Mrs. Pennywhisker!" Paisley said. "It's all our fault she was attacked!"

"And what about Dad?" Argyle cried. "Mrs. Pennywhisker said he had a tredecapus on his face too. He could be anywhere in the house!"

Halle lifted her visor and peered back around the corner. "We need to keep moving."

"Right," Paisley said. "Run and Hide and Don't Come Out Until It's Safe. And I think I know the perfect place to hide out until we figure out what to do next. Come on."

Halle slammed her visor back down and followed Paisley and Argyle down the hall. At the end of the corridor, Paisley ducked through a doorway with Argyle right behind her. But as Halle stepped inside the room, her foot rolled on something, causing her to lose her balance and tumble to the floor.

"Are you okay?" Argyle asked, giving her a hand up.

She took off her helmet to see what had made her fall. A rubber eyeball rolled across the floor. She followed its movement until it knocked into a lifelike replica of an alligator. A very familiar-looking alligator.

"Hold on—is that the alligator from *Swamp Devil of the Deep*?"

Paisley picked up the eyeball and bounced it nervously on the ground. "Probably. This whole room is filled with Dad's old movie props. I figured there would be plenty of places to hide in here."

Halle's jaw dropped. On one shelf, she recognized the crystal ball from *The Masked Mummy*, and hanging on a rack next to that was the actual latex costume of the Fire Kraken! She set her helmet down and ran her fingers over a silken purple cape—the very one that the high priest wore in *Lair of the Demon King*. And that fluffy little stuffed dog in the corner was a model of Mr. Twinkles, the dog that got eaten in *The Manic Marmoset Invasion*.

"This is incredible!" she squealed.

In response, a slurping noise came from the cake carrier in Argyle's hands. He pressed his ear against the chrome lid.

"You brought that thing with you?" Halle said.

"Ugh. What in the world are we going to do with it?" Paisley asked.

Argyle set the cake carrier down on the floor. The tredecapus slapped against the inside of the lid in a frenzy.

"We could take it out and study it," Argyle said. "Find out what it is and what it's doing to people."

Paisley shook her head. "That's against the rules. Avoid Peril at All Costs, and that thing is nothing but peril."

"But maybe we can figure out how to stop it."

Halle crossed her arms. "Absolutely not. We'll leave that to the professionals. Our job is to keep that thing closed so it doesn't get loose. Once help arrives, we can turn it over to the proper authorities."

"What do you think it wants?" Paisley asked.

Argyle double-checked the latch that secured the top of the cake carrier in place. "It wants what any creature wants. To survive. To feed. To reproduce. But it needs a host in order to do it. It's a parasite."

Halle nodded. "Just like the wasps in *Attack of the Mutant Mole Man*. That's how they destroyed the Mole Man after he went on his rampage. They injected the eggs of a parasitic wasp into him. The eggs could only incubate inside another living creature. Then, when the wasp larvae hatched, they ate through the Mole Man from the inside, and his stomach exploded. Of course, that led to the sequel, *Mole Boy's Revenge*. And then that led to another sequel, *Wrath of the Wasp Woman*. But that must be why the tredecapus is attacking people. It's the only way it can reproduce."

Paisley was now juggling several rubber eyeballs. "But what is it doing to the people? Will they be okay?"

"We don't know yet," Argyle said. "We can't get close to them without risking another replication. But at least the creatures aren't replicating inside their hosts like Halle's parasitic wasps."

Argyle sat on the floor and tapped the top of the cake carrier. The creature slapped around inside.

"It must somehow sense when a host is nearby. This one's going crazy. And the tredecapuses on Mrs. Danworth and Mrs. Paprikash seemed to respond immediately when a host was close, by starting the replication process."

"How did they find us?" Paisley asked. "Mrs. Paprikash and Mrs. Danworth can't see anything with their faces all covered up. And the tredecapus doesn't have any eyes."

"Noise," Halle whispered, remembering how the creature inside the cake carrier had started acting up right after she'd squealed.

"But we left Mrs. Danworth way over by the medieval room, and the last time we saw Mrs. Paprikash was in the laundry room. How could they have heard us all the way on the other side of the house?" Paisley asked.

"The grandfather clock," Argyle said. "It's in the entrance hall just outside of the ballroom. When the clock struck noon, it must have drawn them toward that part of the house. And then they heard us talking."

"We need to stop making all noise whatsoever," Halle whispered.

"Hang on," Argyle said, looking around the room. "I think I've got a better idea."

He grabbed one of the remote controls and turned it on. The alligator began snapping its jaws and lashing its tail. Then Argyle ran around the room, turning on all the other remote-control animals. Soon a chorus of squawks, barks, and growls came from every corner of the room.

"What are you doing?" Halle said. "You'll lead them straight to us!"

"That's the point. With all these animals making a racket in here, they'll be confused. And since they can't see, we can sneak off to another part of the house while they're distracted. They'll stay in here where it's noisy."

Halle didn't like this plan at all. They were basically luring the creatures right to them. Aunt Rhona would never approve. "We need to get out of here now before they come," she said.

"No," Argyle said. "They might be in the corridor already. If we leave now, they'll just follow us. We should wait here until they come, and then sneak out quietly."

"Argyle," Paisley pleaded, "this is a complete violation of rule number one!"

"I'm sure this will work. Trust me."

All of Halle's instincts were telling her to run. She glanced at the doorway. There was no sign of anyone yet. Would Argyle's plan really work? And where was her helmet?! She must have set it down somewhere. She scanned the room. It was nowhere to be seen amid the clutter. She had to find something else—and quick! Some other armor or mask or anything that could cover her head.

She began rummaging through the shelves and bins, tossing useless items to the floor. Rubber fingers, plastic ray guns, miniature model spaceships—none of it would do any good. There had to be something useful in this room.

A glint of silver beneath a pile of rusty chains caught her eye. She reached in, wrapped her fingers around the gleaming object, and pulled it out. It was a dagger. The steel blade was pristine, and the ornate, carved handle was encrusted with sparkling red rubies.

She gasped and turned it back and forth in her hand. "This is Raven Rose's dagger!"

She couldn't believe it. Raven Rose had carried this dagger through swamps, forests, and deserts. She had killed countless demons with this very weapon. Holding it in her hands, Halle felt a flutter of something in her heart. A flutter of hope. A flutter of confidence. A flutter of power.

Argyle elbowed her, bringing her back to the present. "Someone's coming," he whispered.

Halle froze—the hope, confidence, and power instantly dissipating.

"Over here!" Paisley dashed behind a fiberglass meteorite in the corner. Argyle and Halle followed.

Over the din of mechanical animals, the unmistakable sound of footsteps shuffled closer and closer. A moment later, Mrs. Danworth appeared in the doorway of the prop room. By now she was entirely coated in slime. She swiveled her head back and forth from one robotic beast to another, disoriented by the barking and squawking. But there was no other way out of the room. She had planted herself directly in front of the only exit.

Halle and the twins eased down behind the meteorite and peeked over the top.

"What do we do now?" Paisley whispered.

"We just need to wait for our opportunity," said Argyle. "And don't make a sound."

Mrs. Danworth stepped into the room, revealing Mrs. Paprikash right behind her. She had two tredecapuses on her—the one that covered her face and a shiny new one perched on top of her head. The second one must have emerged when she was on the other side of the servery door. As the two women moved farther into the room, the creatures on their faces began to bubble and writhe, preparing to replicate once more. The unattached tredecapus scuttled down Mrs. Paprikash's arm. It reached the floor and skittered in one direction, then another, leaving gloppy puddles of slime everywhere as the robotic animals growled and snarled all around it. Finally, with much hesitation, it pounced on Mr. Twinkles.

The robotic dog continued its playful yipping and tail-wagging. Sensing its mistake, the tredecapus released the poor thing, leaving its white, fluffy fur gooey and matted.

Mrs. Danworth took a few more steps into the room. The creature on her face, now very engorged, shifted slightly, uncovering the tip of her nose. Mrs. Danworth inhaled deeply and then turned her head directly to the corner where the three of them huddled behind the meteorite.

Halle's heart stopped.

"Oh no," Argyle whispered. "They can smell us!"

Paisley still had a handful of rubber eyeballs. She tossed them across the floor in the direction of the two women.

Mrs. Paprikash took a step forward and immediately slipped on one of the eyeballs. She tumbled down, knocking into Mrs. Danworth in the process. Mrs. Danworth went down too and landed in a puddle of tredecapus goo. Already mostly covered in slime, she flipped and flopped on the floor, unable to get up, while Mrs. Paprikash struggled to stand amid the scattered rolling eyeballs. At the same time, the rogue tredecapus leaped from one animatronic animal to the next, unable to find a suitable living host.

Seeing their chance, Halle stashed Raven Rose's dagger in her belt and grabbed the twins' hands. She pulled them past the two floundering women, through the doorway, and all the way down the empty corridor. They turned a corner and ducked through another doorway. A large round table with a heavy floor-length tablecloth covered in hearts, diamonds, spades, and clubs sat in the middle of the room.

"Quick—under here!" Halle said, lifting the bottom of the tablecloth. They scrambled under and crouched together, panting for air.

"Don't make any noise," she told the twins.

"It doesn't matter," Argyle said. "They can smell us."

"He's right," said Paisley. "We're not safe here."

"Then we need to find a different room where we can close the door and seal it with something," Halle said. "None of these rooms have doors."

"Upstairs. One of the bedrooms." Argyle started to creep out from under the table, but Halle grabbed him.

"Wait. They're coming!"

They huddled motionless under the table as two sets of shuffling, squelching footsteps drew near. Halle couldn't see a thing from under the thick tablecloth. The footsteps entered the room and paused. She heard several deep inhalations and held her breath. The slightest disturbance in the air could give them away. The footsteps approached the table. Halle covered her face and squeezed her eyes shut, dreading the moment they would be discovered. Unseen noses sniffed at the air, searching like bloodhounds on the trail of a rabbit. This was it. She braced herself for the attack.

But a moment later, the footsteps started up again. They overshot the table completely and passed to the far side of the room. Halle opened her eyes and peered through her fingers at the twins.

Argyle bent his head to the floor and lifted the tablecloth a fraction of an inch. He waved his hand for Halle and Paisley to take a look.

Halle crouched down and cautiously peeked under the tablecloth. The two women had their noses in the air, sniffing impulsively and heading straight toward a strange stainless-steel door that was ajar. When they reached it, they sniffed more urgently at the opening until they managed to push their way inside, nose first. The door eased open all the way, and Halle stared as the two women descended upon shelves and shelves of cigars, knocking down cases and sending loose ones rolling onto the floor.

"We forgot to close the humidor," Paisley whispered to Argyle. "All they can smell are those stinky cigars!"

As Paisley spoke, the two women swiveled their heads in her direction.

Halle gasped and covered her face again, but Paisley scrambled out from under the table, darted across the room at lightning speed, and slammed the door shut, trapping them inside.

20

SLUGGISH CREATURES

Argyle peered through the small window in the door of the humidor. With Mrs. Danworth and Mrs. Paprikash trapped inside, he was able to really observe their behaviors. There were now six tredecapuses wriggling around amid the boxes of cigars in addition to the original two still clinging to Mrs. Danworth and Mrs. Paprikash.

"These things must really like Dad's stinky cigars," he said.

Paisley and Halle stopped their nervous pacing and joined him at the window. The creatures on the two women had enlarged once again and were preparing to replicate for a fourth time.

"Check this out," he said, playing with a small control panel on the door. "When I turn the temperature and humidity up, the creatures get really sluggish, but when I turn it down, they're more active. They must prefer the cold. Some species of fish can only reproduce in certain temperatures."

"So let's make it really warm in there," Halle said. "Maybe that'll stop them or at least slow them down."

"But not too warm," Paisley warned. "We don't want Mrs. Danworth and Mrs. Paprikash to get dehydrated. You know how Mrs. Danworth always insists we drink plenty of water when it's hot out."

Argyle adjusted the temperature to a balmy seventy-eight degrees. As the temperature rose, one of the creatures slid down from a shelf and flopped onto the floor. Its tentacles waved limply from side to side.

"I could watch these things all day," he said, fascinated. "If only it would roll over, I could get a look at its underside. It would be good to know if there's some kind of mouth or appendage on its belly. Hopefully they're not using their hosts as lunch."

Paisley frowned and started pacing again.

But at the mention of lunch, Argyle's empty stomach reminded him of how hungry he was. If only he'd saved that last granola bar he'd found in the snack drawer in the study.

"So . . . now that these two and their tredecapuses are contained, maybe we could think about getting those sandwiches?" Argyle suggested.

Halle turned away from the humidor. "Forget it. We're still in danger. There's a tredecapus loose in the prop room, your governess could wake up and come after us at any minute, and now we know your dad's been . . ."

"Tredecapized," Paisley finished.

"Right. And he could be anywhere in the house," Halle said.

Argyle sat down at the pinochle table. Now that he was aware of his hunger again, he realized his legs had begun to weaken.

"Wait a minute," Halle said. "We could call my mom at the flower shop! Why didn't I think of that in the first place? In fact, that should be in the rules: Call for Help. We could even call 911. Where's the phone?"

"Dad has a phone," Argyle said, "but he keeps it in his pocket."

"Do you have a house phone?"

"No. The staff all have cell phones, but they probably have them in their pockets as well, and I sure don't want to get close enough to search them."

Halle turned back to the humidor window. "So, we've got no one to help us and no way to call anyone."

"And no one to make our lunch." Argyle dropped his head down on top of his arms. His energy was draining by the minute, just like the creatures in the warm humidor.

"This isn't the time to think about lunch," Halle said. "We need to figure out some other way to get help."

But Paisley approached Argyle and rested her hand on his back. "Argyle? Are you feeling okay?"

He moaned in answer.

Paisley looked over at Halle. "We need to get him something to eat."

"Um, hello!" Halle said. "These are Dire Circumstances, remember? Peril doesn't stop and take a lunch break!"

"You don't understand. Argyle needs to eat constantly to keep his energy up. If he goes too long without food, he might pass out."

"Oh, wow," Halle said. "I didn't realize it was so serious. My aunt has anemia and has to take vitamin B12 or else she gets like that too. Has he seen a doctor?"

"No. Dad just says Argyle is . . . delicate." Paisley rested her head down next to Argyle's and looked him in the eye. "Do you think you can hold out for a little while?"

"I don't know. How far are we from the kitchen? Maybe we can sneak back there and look for some cold cuts in the refrigerator before we figure out what our next move is."

"There's no way to get to the kitchen without passing the servery, and that's the last place we saw Mrs. Pennywhisker."

"What if we're super quiet?"

Halle shook her head. "No way. They can smell us, remember?"

Argyle pressed his lips together. Earlier, he'd been so excited about not having any adults around. No adults meant no boring lessons and no lectures on what they could or couldn't do, but now he realized it meant no one to take care of them in Dire Circumstances. He summoned all his energy and forced himself upright again. "I'll be okay. Really. Let's figure out a plan."

21

THE CLOAKROOM

Paisley turned to her notepad once more. It was the one thing that brought her any kind of comfort. She added Halle's new rules—"Never Split Up" and "Call for Help"—then drew a rough floor plan of the house on a clean page. They couldn't get to the kitchen, and Argyle seemed to be holding up okay for now, but they could get to the cloakroom by taking a circuitous route around the manor. That way, they could try to find some shoes to wear. After the incident with the broken china, she didn't want to take any more chances.

They tiptoed through the winding corridors, avoiding hallways that were particularly slimy, and made it to the cloakroom without incident. The cloakroom was especially useful, not only because it was full of outerwear, but also because it was one of the few places on the first floor that had a door. They could close themselves safely inside, making sure to seal off—and *guard*—the crack under the door while they rummaged around.

First, Paisley found her ski boots from last winter when they'd had the ski lift installed on the sloping west lawn. She tried them on but immediately took them off again. They were too small now and way too stiff for running. Next, she found her flippers from when they had scuba lessons in the outdoor swimming pool. Eddie the groundskeeper had thrown all kinds of "treasure" into the water, which they happily retrieved as if they'd sunk in a shipwreck like the *Titanic*—jewelry, coins, and even their sousaphones. But scuba flippers wouldn't be of any use now unless they had to swim away from the tredecapuses, and that wasn't very likely. Then there were her tap-dancing shoes. Those wouldn't do either—way too noisy! She glanced over at Argyle on the other side of the cloakroom, but he didn't seem to be looking for shoes at all. Instead, he was checking all the coat pockets for snacks.

Paisley sat down amid the growing pile of useless

footwear. They were all perfectly suited for something, but none of them were made for running from slimy tentacled creatures.

She flipped open her notepad again and chewed on her lower lip. The cloakroom was a bust. There had to be more shoe options but where? And once they did find shoes, they still had to figure out how to get help without breaking any of the rules.

She turned to Halle, who was on guard duty at the door.

"What do you think we should do now?"

"We have to get to a safer location. Your dad or your governess might hear us in here. Maybe we could try one of the bedrooms upstairs."

Paisley's stomach twisted into a tight knot. "We were supposed to find our dad, not hide from him. He's the one who always makes sure nothing bad ever happens. He's the one who keeps promising us this is The Safest Place on Earth."

Halle's gaze fell to the floor. "I'm sorry. I guess we're on our own."

Paisley returned to her notepad even more determined. "Okay, so we don't have a phone, we have to Avoid

Peril at All Costs, and we can't leave the estate because it's surrounded by the woods."

"We really shouldn't even leave the house," Halle said. "Who knows how many more of those things are out there. The butler and the laundry woman were outside, and they are tredecapus hosts."

"And don't forget Eddie and Brewster," Argyle said, slouching to the floor after giving up his search.

Paisley flipped to her rough floor plan and drew a star where each member of the household staff was last seen outside. "That's one on each side of the house. We're basically trapped."

"If only there was some way to send a Morse code signal with flashlights or mirrors like they did in *The Children of Deadwood Orphanage*. Or we could hang a big SOS sign from the roof like in *Apocalypse VII: Aftermath of the Aftermath*. I can even see the top windows of your house from outside my mom's flower shop all the way in town."

Argyle rested his head against the wall. "It's not like we've got a handy SOS banner lying around."

"Oh my goodness," Paisley said, flipping the pages of her notepad. "It's right here on the Schedule of the Day!"

Argyle grumbled. "If we can't get lunch, we definitely can't do anything else on your schedule. Dire Circumstances and all that, remember?"

"No, I mean the tapestries. If people can see our house from town, then we can use the tapestries to send a message. They're alphabet tapestries."

"Alphabet tapestries?" Halle asked.

"Yes. Mrs. Pennywhisker has us doing a big project where we make a huge tapestry for each letter of the alphabet. We had to plan scale drawings on graph paper first and then convert the measurements to the actual dimensions. It's supposed to teach us about math, art, history, and textiles all at the same time. Each finished tapestry is twelve by sixteen feet. They're enormous. If you can see our house from town, we could hang the tapestries from the attic windows to send an SOS message."

"But we've only made it up to letter R," said Argyle. "And even if we had the whole alphabet, we still couldn't spell SOS because we'd need two S's."

"We can spell something else." She flipped to a clean page in her notepad.

"I guess 'BEWARE OF SLIMY CREATURES THAT

WILL CLING TO YOUR HEAD' is out of the question," Argyle said.

"Don't be absurd. This is serious." She jotted down words in her notepad. "'DIRE CIRCUMSTANCES' wouldn't work but 'PERIL' would. No, that wouldn't get anyone to come save us. That would just keep people away. How about 'SAVE US' or 'HELP US.' Or . . . just 'HELP'! That's it! It's perfect!"

A short while later, Paisley stood with Argyle and Halle at the foot of the rickety, narrow staircase that led to the attic. On the floor next to them was a heavy bundle of woven tapestries rolled together.

"Okay, let's do this," Paisley said. She picked up the front end, Halle grabbed the back, and Argyle supported the middle as best he could. They climbed the twisting staircase, stumbling and bumping into the walls with the large, cumbersome bundle. Finally, after a lot of grunting and adjusting, they reached the top and entered the attic of Price Manor.

A dark hallway lined with doors stretched out in both directions from the stairway.

"This way," Paisley said, pulling them off to the right.

She led them to the doorway at the end of the hall where they could access the windows that faced the tiny town in the valley below. They dropped the tapestries to the floor and shoved them into the dark room.

"Can we seal the door?" asked Halle.

Paisley's eyes adjusted to the dim light. There was a set of solid leather block benches nearby. "Yes. Help me bring these over."

They pushed the heavy benches against the door, closing off the crack completely.

"Is there any other way in here?" Halle asked.

"No," Paisley said. "We're absolutely safe."

"What is this place?"

Argyle flipped the row of light switches by the door, and the room lit up with color. Disco lights flashed across the walls and spinning mirrored balls hung from the rafters, sending tiny dots of reflected light all over the wide room. In the center was a waist-high, oval-shaped wall with nothing but a smooth, polished wood floor inside it. Against the back of the room were shelves and shelves of roller skates in every color of the rainbow.

They were inside the roller-skating rink.

22

QUEENS OF THE RINK

When Halle was ten years old, she started spending every Saturday night at the noisy, crowded roller-skating rink in Bedlamton. It was where all the kids from her class went. Her mother was usually exhausted from a busy day in the shop, so Aunt Rhona would take her. Halle loved the loud, pulsing music and the colorful flashing lights, but most of all she loved talking and laughing with her two best friends, Georgia and Dakota. But at some point during that year, Georgia and Dakota started to change. They got sleek new skates instead of using the plain brown ones with frayed laces that the rink provided.

They started getting together before skating so they could do their hair and pick out matching outfits. And there were changes happening with them at school too. They huddled over fashion magazines at recess instead of playing on the swings, and they would bat their eyelashes at one of the seventh graders whenever they saw him. On Fridays after school, Georgia's mom started taking them to the mall. Halle went with them a few times, but she always complained that shopping was boring and that she'd rather go to the ice cream parlor or the miniature golf place, so they stopped inviting her. She never had a lot of spending money anyway.

At the roller-skating rink, Georgia and Dakota got better and better at skating, but every time Halle stepped onto the rink—disaster. She had no sense of balance. The skates would slip out from under her, leaving her splayed out on the polished floor while other kids zoomed around her without missing a step. Georgia and Dakota got tired of helping her up all the time. They would cruise around the rink, hand in hand, giggling and whispering as they passed by her. Aunt Rhona tried to help Halle with her skating, but by then Halle was already taller than her tiny aunt. Every time she held her hand, Halle would end up pulling Aunt Rhona right down with her when she fell.

Sometimes a big group of kids would link arms and glide around the rink in a long line. Halle was sure that if she could just latch on, all those kids together would be able to keep her up and pull her right along with them. So one Saturday night, a line formed with Georgia and Dakota at one end. She saw her chance. As they skated by, Halle grabbed hold of Dakota's hand and held on tight.

"Ouch! You're crushing my fingers. Get off!" Dakota said.

"Sorry!" said Halle, but she didn't dare let go. They were going faster than she had ever gone before. Her body went rigid as the line pulled her along. If she let go, she'd be done for.

Then someone—it sounded like Georgia—shouted, "Crack the whip!"

The girl at the other end of the line slowed down and started turning in a circle. The rest of the line pivoted around her, gaining speed. Halle was at the very end, going so fast that everything was a blur. She gripped Dakota's hand tighter.

"Faster!" someone yelled, but they were already going impossibly fast.

Halle flew around and around. Her hand grew slippery from sweat, and before she knew it, Dakota wrenched her hand free. Halle was propelled from the line. Her skates twisted under her, and she crashed against the wall. Pain seared through her ankle.

Halle ended up having to wear a cast for two months. No more roller-skating. But while her broken ankle would heal, she wasn't sure her friendship with Georgia and Dakota ever would. They never even said they were sorry.

The weekend after she got the cast, Aunt Rhona took her to her first horror movie to get her mind off her misery. It was a revival of the old Maximus Price film *Raven Rose and the Curse of the Soggy Bog Beast.* As they sat in the theater, waiting for the movie to begin, Aunt Rhona told Halle all about Maximus Price, including the fact that Halle's own mother's flower shop had been sending a weekly delivery to Price Manor for years.

Halle was only half listening. She was fantasizing about being back at the roller-skating rink, surrounded by a ton of kids lining up to sign her cast while Georgia and Dakota sulked in a corner. But once the film started, Halle was enthralled. All the movies she had seen before had been silly kid movies and lighthearted tales of pretty people falling in love while singing and dancing. The actors were always too perfect. They didn't seem real at all. This was different. Instead of fancy costumes and sparkly engagement rings, the actors wore tattered clothes streaked with dirt. They stumbled through the woods, tripping and falling over rocks and tree roots. And boy, was there a lot of screaming. To Halle, the characters seemed more like real people. Plus, there were gruesome monsters and fierce battles. She sat through the whole movie on the edge of her seat, her heart pounding, her eyes gleaming, her hands gripping her bag of uneaten popcorn.

After that, Halle wanted more. For the next few weeks, while her ankle healed, she and Aunt Rhona watched every Maximus Price movie they could get their hands on. Halle also volunteered to help her mom with Maximus Price's weekly flower arrangements. Her mother's standard arrangement of red and white roses simply would not do. The roses needed to be red and black, and she researched the best methods for dying flowers.

When Halle got her cast off, the doctor told her she still wasn't ready for roller-skating, so Aunt Rhona introduced her to classic horror films—*The Wolfman*,

Frankenstein, *Nosferatu*—the scarier, the better. They even watched foreign horror films from places like Japan and Argentina. But Maximus Price movies were her all-time favorites. She covered her bedroom walls with posters of Scream Queens and studied their faces, reenacting scenes of being chased by monsters, falling, fainting, even dying!

At school, though, every time Halle brought up horror movies, Georgia and Dakota would run off making gagging noises. The other kids were not much better. They all preferred skating and going to the mall and the singing-and-dancing movies.

Finally, after months of recovery, her doctor announced that her ankle had completely healed. She could even go back to the roller-skating rink. But Halle had no interest in roller-skating anymore. She wanted to be an actress. She wanted to be in horror movies. She wanted to scream and run from terrifying monsters on the big screen.

But she never imagined she'd be screaming and running from any real monsters, and certainly not hiding from them in a roller-skating rink of all places.

She looked around at the huge empty rink in the attic

of Price Manor. “Have you guys ever gone skating in Bedlamton?” she asked.

“There’s a roller-skating rink in Bedlamton?” asked Paisley.

“Yeah, but it’s really crowded, and everyone skates super fast.”

“I wouldn’t mind that,” Paisley said, gazing across the vacant rink. “But we wouldn’t be able to go anyway. We have to stay here. It’s The Safest Place on Earth.”

“Why do you keep saying that? That this is ‘The Safest Place on Earth’?”

“Because it is,” Paisley said. “Dad makes sure of it. He says we’ll always be safe as long as we stay within the stone walls of the estate.”

“You mean . . . you never leave?”

“Dad says it’s too dangerous after what happened to our mom,” Argyle said.

“So, you’ve never even been to the movie theater? Or the ice cream parlor? Or the miniature golf place with the exploding volcano?”

“We did have miniature golf here one summer,” said Paisley.

“It didn’t have an exploding volcano, though!” Argyle said.

Paisley ignored his complaint. "And we can watch movies and eat ice cream here whenever we want."

"But that's not the same. The Bedlamton Ice Cream Parlor has a 1950s-style jukebox and over a hundred toppings! You can get gummy bears or fresh strawberries or even cereal on your ice cream. And movie theaters are awesome. You sit in the dark in front of a huge screen with the sound coming from speakers all around you, so it feels like you're actually part of the movie."

"That doesn't sound any different from the movie screening room we have here," Paisley said. "Dad always makes sure we're not missing anything."

"But in a movie theater you get to see previews about what movies are coming out next. And the whole place smells like buttered popcorn. And you're surrounded by tons of other people. Everyone screams at the scary parts and laughs at the funny parts all together, as if you're in a huge room filled with friends."

"You must have a lot of friends," Paisley said. She gazed at the empty rink again. "I guess that's one thing we're missing."

Halle sighed. "Well, after the movie ends, everyone kind of goes back to being strangers again."

For a moment, no one said anything.

Finally Paisley broke the silence. "We should get these tapestries up. The sooner the better."

The tapestries were like nothing Halle had ever seen before. The first, an ornate golden H intertwined with green vines on a deep crimson background, would stand out well against the gray stone exterior of Price Manor. They hefted it onto the first windowsill, straining with the weight and sheer breadth of the heavy woolen hanging. Gripping the top tightly, they let the bottom drop down over the side of the house. While the twins held the top of the tapestry taut across the edge of the windowsill, Halle brought the window down on top of it, securing it in place. They quickly moved on to the second letter, a brilliant blue E on a pale-yellow background. They hung it from the next window, followed by the last two tapestries at the remaining windows.

"And now we wait," Halle said. At least they were safe up in the roller-skating rink with the door barricaded. They could even keep a close watch through the windows for help to arrive. Halle stared down into the valley at the town below, but all she could see were treetops and the unrecognizable roofs of buildings. Somewhere down there was the corner outside her apartment and the flower shop, where she had stood many times, looking up at these very windows. She'd never imagined she'd ever be inside Price Manor looking down.

Halle shook herself from her thoughts. She turned from the windows to see Paisley pulling on a pair of roller skates.

"What in the world are you doing?"

"It's better than sitting around doing nothing. Can you help me with the laces? Mrs. Pennywhisker usually ties them for us."

"You guys can weave beautiful, intricate tapestries, but you can't even tie your own shoelaces?" As much as she admired Maximus Price, she had to admit, his parenting skills were seriously lacking.

23

PAISLEY SKATES HER TROUBLES AWAY

Paisley attempted to follow along as Halle showed her how to lace up the skates and tie a bow with a double knot. But there were so many loops and twists, and the laces were way too long. Half the time, she felt like she needed a few extra fingers, and the other half, she felt like she had too many—like the tentacles on a tredecapus. Her mind was clouded with worry and anxiety. She needed to focus on something else. She needed to release her nervous energy. She needed to skate. Eventually, she gave up trying to learn how to tie the laces, and Halle just tied them for her.

"I'm still not sure you should be roller-skating at a time like this," Halle told her.

"Well, we can't just sit around!" she replied.

"Actually, that's exactly what I intend to do," said Argyle, who had collapsed on one of the cushioned benches after exerting his last bit of energy on the tapestries.

Paisley shook her head and skated over to the rink opening. She couldn't just wait around until help arrived. It would only lead to panic. She launched herself onto the smooth surface of the rink with a burst of energy, hoping to bring relief to her muddled mind. She did a few warm-up laps and then swung her arms to one side and spun in a circle.

But even with her muscles at work, her thoughts kept returning to one glaring truth: this was no longer The Safest Place on Earth. She looked over at the others, desperate for some company. Skating by herself wasn't enough. "Who wants to come join me?" she called.

Argyle slumped deeper into the cushioned bench while Halle pointed to the windows and went back to keeping watch.

Paisley sighed and resumed skating, this time going

backward. Help would come soon. They were safe in the roller-skating rink for now. She picked up speed, jumped, did a spin in the air, and came out of the turn going forward again. Concentrating on her tricks was helping. When she focused on the moves, her worry slipped away. She glided on one leg with her other leg stretched straight behind her. Then she pulled her leg in, tucked her arms, and did a quick pirouette on the tip of her skate. As she came out of the spin, she noticed Halle watching her with her mouth hanging open. Paisley smiled. Maybe if she did a few more tricks, Halle would come skate with her, and they could both feel better. She set off across the rink again, crisscrossing her legs in fancy figure eights. She squatted down and stuck one leg straight out in front of her, angled her other foot to the side, and spun in a three-hundred-sixty-degree wheel spin.

When she came to a stop, she looked over at Halle. "Are you sure you don't want to skate with me?"

Halle quickly shook her head as if she'd been caught spying and pointed at the windows again.

"It will make you feel so much better," Paisley said. "Argyle can watch the windows. Plus, I never get to skate with friends."

Halle bit her lip. "Uh . . . actually . . . I'm not that great a skater."

"Well then, I'll give you some pointers!" Paisley skated across the carpeted area to where Halle stood. This would surely keep their minds off the looming danger until help arrived. "Let's get you some skates. You can pick whatever color you want. I can even teach you a few tricks to show all the other kids in town."

"I'm definitely not ready for tricks. I can barely let go of the wall without falling."

"Well, there's your problem. You'll never learn to skate if you hang on to the wall. Come on. It's not as hard as it looks. We'll start slow and simple, and I'll help you stay up."

"Really? Do you think I can learn?"

"Sure." Paisley took her by the hand and led her to the wall of roller skates. "What size do you wear?"

24

THE KITCHENETTE

Argyle took Halle's place at the window. Thick mist hovered over the grounds below. The fog machines usually only ran from dusk until dawn, but with Mr. Belloft, Eddie, and everyone else unable to turn them off, they'd been running nonstop since last evening. His eyes traveled over the foggy estate, past the scarecrows that flanked the driveway, all the way to the stone wall and iron gate that marked Price Manor's boundaries. How long would it be before someone came to help them?

He peered past the gate to the road winding down the mountain and into the town below. Somewhere in that town was a roller-skating rink like this one except filled with people. And a movie theater. What would it be like to have all those people around? Probably pretty exciting. He supposed there were playgrounds and restaurants too. Restaurants where you could order anything on the menu instead of eating only what was put in front of you. Argyle's mouth watered. If he were in a restaurant now, he would order a big tuna sandwich with arugula and tomatoes. Or perhaps a grilled panini with salami, brie, and pickles. At this point, he'd even take something simple like peanut butter and jelly.

Wait a minute—what was he thinking? He must have been really out of it. There was a little kitchenette right here in the roller-skating rink! Mrs. Pennywhisker usually prepared snacks while they were skating, and when they were ready, she'd call them over and they'd sit on the swiveling stools of the snack counter to eat. Just last week, Mrs. Pennywhisker had made them hot dogs and French fries.

Argyle hopped off the windowsill and headed straight for the snack counter. When he entered the little kitchenette, he took a quick survey of the layout. It was much smaller than the main kitchen downstairs and much more manageable. In addition to a small stove and mini fridge, there was a toaster oven, a blender (which he shuddered at), and—oh wow—a panini press! He ran his hand over the shiny chrome lid. How hard could it be? There was only one dial that said on and off.

25

HALLE LEARNS TO ROLLER-SKATE

Halle hugged the wall at the opening of the rink and carefully placed one skate and then the other onto the polished wooden floor. Immediately her legs slipped and slid under her as she clung to the side and tried to make her way forward.

Paisley skated over and took her hand.

"I'm totally going to fall," Halle said.

"Don't worry. I've got you. Just let go."

Halle wobbled as she let go of the wall.

"Now, center your weight over both feet and try to relax your legs," Paisley said. "You're very tense."

Halle tried to relax, but she was bracing herself for the fall she knew would come at any moment. "Maybe this isn't such a good idea."

"Don't worry," Paisley said. "We'll go slow. You don't have to pull on my hand—just use it for balance." She took hold of Halle's other hand and skated backward as Halle teetered forward. "Keep your feet pointed in the same direction. If you feel your legs drifting apart, just shift your weight to one foot and adjust your other one until it's going in the same direction."

Halle followed Paisley's instructions, and to her surprise, she didn't fall. With Paisley holding both of her hands, she was more balanced. They progressed forward, not too fast but not too slow, and Halle began to relax. Colored lights danced over the floorboards and walls around them as Paisley guided her. They glided past a giant mirror along the back wall, and Halle couldn't help taking a peek at herself. A huge grin covered her face. Now there was an expression she didn't see much in her bedroom mirror. It was more than joy. More than happiness. It was Elation. The only thing missing was music, but she definitely didn't want to attract any of those creatures up here with unnecessary noise, even with the door barricaded.

As they started moving faster, Halle faltered. They were coming to the curve, and she didn't know how to turn. She tensed her legs as her expression shifted to wide-eyed fear. "We're going to crash!"

"Relax," Paisley said. "I'll hold you steady as we go around."

But they were going too fast for Halle. As she lifted her right foot, her left one slipped forward, catching against Paisley's skate. They both went down and landed in a tangled heap on the floor.

"I'm so sorry!" she said as she struggled to unhook her skate from Paisley's. But the uncooperative skates stayed locked together.

Paisley started to giggle. It was not the same kind of giggle that Georgia and Dakota used to do when they skated by her at the rink. This giggle was friendly. They tugged and tugged to get their skates apart, and soon they were both rolling on the floor, laughing.

Once they untangled themselves, Halle sat up. Paisley was still sprawled on the floor, laughing uncontrollably. Halle took a good look at her in her roller skates and pajamas and began to crack up even more.

Finally, they got ahold of themselves, and Halle said, "I'm sorry. I'm a terrible skater."

Paisley propped herself up, a humongous grin on her face. "No, you were doing great! Let's try again."

They set off around the rink again. Halle still wobbled a bit, but she could tell she was getting better with Paisley's strong hands leading her along. They even managed to get around the curve a few times without falling, though when they did fall, they just laughed and squealed and helped each other up again.

She had almost forgotten all their troubles when a burning smell reached her nose.

"What's that?" she asked, sniffing at the air.

Paisley turned her head toward the smell. "Oh no! Argyle!"

A plume of smoke rose from behind a counter as Argyle waved frantically at the air with both hands. Paisley led Halle quickly out of the rink and across the carpeted section of the floor. They peered over the counter to see a sizzling, smoking panini press with juices dripping all down the sides. Argyle lifted the top and revealed a burnt, oozy mess of charred bread, soggy pickles, pastrami, tuna, peanut butter, coleslaw, and mozzarella cheese.

"My sandwich!" He hacked at the mess on the griddle with a fork until blackened, crusty clumps of tuna fused with cheese and peanut butter littered the countertop. With a sigh, he scraped all the pieces onto a plate and mushed them together to form a ball.

"Anybody want some?" he asked, holding the plate up.

Paisley eyed the burnt monstrosity. "Uh . . . no thanks."

"Why don't we start over?" Halle said as she scooted around to the other side of the counter, not even needing to hold Paisley's hands.

26

SANDWICHES

Argyle watched Halle spread a thin layer of mayonnaise on one piece of bread and some mustard on another. She placed two slices of ham and one slice of cheese on the mayonnaise side and topped it off with a leaf of lettuce before placing the other slice of bread on top, mustard side down. He mirrored her steps exactly, not wanting to admit how simple it really was. When he'd first seen all the options inside the refrigerator, he had wanted everything! Paisley wrote down the steps in her notepad, though Argyle didn't really think that was necessary, and then they cut the sandwiches into halves and sat at the counter to eat them.

Argyle took a bite. The soft bread was like velvet on his tongue, and the flavors of the ham, cheese, and mustard melded into a heavenly combination of deliciousness. "Mmmm. This is the best sandwich I've ever had."

"That's because you made it yourself," Halle said. "I feel like food always tastes better when you're the one putting in the effort. Pickle?" She offered him the pickle jar.

Argyle fished out one of the spears and then passed the jar to Paisley.

All too soon, their sandwiches were devoured. Energy seeped back into Argyle's body and his head cleared. He swiped the last of the crumbs from his plate with his index finger and popped it into his mouth.

"And now we've got to clean up," Halle said. "My mom never lets me leave the kitchen when there are dirty dishes."

Paisley put the ingredients away while Halle showed Argyle how to wash the plates and stack them in the drying rack. They were even able to scrub the burnt surface of the panini press, which Argyle had been sure was ruined. When all the washing was done, Paisley went to put the skates away while Argyle dried the dishes and Halle wiped the countertops.

"How do you know how to do all this?" he asked.

"I don't know. I guess my mom taught me. And I've had lots of practice. I make my own breakfast and pack my lunch for school. Sandwiches are pretty easy, but I also cook eggs and heat up soup. And I can make chocolate chip cookies."

"Chocolate chip cookies?! All by yourself?! That's amazing!"

"It's not so hard. I make them all the time."

"Do you think you can show us sometime?"

"Of course. All you have to do is follow the directions on the chocolate chip package. Have you ever baked anything before?"

"Dad doesn't want us near the oven. It's too dangerous. We could get burned or start a fire—like I almost did with the panini press!"

Halle was quiet for a moment as she wiped a glob of peanut butter off the counter. Then she looked up at Argyle. "Maybe you wouldn't if someone showed you how to do things *safely*. Maybe it's actually more dangerous if you're *not* taught how to use things like ovens and panini presses properly."

Argyle sighed. "That makes a lot of sense, but our dad would never go for it."

"Have you ever talked to him about it?"

"No. It's just always been like that."

"Well, maybe you should. Maybe he needs to realize that you aren't little kids anymore. You can't have people tying your shoelaces and making all your meals for you forever. You'll be thirteen tomorrow. What happens when you're fourteen? Or sixteen? Or twenty-three? Will he still not let you use the oven then? Will you still never be able to set foot outside of Price Manor?"

Argyle widened his eyes in horror. "I never thought about that before!"

Halle laughed and rinsed out the dishcloth. "I would be more than happy to show you how to bake chocolate chip cookies, but I think you need to talk to your dad first." She wrung out the cloth and hung it over the faucet, then dried her hands with a fresh dish towel.

Argyle marveled at how comfortable Halle was in the kitchenette. She maneuvered around as if she'd been making sandwiches and washing dishes for years. And that wasn't all. She'd ridden her bike here all the way from town, completely on her own. If she could do those things, why couldn't he and Paisley? Argyle puffed out his chest and grinned. "You're right. I will talk to him." Then he sagged. "As long as we can find him and get that tredecapus off him."

"Hey!" Paisley cried from the window. "I think help is on the way!"

A police siren off in the distance grew louder. Argyle and Halle rushed over to join Paisley at the window.

27

HELP ARRIVES

Halle watched through the window as the police car cut its siren and turned through the gate of Price Manor. Its headlights pierced the thickened mist while the car wound its way up the gravel driveway toward the house.

"Finally," Paisley said. "Let's go down and explain what's going on."

Halle grabbed her arm. "We can't go down there! It's too dangerous!"

"Then how are we supposed to tell the police what's happening?"

The car came to a stop in front of the main entrance.

Two police officers, a man and a woman, got out.

Halle banged on the attic window and shouted, “Hey! Up here!”

The officers didn’t hear her. They walked toward the front steps and peered wide-eyed through the mist. Halle pounded at the window again, and Paisley and Argyle joined in, knocking on the glass and calling to get their attention. It didn’t work.

Halle’s mind raced. They hadn’t thought this through at all. These police officers were in just as much danger as they were. Even more since they were out in the open with no idea what was lurking in the mist and the shadows.

What would Aunt Rhona do now?

Halle threw open one of the windows, causing one side of the letter P tapestry to slip down.

“We’re up here!” she shouted.

Argyle and Paisley joined in as well, yelling to get their attention.

Finally, one of the police officers looked up. She nudged her partner and pointed to the attic windows. Halle and the twins waved their arms and shouted as the officers jogged over toward the east wing. That was when

Mr. Belloft stepped into their path. The officers stopped in their tracks. As Mr. Belloft staggered toward them, they shouted and gestured for him to stop, but he kept advancing on them.

"Watch out!" Halle called down to them.

The officers looked up. The man called out, "Stay there, kids, we've got everything under control," while the woman pulled out her baton and tried to reason with Mr. Belloft. The thick mist swirled around the butler's waist as he came to a stop in front of the officers.

"Oh no," Halle said under her breath. In the next instant, Mr. Belloft's tredecapus split open.

The officers looked at each other, then back at Mr. Belloft just as the new creature emerged, pink and shiny. It quivered, then leaped through the air. The officers jumped back in surprise, and the tredecapus missed, landing somewhere on the ground, obscured by the dense fog. The two turned to one another, confused at first, but then they smiled and shook their heads in amusement. The male officer wiped his forehead and laughed, while the female officer poked her baton around in the mist, searching for the creature.

"Don't worry, kids. It's just a movie prop!" the man

shouted, looking up toward the attic window. "Maximus Price is up to his old tricks again." Then he pulled his radio out and gave the all-clear to his dispatcher.

Halle's heart sank.

"No!" Paisley yelled. "You need to get away from it!"

"What's that, sweetheart? I can't hear you." He put the radio back on his belt and walked toward the east wing windows.

"It's not a prop!" Argyle shouted. "It's real, and it's dangerous!"

The officer came closer. "You'll have to speak up."

Behind him, the other officer poked at the misty ground again and again until the creature launched itself up onto her face. She reeled back, grabbing at the slimy tentacles while her partner sauntered obliviously toward the east wing.

Halle and the twins gasped.

"It's all right, kids. My name is Officer Munrap. Everything is fine."

"No! Look behind you!" Argyle shouted. He pointed at the other officer, who staggered back and forth before collapsing on the ground, enveloped by the mist.

Halle lowered her head into her hands and backed away from the window.

"Why don't you come on down so we can talk?" Officer Munrap called up to them. He started to head back to the front door but then stopped and swiveled his head around, looking everywhere for his partner. He turned a full circle, confused, then checked inside the patrol car. With no sign of his partner, he approached Mr. Belloft and tried to question him, but of course Mr. Belloft was no help. Then Ms. Tuttle, the laundry woman, appeared from around the side of the house, drawn by the noise and commotion.

The twins yelled for Officer Munrap to take cover, but he was too busy trying to question Mr. Belloft. Ms. Tuttle came closer and closer, and the tredecapus on her face grew larger and larger, while the policeman was getting more and more frustrated with Mr. Belloft's lack of cooperation. When Officer Munrap took out his handcuffs and shook them threateningly at Mr. Belloft, Ms. Tuttle was directly behind him. Her tredecapus split open with an explosion of slime that splattered the officer all over his back. As he turned to face the source of the unexpected wetness, the fresh new tredecapus leaped onto his head and sent him sprawling to the ground.

"We have a big problem," Halle said.

She walked from window to window and threw each

one open, letting the tapestries fall to the ground below. "We can't let anyone else come here. The more people who come, the more creatures there will be. We've got to seal off this whole place, or we'll have an apocalypse on our hands. Just like in *Zombie Chickens from Mars*. And then we need to get as far away from here as possible."

"You think we should leave?" Paisley said in alarm.

"Every time someone comes near those creatures, they replicate. We need to stop them from spreading. And to do that, we've got to get out of here and warn everyone. We've got to call in the National Guard. Or some kind of secret government pest control team."

"But what about the rules?" Paisley said, pulling out her notepad. "What about 'Run and Hide and Don't Come Out Until It's Safe'? What about 'Never Go into Spooky Places,' like the woods? What about 'Protect Yourself'?"

Halle gazed out the windows at the town of Bedlamton in the valley below. It would be bustling with people at this time of day. No one knew there was a potentially lethal creature just a few miles away from them. Even the police had given the all-clear. She and the twins were the only ones who could do anything. And they had to act now before it was too late.

She turned back to them. "This is about more than just saving ourselves. This is about saving the town. Maybe even saving the world."

The twins blinked. Argyle chewed on his lip, and Paisley wrung her hands together. Finally, Paisley flipped back to her Schedule of the Day and took her pencil out. At the bottom of her list, in big fancy letters, she wrote:

SAVE THE WORLD

28

THE PIED PIPER

Paisley turned to a fresh page in her notepad and titled it "The Plan." She gave Halle her full attention as Halle launched into another riveting lesson.

"First, we need to eliminate the danger to innocent civilians."

"The stone wall surrounds the whole estate," said Paisley. "The only way in or out is through the gate. If we close the gate, Mr. Belloft and the others can't leave."

"But those critters can," Argyle said. "If they're not attached to a host, they could slip right through the bars or even climb over the wall."

"Which means we need to lure all the creatures into the house and lock them inside. And the hosts too," said Halle. "That'll be the first step."

Paisley wrote it down, though she wasn't sure she liked the idea of her home—The Safest Place on Earth—being overrun with tredecapuses. "How would we even lure them all in?" she asked.

"The same way we did in the movie prop room," Argyle said. "We need to make a lot of noise."

"That kind of reminds me of *The Pied Piper of Hamelin*," Paisley said.

"The Pied what?" Halle asked.

"The Pied Piper. It's an old fairy tale. We studied fairy tales and their origins last year with Mrs. Pennywhisker. In that one, the town of Hamelin was infested with rats, and the townspeople called in a rat catcher who led the rats away by playing his magic pipe."

Halle's eyes lit up. "Just like in *Bloodthirsty Badgers of Baton Rouge*! They were able to lure all the killer badgers over a levee and into the Mississippi River by playing accordion music from a riverboat. We could totally do that here. Then we can escape into town and get help."

Paisley's stomach turned. "Wait a minute. You can

escape, but Argyle and I can't leave. We're not allowed to. I thought the plan would just be to lure them inside the house—and then wait outside."

"It's too risky," Halle said.

"But what if the tredecapuses are already outside of the estate? What if they came from somewhere else and are already all over town?"

"I don't think so," Halle said. "If they were, the police officers would have known about them already."

Paisley tapped her pencil nervously against her notepad. She was starting to regret bringing up the Pied Piper. The more they talked about the idea, the more it seemed like they were actually going to go through with it. She didn't know which was worse: having the safe haven of her home become infested with tredecapuses or leaving it entirely. She pocketed her notepad and started pacing back and forth. "I still don't like the idea of leaving. Something terrible could happen to us out there."

"It's here we need to worry about. I've lived in Bedlamton my whole life. Nothing bad is going to happen."

Paisley turned to Argyle. "You're not thinking of going, are you?"

"We can't stay here forever," he said. "Especially if

there's a face-eating creature attacking people. Besides, Dad travels the world all the time and nothing happens to him. Why should he keep us locked up here?"

Paisley came to a sudden halt. She had never heard Argyle talk like this before. Sure, he was grumpy when he got hungry, but he always did what was expected of him without question, like she did. "We're not exactly locked up," she told him. "Look around. We have an enormous estate and everything we need. Why would we want anything else?"

"Think about what we're missing, though! Mini golf with an exploding volcano, ice cream parlors"—he turned to Halle—"and friends!"

Paisley looked over at Halle. She'd never laughed so hard as when they were falling on top of each other in the roller-skating rink. Friends were one thing she never realized she needed before, but skating with Halle had been so much fun.

"Besides," Halle said, "we also have to stick together. It's in the rules, remember?" She smiled and pointed to the notepad in Paisley's pocket.

Paisley couldn't help chuckling at Halle's playful teasing. Maybe Argyle was right. Maybe there were things outside of their home that she'd been missing out on. Things she'd never even imagined. Their father was so focused on all the dangers of the world that he'd never shared any of the world's wonders with them. Even the things Mrs. Pennywhisker taught them—things like old fairy tales and native rocks of East Anglia and calligraphy—had nothing to do with what was really just outside the gate. What was the point of calligraphy anyway? It sure wasn't helping them now.

And if she could have fun roller-skating with Halle, surely there were more kids like her who enjoyed doing other sports. Whole teams of them! Over the years she'd learned all the skills of basketball, volleyball, soccer, and hockey, but those had all been fleeting interests. She'd never stuck with them because she never had anyone to play with. And all that sports equipment now sat unused and forgotten in the—

"Oh my goodness!" she said. "I just remembered something."

She ran to the other side of the room to a huge storage closet. She flung open the door and started throwing objects out. Shin guards, football pads, baseball gloves, and more. Finally she found what she was looking for.

Two pairs of tennis sneakers—with Velcro fastenings.

Halle laughed. "One of these days, you two have got to learn to tie shoelaces."

"No problem," Argyle said. "We'll get to it right after we save the world."

29

THE PORTRAIT

Halle and the twins arrived back in the entrance hall dressed for battle. Well, not really battle. More like some kind of weird triathlon. They'd raided the attic storage closet for whatever head protection they could find. Paisley wore a hockey helmet with a metal cage that covered her face, Argyle had on a snorkel mask, and Halle wore a fencing mask that had a wire mesh front and a padded bib that went down past her neck. The twins' two shiny sousaphones were propped up by the front door.

Halle took a long look around the entrance hall. She wasn't sure she'd ever get to come back again. Marble columns rose on both sides of the massive room, supporting a grand balcony on the second floor that circled the entrance hall. Under the balcony, several arched doorways led to the magnificent rooms and corridors that stretched to the other parts of the house. Suspended from the ceiling was an ornate chandelier made of what looked like interlocking bones. *What an incredible place to live in*, Halle thought. Paisley and Argyle had no idea how lucky they were. Each room, each piece of furniture, each decoration was an exquisite work of fantastical art. The chandelier's lights reflected across the polished-marble banisters and down the gleaming staircase until they reached the awful puddle of slime where the housekeeper had lain earlier. Halle shuddered. Perhaps they weren't so lucky after all.

She turned away from the puddle of pink goo and paced back and forth in front of the door. In a few minutes, they'd be walking through that door and straight into harm's way. She stopped to rub her sweaty hands on her pants. That's when she spotted the elegant marble-topped table on the other side of the hall. On it was a vase of drooping roses. Her roses. The ones she had arranged for Maximus Price the week before. A pang of guilt spread through her once more as she thought of the fresh roses she had dropped outside. But of course, none of that mattered now.

Then she remembered what Paisley had said about the flowers. They were a tribute to their mother. Halle's eyes traveled up to the portrait hanging just above the flowers. Her jaw dropped. She pulled the fencing mask off her face for a better look.

The woman in the painting was clad in armor and chain mail and stood at the top of a rocky cliff. Her long black hair blew dramatically to one side with a few dark tendrils waving across her face. One hand rested on her hip, while her other was high in the air, proudly raising the grotesque severed head of a demon.

"That's Raven Rose, the demon huntress!"

Paisley looked up. "Yes, that's our mother. She was Raven Rose. Well, you know, she was the actress who played her."

Halle could not take her eyes off the portrait. "I can't believe that's your mom."

Argyle shrugged. "We don't really remember her. We were only two when she died."

"You guys, she is amazing! She's brave and strong and

she fights all kinds of horrible monsters." Halle beamed up at the portrait as she gushed on. "Plus, she's smart and loyal. I can't believe you've never seen her movies. She's incredible! She's practically invincible!"

Paisley squinted at the portrait. "Well, it turns out she was just a regular human being after all."

Halle's eyes dropped to the floor. "I'm sorry. I didn't mean . . ."

"It's okay," said Paisley. "Dad doesn't talk much about Raven Rose. It's kind of neat to hear about her."

"You have got to see her movies. Raven Rose could take on anything—demons, evil wizards, monsters" Halle pulled the fencing mask back over her face and puffed out her chest. "And so can we. Are you guys ready?"

Argyle handed Halle the four loose cigars they'd retrieved from just outside the humidor.

"Ready," he said.

He and Paisley picked up their sousaphones, and the three of them turned to the massive front door.

30

THE ERRANT CIGAR

When Argyle opened the front door, a thick cloud of fog wafted in from outside.

He waved at it with his hands. "This mist is getting worse by the minute. The fog machines have been running all day."

"It's going to be impossible to see anything," Paisley said.

"I could probably turn them off from the garage, but it would take a while for the fog to dissipate."

"We don't have that kind of time," Halle said. "We've got to get these creatures into the house. If any of them get outside the grounds, they'll spread through Bedlamton like an alien plague."

"Or like any invasive species in its non-native habitat," Argyle added. "It happens all the time. The giant African land snail. The spotted lantern fly. The northern snakehead fish. When an invasive species is introduced to a region, it has no natural predators. They can reproduce uncontrollably and wreak havoc on an ecosystem if they're not stopped early enough."

"Then, I say we move forward with the plan. It's now or never," Halle said.

He peered into the fog. He could see only about ten feet in front of him. Hopefully that would be enough to get out of the way when they needed to.

"Is everyone set on what we're doing?" he asked.

Halle nodded under the fencing mask.

"Paisley?"

His sister bit her lip and fidgeted with her sousaphone. "I still don't like it, but I guess it's the best option we've got."

"Okay then." He turned to Halle and extended his palm toward the open door. "After you."

Halle crept down the front steps and took her place

behind the banister.

Argyle turned to Paisley. "Ready?"

She lifted her hockey mask from her face and rested it on top of her head. "Ready as I'll ever be."

"Okay. Let's put those sousaphone lessons to good use." He pulled his snorkel mask up against his forehead and got ready to play his heart out.

Paisley set her feet in motion and initiated a steady marching beat. Argyle took his place next to her and marched in time. They descended the steps in unison. Once they reached the ground, they brought the instruments to their lips. In a moment, the air was filled with the joyful um-pah um-pah sounds of Mozart's "Turkish Rondo."

They marched back and forth in front of the steps with the mist swirling around them. Argyle kept his eyes on the banister, waiting for Halle's signal. After a few minutes, she held her hand up to silence them. They stopped marching and cocked their ears.

Footsteps scraped across the gravel somewhere in the distance.

"I wish we could see who was coming," Paisley whispered.

They listened some more until the footsteps stopped.

"They're not close enough yet," Argyle said. "Keep playing."

They brought the sousaphones back to their lips and played a few more measures. Finally, Argyle could make out shadowy figures moving in the mist several yards away.

"Is it time?" Halle whispered.

"Wait until they're a little closer," Argyle replied.

"If we wait much longer, it'll be too late!"

"We need to make sure they're close enough to smell the cigars," he replied.

Four figures shuffled like zombies toward them through the mist, closing the distance between them until they could even see Mr. Belloft's purple bow tie through the fog.

"Okay . . . now!" Argyle whispered as he and Paisley joined Halle on the other side of the banister.

Halle shot up and threw the first cigar into the open door of the house and then ducked back down. Argyle watched the figures in the mist. Now that the sousaphones were silent, the hosts had stopped moving. The tredecapus on Ms. Tuttle shifted, exposing her nose. She inhaled deeply and took another step forward.

"Throw them all or else they'll smell them on us," Argyle said.

Halle stood up again and threw the second cigar. Then the third. As she drew the last cigar back, Paisley caught her arm.

"Wait!" she said. "They're not all here. I don't see Eddie or Brewster."

"Brewster had his seat belt on when we spotted him through the window," Argyle whispered. "He's probably stuck in the car. And who knows where Eddie is. He could be wandering around the hedge maze for all we know. Throw the cigar before they come after us!"

Halle tossed the last cigar. Argyle held his breath as the cigar arced through the air toward the open front door. But the angle was too high. The cigar smacked against the top of the doorframe and bounced back to the landing before slowly rolling down the stone steps. His heart sank lower with every drop until the cigar came to a stop somewhere on the foggy ground.

Instead of staggering up the steps and into the house like they were supposed to, Mr. Belloft, Ms. Tuttle, and the two police officers stopped at the foot of the stairs where the stray cigar had fallen. At once the tredecapuses began to bubble and pulsate.

Argyle met Halle's and Paisley's eyes. "Now what do we do?" His voice was barely audible.

Halle didn't hesitate. "We run."

31

INTO THE MIST

Running was not a problem for Paisley, even with the cumbersome sousaphone draped over one shoulder. Only, usually, she was running to some place. Now she was running away. The hockey mask loosened and bounced against her head with the pounding of her feet until it eventually bounced off, but Paisley kept moving, her fear pushing her faster than she'd ever run before.

By the time she stopped to figure out where she was, there was no sign of Halle or Argyle. She peered through the mist, not daring to make a sound. Who knew how many tredecapuses were scuttling around after finding that stinky cigar? She focused her mind on the rules. Avoid Peril. Protect Yourself. Don't Investigate. But the more she thought about the rules the more anxious she got. She had to keep moving. She had to get somewhere safe. But was any place safe anymore?

The ground sloped downward. That meant she was nearing the west lawn. If she kept heading down, she might get far enough away from the fog machines that she'd be able to see her surroundings and hopefully meet up with the others. She broke into a run again.

32

LEFT BEHIND

Argyle had gotten a slow start when Halle gave the order to run. His legs had fallen asleep while they were crouched behind the banister, and at first all he could manage was a stiff, painful stagger. The fact that he was lugging a twenty-pound sousaphone with him didn't help.

As the numbness in his legs subsided, he was gradually able to pick up speed. His breath quickened with the effort, but he pushed on. He kept close to the house as he fled from the front entrance. He had no idea if the others had done the same, but it was the most sensible choice. There was no point running out into the thick mist and getting lost.

When he was sure he was a safe distance away, he leaned against the wall to catch his breath. He removed the snorkel mask and wiped the sweat off his brow with his pajama sleeve. The cool spring air and the damp mist felt good against his skin, but he couldn't linger. He had to catch up with the others. They needed a new plan. The cigars might be enough to keep the hosts occupied, but any new tredecapuses would be roaming free, attracted to any sound or vibration.

He ditched the sousaphone, pulled the snorkel mask over his face once more, and kept moving.

33

SILENCE

Halle had kept up with Paisley pretty well until she tripped over a root and fell—just like they always do in horror movies. The fall surprised her more than it hurt, and she scrambled up again, but by then Paisley was gone, somewhere off in the mist. Halle spun around in a circle. There was no sign of Argyle either. She listened for their footsteps, but the fog seemed to swallow up all sound except for her own heavy breathing inside the fencing mask. In the eerie silence, she crept forward, concentrating on not tripping again. She lifted the fencing mask off, just for a moment, but her visibility was no better. The mist was the problem, not the mask. She quickly covered her face again.

It was too quiet. In movies, when things were this quiet, it usually meant something was about to pop out from nowhere.

She waited in the silence, her body rigid and tense. Something rustled nearby, and she whipped her head toward the noise. The mist whirled around her, and then the rustling died down. Just the wind blowing some dried leaves across the ground. She listened intently for any other sounds. Surely she'd hear a tredecapus if it were scuttling over the dead leaves. But there was nothing. Not even birds singing.

34

THE SKI LIFT

Paisley, in her haste, nearly crashed into one of the stone pillars that supported the ski lift. She was farther down the hill than she had realized, and the fog was still as thick as anything. How would she ever find Halle and Argyle in this? The mist hadn't looked so bad from the attic windows, but in the middle of it, she was nearly blind. Panic filled her heart as she strained her eyes to see. She hadn't meant to leave the others behind.

If she kept running down the hill, she'd be even farther away from them, yet if she turned back, she'd be heading right toward those creatures again. But she couldn't just stand here and wait. She had to get somewhere safe.

She peered up at the pillar. It was a rectangular column constructed from the same type of stone as the manor itself so that it resembled a giant chimney. Closer to the top, the mist thinned out. She could see the ski lift cables stretching out in both directions high above her, and beyond that the hint of clouds against a blue sky. If she could get to the top, she'd be able to see out over the whole west lawn.

Without wasting another moment, Paisley grabbed hold of two of the stones that were embedded into the pillar just above her head. She gripped them tightly, then slid her sneaker into a sizable crevice and hoisted herself up.

After years of rock climbing on her bedroom wall, Paisley's arms, legs, and fingers were perfectly conditioned for this task. She moved carefully and methodically up the pillar. But it would take some time for her to chart a course to the top. Unlike the mapped-out plastic grips mounted to her wall, which were specifically designed to provide a secure yet challenging climbing experience, these stones were placed haphazardly. Before each move, she had to test out various grips to ensure she could hold on. About a quarter of the way up, she noticed the added weight of the sousaphone. She'd been so eager to get off the ground, she hadn't even considered setting it down. Now, with both hands clinging to the stones, there was no way to take it off. Paisley pressed on, determined to make it to the top.

About halfway up, her foot slipped, and she tightened her grip with her fingers. The sousaphone pressed hard into her shoulder. Adrenaline pumped through her as she swept her sneaker blindly over the stones, trying to locate a better toehold. Her arms trembled with the effort of holding on. Finally, she found a suitable crevice to wedge her foot into. With her weight evenly distributed once more, she closed her eyes and pressed her face against the cool stones. She just needed to rest for a minute. Then she'd keep going.

The soft thud of footsteps caused her eyes to shoot open. She glanced toward the ground.

"Argyle? Is that you?" she whispered.

A shadow moved toward her through the fog.

"Halle?"

The figure drew closer until it was directly under her. Through the mist, she could just make out the top of a tentacle-covered bald head. Paisley's heart sank. She knew that head. It was Eddie, the groundskeeper. But it was the unmistakable sound of squelching and bubbling from the tredecapus that gave her the burst of energy she needed to continue scrambling toward the top of the pillar.

The sousaphone dug into her shoulder, and her sneakers slipped several more times in her haste. For the first time that day, she desperately wished she was not wearing shoes. When she climbed the wall in her room, it was much easier to do it barefoot. She could slide her toes into tiny nooks and crannies. Mrs. Danworth didn't know she did this, of course, and Paisley didn't see any harm as long as she wore shoes while on the ground.

She hugged the pillar as hard as she could and raked the heel of her sneaker against the rough stones. After a few tries, the sneaker slid off and fell to the ground. Straining with the effort to hold on, she lifted her free foot higher, feeling along the surface with her bare toes until she found the perfect crevice to slide her foot into. She pushed her body higher and then scraped her other shoe off so that both feet were free. Her forearms burned. Just a little bit more, and she'd be able to reach the top. If only she could get rid of the sousaphone like she'd gotten rid of her shoes. But it encircled her body completely, and she was in such a precarious position that, if she attempted to take it off, she'd surely lose her grip and fall. There was no way around it. The sousaphone was coming with her.

A sound like a water balloon bursting made Paisley look down again. A slick, shiny new tredecapus had emerged and was slithering over the top of Eddie's bald head. How high could it jump? Paisley grunted as she scaled the last few feet and finally got her hands on top of the column. She gripped the ski lift cables, pulled herself up, and slumped over the top of the pillar. Then she drew her legs up and twisted herself around to perch at the top. She peered down just in time to see the tredecapus spring through the air and wrap its tentacles around one corner of the pillar. It slid its fleshy tendrils between the stones like the ivy vines that snaked up the manor walls. Paisley's heart pounded as the creature slithered upward, slowly reaching higher. But as its slimy secretions oozed from its pores, it began to slip back down again. It was losing its grip. It wriggled and stretched, slapping its arms against the pillar, making it even more slippery. In mere moments, it could hold on no longer and flopped to the ground.

Paisley let out a great sigh and extricated herself from the sousaphone at last. She was stuck twenty feet in the air, but she was safe.

35

THE ROLLS-ROYCE

Argyle glimpsed the northwest corner of the house several feet in front of him and stopped. If he rounded the corner, he'd come to the garage, where he could turn off the fog machines and shut himself safely inside. But between Argyle and the garage was the Rolls-Royce with its door wide open and Brewster and the tredecapus still inside. Brewster couldn't get out—he was trapped in his seat belt—but Argyle had no doubt the tredecapus would replicate if he got too close.

A cool breeze blew over him. He shivered, wishing they had stayed inside the warm house. From what he had seen in the humidor, cold temperatures and moist air were the perfect conditions for tredecapus replication. The garage was his best bet. But he'd have to make a run for it.

He shook out his arms and legs and took a big breath in and out. No time to waste. He bolted forward and sped around the corner, straight toward the garage door. He grabbed hold of the handle at the bottom and yanked at it, only to find that it wouldn't budge. He ran to the other door and pulled at that too, but it was no use. Also locked. Panic flooded his body, but he pushed it away. His eyes darted around. There was another door on the side of the garage. A people door, not a door for cars. He sprinted over and twisted the knob, but it didn't give.

Argyle deflated. There was only one other way to get into the garage. He didn't want to look, but he had no choice. He turned and faced the Rolls-Royce. Over the rearview mirror, just above Brewster's head, above the slimy, pulsating tentacles of the tredecapus, was the remote-control garage door opener. Brewster used to let him play with it when he was younger. It was his last hope.

Argyle sprinted around the car to the passenger side door and flung it open. Brewster turned his head toward him. Argyle froze, paralyzed by the sight of the oozing, bubbling creature. But there was no time to lose. He scuttled all the way into the car, reached over Brewster's head, and jabbed the green button on the remote control just as the creature's back split open with a gush of pinkish liquid. Argyle reeled backward, falling out of the car and onto the ground as the garage door began to rise with a grinding, motorized hum. The newly born tredecapus wriggled free of its parent and flopped down onto the pristine leather seats of the car. Argyle kicked the passenger door shut with a loud clunk and scrambled to his feet. If he could close the driver's side door in time, he'd be safe. He dashed around to the other side of the car and threw himself against the door, slamming it shut.

He'd done it. It was easy. He leaned against the car, panting with relief.

But something wasn't right. He could still clearly hear the slurping, slapping sounds of tentacles against leather—not muffled as it would be in a closed car. Argyle recoiled in horror. The driver's side window was wide open. The tredecapus slithered over Brewster toward the gaping hole. Toward Argyle.

Argyle sprinted into the garage and pounded the

Door Close button inside. The garage door, which was nearly to the top, ground to a halt, then paused for an excruciatingly long moment before it slowly reversed back the other way. He jabbed the button over and over again, desperately trying to speed the door's descent. The garage grew darker and darker as he stared at the shrinking gap. He abandoned the button and felt along the wall, frantically searching for a light switch. His fingers brushed cobwebs and rusty nail heads but no switch.

A shadow flickered under the door as the last of the daylight disappeared, and then the door thunked against the ground. Argyle stood in the darkness, not moving, not making a sound. All was silent.

Had he just imagined it, or was he trapped in the garage with a tredecapus?

36

THE GATE

Halle crept blindly through the mist. She needed to get to the gate and out to the road. That had been the plan. Although she wasn't sure they were even still following the plan since she had totally messed up the whole cigar-throwing part. Their heroic idea of trapping the creatures inside in order to save the town hadn't taken into account her total lack of coordination.

But who was she kidding? She wasn't a hero—she was a Damsel in Distress, just like the Scream Queens in the posters all over her room. The only way she'd make it would be to do what Aunt Rhona would do: get away from here as fast as possible. Then at least she could warn people. She only hoped that Paisley and Argyle remembered what she had taught them and were getting out of there too.

The mist hid everything. She swiveled her head in all directions, staring through the metal mesh of the fencing mask, hoping to see the gate or the stone wall or any kind of familiar landmark. She wasn't even sure she was heading the right way. And she was alone. According to every horror movie she had ever seen, something bad always happened to the Damsel in Distress any time she was alone.

Her heart pounded against her chest as she slunk through the mist, trying not to make a sound. Dark shadows of gnarled trees and brambly bushes appeared with every step she took. She stumbled over sticks and stones, nearly falling a few times. And then her feet hit gravel. The driveway! The gate had to be somewhere up ahead. She just needed to stay on the gravel. She quickened her pace.

As she surged ahead, a dark figure on her left made her jump. But it was just one of the creepy scarecrows that lined the driveway. She veered toward the center of the path and continued on.

Finally, through the mist ahead, she could just make it out. Past the last large scarecrow was the hazy silhouette of the iron gate, wide open to the road.

Halle raced toward the opening, ready to run the six miles back to town. The stone gargoyles flanking the gate seemed to sneer down at her as she approached them.

Goodbye and good riddance, she thought.

But before she stepped through the opening, she stopped in her tracks and gasped. Slithering over one of the gargoyle's shoulders was a glistening tredecapus. She could almost hear Aunt Rhona yelling, "Get out of there!" But if she left, then what would happen? This creature—and who knows how many more—would follow her back into town. She'd put everyone in danger. Aunt Rhona, no doubt, would be able to save herself, but what about everyone else? What about her mom? What about Lenny the delivery driver and all the kids from school? Even Georgia and Dakota, whose faces she kind of wanted to see plastered with a tredecapus, if she was being honest. They would all be helpless. None of them were prepared for Dire Circumstances. Halle had to keep the creatures and Mr. Belloft, the police officers, and all the other hosts inside the grounds of Price Manor—and keep everyone else out!

The first thing she needed to do was close the gate. She adjusted her fencing mask and took a cautious step forward, keeping her eyes on the tredecapus perched on the gargoyle's shoulder. The breeze rustled through the trees, hopefully masking any sounds her footsteps made. When she reached the gate, she grabbed hold of the iron bars and pushed.

A loud groan from the gate's hinges erupted into the air. Halle recoiled as the tredecapus slithered up the gargoyle's neck in response to the sound. She needed to run—now! But the gate still wasn't closed all the way. Paralyzed with fear, Halle stared at the tredecapus. It seemed to be waiting for her to make another move. Another sound.

On the other side of the gate, a squirrel scampered through the mist. The tredecapus shifted its attention toward the unsuspecting animal. No, no, no! If the tredecapus got out, all would be lost. Halle swallowed as she realized what she needed to do. Sometimes, in the movies, the Scream Queen had to be the bait.

She leaped forward and threw her weight against the

iron bars. The gate gave a final groan and clanged shut. In the next instant, the tredecapus sprang through the air. Halle ducked. The creature sailed over her head and landed at the base of the gate. The squirrel on the other side rustled innocently through the leaves, and the tredecapus immediately rose onto its tentacles, preparing to pounce through the iron bars.

"Oh no you don't!" Halle said as she stomped loudly on the gravel driveway.

The tredecapus refocused its attention back on Halle. It sprang toward her as she backed away. Bit by bit, with soft, crunching footsteps, she led it away from the gate, up the driveway toward the manor.

37

THE SCHEDULE OF THE DAY

At the base of the ski lift pillar, the tredecapus attempted to scale the stone surface for a third time. It clung to one corner of the column and creeped its way up a few feet before losing its grip again and sliding back down in a puddle of slime. Each time it made a new attempt, Paisley's whole body filled with dread. There was nothing she could do but wring her hands and pray that the slime would stay slick enough to keep her safe at the top of the ski lift. From her perch, she peered around the grounds. The fog was definitely thicker in some areas than in others, probably because it came from multiple machines spread out along the walls of the estate, but she couldn't see any signs of life aside from Eddie, who remained standing at the foot of the pillar, swaying slightly. Where were Argyle and Halle?

The creature on Eddie's head showed no indication that it would replicate again. No bubbling. No bulging. According to Argyle's theory, the tredecapuses relied on sound, and she wasn't making a peep, but the hosts could find more victims through smell. She sniffed her armpits. She had definitely built up a sweat with all that running and climbing, but perhaps she was too far out of range for Eddie to catch her scent.

The more Paisley peered through the mist, looking for any sign of movement, the more she began to fidget. She needed to let off her nervous energy somehow, but there was certainly no room for sit-ups or push-ups or jumping jacks at the top of the pillar. She had no idea how long she would be up there. With no other outlet for her energy, she brought her knees up and hugged her legs tightly against her chest, squeezing with all her might.

Inside the pocket of her pajamas, her notepad shifted. She took it out. Perhaps there was something in her notes that could help her. Of course, there was nothing about what to do if you were trapped at the top of a ski lift pillar. She flipped through the pages, desperate for some clear directive. She needed instructions and directions and schedules. With them, she had order and focus and calm. And completing her tasks gave her a sense of accomplishment and purpose. Without them, she was lost in a maze of chaos and uncertainty. She turned to her makeshift Schedule of the Day and read through the list. A sad smile spread across her face. They had managed to hit almost everything on the list.

Sousaphone practice—*check.*

Calligraphy—*check.* (She had written most of the list in her fancy, loopy script.)

Make sandwiches—*check.*

Tapestries—*check.*

Roller-skating—*check.*

She had even managed to get in some rock climbing.

There were just two things left on the list. "Find Dad" and "Save the World."

38

THE GARAGE

As Argyle's eyes adjusted to the dark, he scanned the cluttered garage. Along with a vintage red Plymouth Fury and a junky yellow Oldsmobile, there were shelves lining the walls full of tools, old paint cans, and other odds and ends. In front of the shelves and along the other walls were cardboard boxes, wooden crates, garden implements, broken lawn chairs, a riding mower, and many, many other dark shapes he couldn't quite identify. Plenty of places where something could hide. He felt along the wall until he found what he needed. A control panel. He flicked the first three switches, and the garage lit up with fluorescent light. As he squinted in the brightness, he thought he saw something scuttle across the floor. He turned toward the movement, but nothing was there. Had it darted under the Plymouth? He waited, his eyes fixed on the floor in front of the car. Nothing moved. Nothing made a sound.

Convincing himself it was just his imagination, Argyle turned back to the control panel and searched the rows of switches. On the bottom, he found the section labeled "Fog Machines" and flipped them all to their off position. It would take a while for the fog to dissipate, but at least it wouldn't get any thicker.

A wet, slurping noise forced Argyle to spin around and face the Plymouth once more. A dark splotch stained the floor directly in front of the car. Had that stain been there before? Maybe an old oil stain? Or was it fresh tre-decapus slime?

As he stared at the spot, a gooey gray tentacle slid out from under the car. Argyle's heart dropped. He hadn't imagined it after all. He scanned the nearby shelves, hop-ing to see an empty box or something he could use to catch the thing, but there was nothing within reach. Two more tentacles emerged from under the car. Now he had to be very careful. The blind creature couldn't see him, but any sound he made might alert it to his location. How long could he stay quietly in one place? He gulped and shifted the snorkel mask back down over his face.

Desperately Argyle scanned the garage once more until he spotted the huge vents on the ceiling. He turned back to the control panel. There was a dial labeled "Blower." These creatures didn't like heat, but fortunately, his dad did. He'd installed an industrial-strength, forced-air heater in the garage so he never had to get into a cold car in the wintertime. Argyle turned the dial as far as it would go. Hot air blasted from the ceiling. The tredecapus immediately retracted its tentacles back under the car to escape the sudden onslaught of heat.

Smiling, Argyle planted himself directly under one of the blowers. The hot air swished against his back, rippling his pajamas. There was no way that thing would come anywhere near him now. He pulled over an old lawn chair and set it up directly under the vent. It was the kind of chair he could stretch out on. He plopped himself down, crossed one leg over the other, and laced his fingers behind his head as he reclined back into the seat. The heat blowing over him was like a warm summer breeze. If only he had a book, a sandwich, and a tall glass of icy lemonade, it would be almost like lounging on the patio on a hot day. His mouth watered and his stomach growled at the thought of it.

He kept his eyes glued to the spot under the car, though, where the dark lump of the tredecapus sat undulating in the shadows.

39

HALLE IS NOT A DAMSEL IN DISTRESS

Halle's heart thudded as she lured the lone tredecapus *backward* up the gravel driveway.

Away from the gate, away from the road, away from safety.

The creature followed, now jumping from scarecrow to scarecrow, while Halle kept a safe distance between them. She was breaking just about every rule she could think of, but she couldn't risk letting that thing get out.

The mist was thinning—thank goodness—and in addition to keeping tabs on the advancing creature, she could turn and see the manor looming in the distance through the haze. She adjusted her fencing mask—the only thing giving her the courage to keep going. But she needed a plan. The closer she got to the house, the more creatures there would be on account of that smelly cigar. Would she be able to lure them all inside the house on her own?

Aunt Rhona would lose her mind if she knew what Halle was doing. Usually, when they went to the movies, Aunt Rhona would mutter things under her breath like, "She'd better not go into those dark woods," or "She did not just go and check out that pile of human bones!" Halle would giggle and nod in agreement. Those people in the movies always got themselves into trouble.

Except Raven Rose. Even when Raven Rose marched into the darkest, densest woods to meet the biggest, baddest demon, she always came out alive. The rules didn't apply to her. She was the one who always saved the day.

Halle had no idea where Paisley and Argyle were, but what if they needed saving? Maybe she was the only person who could help them. Maybe she had to stop thinking like Aunt Rhona and start thinking like Raven Rose.

She searched through the haze for any sign of the twins. The breeze had died down, and everything was still, including the trees and scarecrows, which rose like bony fingers from the ground. But the stillness didn't mean that danger wasn't lurking nearby, ready to pop out at any moment. She softened her footsteps just a little.

As she continued toward the house, the clouds parted overhead, allowing the glaring sun to fully illuminate the grounds. Suddenly, her blood ran cold. There, in the shadow of Price Manor, the ground was crawling. There were maybe forty—no, fifty—tredecapuses scurrying away from the small group of hosts gathered around that lone cigar. Even out here in the open air, the smell of that stinky cigar was causing them to replicate. But the creatures, once born, had no interest in the cigar. They roamed free across the grounds in search of more hosts.

As Halle took her next step, a twig snapped under her foot. Every tredecapus as far as she could see stopped in unison and changed course.

They were coming for her.

At the same time, the one pursuing her from behind leaped to the top of the nearest scarecrow not five feet away. With no real plan, Halle lurched off the gravel driveway and ducked into the shelter of the trees. Into the woods.

"I am not a Damsel in Distress," she whispered to herself as her limbs turned to jelly. She wove this way and that through the wooded grounds, adjusting her mask and swiveling her head in every direction. Something flashed across her path. She darted behind a cluster of trees, trying not to make a sound. Dried leaves rustled all around her as the breeze picked up again. She peered through the wire mesh mask at the undulating ground. The creatures were everywhere, coming from all directions.

Halle trembled and her breath came in short gasps. "I am not a Damsel in Distress," she said again.

She reached for a branch to steady herself, but her hand landed on something cold and wet. Horrified, she pulled away just as a tentacle coiled around her arm. Several tredecapuses clung to the lower branches of the tree and slithered toward her. Shaking her arm frantically, she flung the creature away just as another one dropped onto her shoulder. Halle tried to pluck the slimy creature off, but its slick body slipped right through her fingers. Another one leaped onto her back, sliding one gooey tentacle under the rim of her mask. She thrashed and flailed, but the creatures snaked their tentacles into her clothes and around her limbs. One leaped from the ground and landed squarely on the front of the mask. She staggered back.

Through the wire mesh, Halle could see nothing but the grayish-pink flesh of the creature's belly, covered with tiny undulating suction cups that oozed slime. Blindly, she threw her hands out in front of her and staggered forward as mucus seeped through the mask and dripped onto her neck. More creatures wrapped themselves around her legs as she stumbled across the ground. She couldn't see a thing. She had to get the mask off, but the mask was the only thing protecting her.

She blundered forward, desperate to keep moving, but her foot caught on a twisted root, and down she went. Again. As her knees thudded against the ground and her hands scraped across loose twigs and stones, the mask flew off and rolled out of reach. In a panic, she buried her head in her arms and pressed her face to the ground. Dozens of slimy bodies pounced onto her back. She tightened her arms around her head, hardly daring to breathe. The creatures wriggled across her body and weaved their tentacles into her hair. She had to get them off.

Shielding her face with her hands, she rolled sideways,

turning over again and again. The weight of her body squashed the beasts, causing them to drop away.

Then she hit a tree.

Something sharp jabbed her back. Thinking it was a twig, she reached one hand behind her to shove it away while still covering her face with her other hand. But instead of a twig she felt something smooth, hard, and cold. She took hold of it and peeked between her fingers. It was the dagger she'd tucked into her belt hours ago. Raven Rose's dagger. Halle's eyes gleamed as she raised it in front of her. The rubies sparkled in the sun.

"I am not a Damsel in Distress!" she shouted.

As if to emphasize her point, the low notes of Mozart's "Turkish Rondo" began to echo across the grounds. The few tredecapuses that still clung to her legs suddenly dropped away. Halle scrambled up. Every single tredecapus—the ones that had been clinging to her, the ones she had squashed while rolling, and all the others that had been following after her—scuttled off in the direction of the music. Except for one. The one that was still splayed across the face of the fencing mask. It disengaged itself from the mask and leaped straight at Halle, determined to finish what it had started. But she was ready for it. She ducked out of the way and the tredecapus missed, landing behind her at the foot of a scarecrow. She turned to face the creature, brought the dagger high in the air, and then thrust it down into the fleshy blob.

"Take that, vile fiend!"

But the dagger did nothing. Upon contact, its blade retracted into its hilt, which pressed ineffectively into the creature's squashy body like a fist punching a pillow. The dagger was just a movie prop.

The creature quivered on the ground by her knees, stunned but recovering. Halle flung the dagger aside and collapsed in on herself. She buried her head in her hands and squeezed her eyes shut. The slurping, slapping sound of tentacles drew nearer, and she cringed, waiting for the attack.

Halle waited . . . five seconds . . . ten . . . but nothing happened. The sounds had stopped. There were no slimy tentacles coiling around her body. At last, she mustered up the courage to peek through her fingers. There, not three feet away, the tredecapus lay squashed under the foot of the scarecrow. Halle stared, unable to understand what she was seeing.

Then the scarecrow took a step.

Her eyes traveled slowly up. It wasn't a scarecrow at all. It was a man in a dark blue uniform, soaked with slime. A large bag with a strap hung over his shoulder, spilling envelopes and magazines all over the driveway. It was the mailman! And on his face was another bubbling, pulsating tredecapus just beginning to split open.

Halle screamed.

She screamed a scream ten times more intense than the one she had produced in front of her bedroom mirror that morning.

40

"TURKISH RONDO"

Paisley blew the last few notes of the "Turkish Rondo" then looked down from her perch as she rested her weary lips. The crowd of tredecapuses gathering at the base of the column was much, much bigger than she'd imagined it would be. And still more were creeping their way toward her through the thinning mist. She shuddered. But if she could get them all here, and *keep* them all here, they wouldn't wander past the walls of the estate. They wouldn't be able to replicate. And as long as she stayed on top of the ski lift column, she was safe. But she had to keep playing. She'd play all afternoon and all night if she needed to. She brought the sousaphone to her tired lips again.

41

THE SCREAM QUEEN

A funny thing happened after Halle screamed. The creature clinging to the mailman's face mere feet away from her trembled, drew in its tentacles, and dropped to the ground, quivering.

Halle gaped at the fallen creature. It convulsed once more and then withered into a lifeless pile of skin and slime and tentacles on the ground. Had she done that with her scream?

The mailman, now free of his parasitic pest, doubled over, sputtering and coughing and wiping slime from his face.

Halle jumped up. “Are you okay?” she asked, patting him on the back.

He straightened and looked around in a daze. “Wh-what happened?”

“I think I just saved you.”

“Where am I?”

“You’re at Price Manor.”

He blinked. “Price what? Where’s my hat? And why am I all wet?”

“I think you’d better sit down.” Halle took him by the elbow and led him over to a bench. The fog was now only a thin mist, and she could see most of the grounds in front of the manor. The tredecapuses and the hosts were nowhere in sight. They’d been drawn away by the sousaphone music. But whoever was playing it would need help. And Halle might just be the only one who could do it.

She looked at the mailman again. He was a little confused, but he seemed okay physically. “Stay here and rest. I’ll be back as soon as I can.”

“Who are you?” he asked.

Halle smiled and shrugged. “I’m just the flower delivery girl.”

42

ARGYLE'S DREAMS

Argyle opened his eyes with a start. He'd been dreaming about a sandwich, but when he had bitten into it, gelatinous slime dripped out and oozed down the front of him. His snorkel mask was completely fogged up, so he yanked it off and dropped it to the floor. He looked down at his pajamas. They were plastered to his body with sweat as the blower continued to blast hot air onto him. Had he passed out?

His eyes quickly darted to the spot under the car. With relief, he saw that the blob was still there, its tentacles splayed apart as if it were wilting into the floor from the heat of the garage. There was still the hint of life in its gentle undulations, but it had definitely lost all its pep.

Argyle tried to swallow, but his throat felt like sandpaper. The garage was like an oven now. He had to turn off the blower. As he sat up from the reclined lawn chair, all the blood in his body rushed to his head and dizziness washed over him. He propped himself up, but his arms trembled with weakness. How long had it been since he'd eaten? Or had anything to drink? He lay back and closed his eyes again. He'd just have to stay here until help arrived.

As he slipped back into semiconsciousness, visions swirled in his mind: the smiling faces of Mrs. Danworth holding a pitcher of ice water, Mrs. Paprikash offering a tray piled high with sandwiches, and Mrs. Pennywhisker handing him a book about aquatic creatures of the Arctic Ocean. Argyle reached for the sandwiches, but they turned into cigars as the women's smiles faded and giant tentacles rose out of the water pitcher.

His eyes shot open again. What was he doing? He couldn't just wait for someone to come help him. There was no one left. Mrs. Pennywhisker and Mrs. Paprikash and everyone else were the ones who needed help, not him. And he had no idea what had happened to Paisley and Halle. They could have been tredecapized for all he knew.

Argyle drew himself up with great effort. He stumbled over to the garage door and leaned on the button to raise the door panel. Cool air wafted in as the garage door lifted, and without a plan, without any protection, he ducked under the door and out to the refreshing outdoors.

The first thing he noticed was Mozart's "Turkish Rondo." It had to be Paisley, but the music was much slower than her usual peppy tempo. How long had she been playing? The next thing he saw was the sluggish tredecapus dragging itself past him from the open garage. It bypassed him completely and headed in the direction of the music. He had to get to Paisley.

On shaky legs, he rushed ahead of the tredecapus, following the sound of the sousaphone down the sloping lawn. There, he stopped short. The mist had lifted, and scores of tredecapuses, along with Mr. Belloft, Ms. Tuttle, Eddie, the two police officers, and even Mrs. Pennywhisker, were all gathered around one of the columns supporting the ski lift. At the top sat Paisley, puffing away on her sousaphone.

Footsteps thudded behind him, and he reeled around, only to find Halle running toward him from the driveway. Her eyes were bright, and a wide grin spread across her face.

"I think I know how to stop them!" she yelled. "Come on!"

43

PAISLEY CLIMBS DOWN

Paisley finished Mozart's "Turkish Rondo" for the tenth or eleventh time—she wasn't quite sure now. Her lips were numb, and she was growing more and more light-headed. A crowd of hosts and creatures gathered below her. Even Mrs. Pennywhisker had joined them. She must have been drawn by the music out through the open front door of the house. Paisley scanned the tops of their heads and their clothing, hoping to see her father among them, but he was not there.

While looking down, she felt herself sway and quickly grabbed on to the edge of the pillar for support. This was a lot higher than the loft in her bedroom, and there were no springy mats underneath to break her fall. Not that she would fall. She just needed to rest for a minute. She was so tired.

No longer drawn by the music, the creatures began to turn away. Paisley couldn't rest. She had to keep them here, away from everyone else. She had to save the world.

With great effort, she brought the sousaphone to her lips again.

But before she blew her first note, the quiet of the afternoon erupted with a shrill scream that reverberated across the entire grounds. Paisley clapped her hands over her ears as the sound crescendoed, held in the air, and then faded out. The ground below writhed as dozens of tredecapuses squirmed and trembled. Then the sound came again—another scream, equally loud and long—sending the tredecapuses into frenzied convulsions.

The human hosts dropped to their knees, holding their ears, as the creatures slid from their faces and quivered on the ground. Paisley searched for the source of the sound. Her eyes traveled up the sloping lawn until they rested on Halle standing tall, her head raised and her fists balled, holding the scream as long as she could. Argyle danced around excitedly behind her.

The scream died out, and almost as if a wave passed

through them, the creatures became still. Mrs. Penny-whisker and the other hosts collapsed to the ground, coughing and rubbing slime from their faces. Dozens of tredecapuses lay lifeless on the ground around them.

Paisley wasted no time. She scrambled down the stone column, lugging her heavy sousaphone along with her. "Mrs. Pennywhisker! Mr. Belloft! Are you all right?"

One by one, the adults sat up, blinking as if waking from a deep sleep. Officer Munrap tried to stand but wobbled and fell back to the ground. The others began muttering words of confusion, such as, "What am I doing on the ground?" and "Why am I covered in slime?" and "Where's my knitting?"

Paisley set her sousaphone aside and checked on each of them while Argyle and Halle ran toward her, leaping over gelatinous puddles and shriveled tredecapus skins.

"That was amazing!" Paisley said when Halle appeared by her side. "How did you figure out what to do?"

"Luck, I guess. I came across the mailman—he had one of these things on him too—and I just . . . screamed."

"That was no ordinary scream," said Argyle.

Halle smiled. "Well, I have been working on that a little bit at home."

Paisley inspected one of the fallen creatures. "So . . .

the scream killed them?"

"They must be extremely sensitive to noise or sound vibrations," Argyle said. "They can't see at all, so their auditory system is probably highly developed. That's why they were drawn to sounds all day."

Paisley picked up a stick and poked at one of the dead creatures, turning it over. "I don't see any ears on this thing."

"They probably have internal ear organs. Like fish. And Halle's scream was so intense, it must have caused irreparable damage."

"But I heard Mrs. Pennywhisker scream this morning. Why didn't that stop them?" Paisley asked.

"Her scream is nothing like Halle's." Argyle turned to his governess. "No offense, Mrs. Pennywhisker."

Mrs. Pennywhisker coughed. "No worries, dear," she replied, patting him lovingly on the cheek with a slime-covered hand.

The adults started coming to their senses, though they remained on the ground, too weary to stand on their own.

Paisley took hold of a not-so-slimy part of Mrs. Pennywhisker's arm. "We've got to get back inside," she said. "Mrs. Danworth and Mrs. Paprikash are still in the humidor. And we need to find Dad." She helped the governess to her feet. "Where did you say you saw him?"

"Who?"

Paisley gave Argyle a worried look. Mrs. Pennywhisker was always a bit forgetful, but what if the tredecapus had damaged her brain? Paisley tried again, speaking in a loud, clear voice. "We need to find our dad. Do you remember where you saw him?"

"Oh . . . I haven't seen him for quite a while. He's been out of town."

Paisley bit her lower lip. "Earlier you said you saw him. He had one of these things covering his face. I heard you scream."

The old woman furrowed her brow. "I'm not sure."

"Come on," said Argyle. "Let's get everyone back inside the house. They probably just need something to eat."

Paisley helped the others to their feet, hoping Argyle was right, but her insides were twisted with worry, and this was not the time for sit-ups, push-ups, or jumping jacks.

44

MAXIMUS PRICE'S JOURNEY

Argyle passed steaming mugs of hot cocoa around to everyone in the drawing room.

"My goodness, Argyle! Where did you learn to do this?" Mrs. Paprikash asked as she took a sip.

"It was easy," he replied modestly. "The directions were right on the package." He winked at Halle.

After they had extracted Brewster from the Rolls-Royce, they'd circled back around to pick up the mailman, then rescued Mrs. Paprikash and Mrs. Danworth from the humidor with another one of Halle's incredible screams. Paisley and Argyle then led everyone, including the two bewildered police officers, into the drawing room, where the twins doted on them, fluffing their cushions and propping their feet up on ottomans.

The adults were starting to remember bits and pieces of their morning, much to Argyle's relief, but no one could remember anything after they'd been attacked by the tredecapuses. Aside from some temporary memory loss, they all seemed to be fine.

"Okay," Argyle said, "we need to find our dad. Does anyone remember where they saw him last?"

Brewster spoke up first. "I drove him home from the airport this morning. His flight got in very early. Along with all his luggage he had an insulated box that he said he wanted to keep on his lap so it wouldn't get knocked around in the trunk. I didn't ask him what it was. When we got home, I remember pulling up to the garage. Mr. Price had been unusually quiet, so I figured he was tired from the long flight and had fallen asleep. I turned around to wake him and that thing was on his face. The box was lying open, and there was ice melting all over the seats. I don't remember anything after that."

"That's when you must have been attacked," said Argyle. "How did Dad get out of the car, though?"

"That was me," said Eddie. "I was pruning some bushes and saw the car just sittin' there with no one gettin' out of it. When I went over and saw those creatures on their faces, I just about fell on the ground laughing. Thought they were trying to pull one over on me. I opened their doors. Brewster was playin' dead—at least, that's what I thought—but Mr. Price was stirrin'. I helped him out of the car. He was havin' trouble seein', ya know? I told him he'd give everyone a right good scare with that thing. That's when the other little bugger jumped out at me. I hadn't seen that comin'. Sent me runnin' off around the other side of the house." Eddie scratched his head. "I guess it caught up at some point and got me."

"I went out to help with the luggage," Mr. Belloft said. "That's when I saw Mr. Price. He had come around to the front of the house. That must be where it got me."

"Well, you left the front door wide open," Mrs. Danworth scolded. "I was doing my dusting. The pollen's starting up, you know. When I saw the open door, I had a sneezing fit. Mr. Price was right out there on the path. I didn't want to let any more pollen in, so I hurried him inside and shut the door. I suppose I was so busy sneezing that I didn't even properly look at him.

That's the last thing I remember."

"He must have made his way to the kitchen after that," said Mrs. Paprikash. "I was getting ready to make breakfast when he came in."

Argyle turned to Mrs. Pennywhisker. "What about you? Where did you see him?"

Mrs. Pennywhisker furrowed her brow. "I'm still having trouble remembering."

"Do you remember seeing us in the servery?" he asked.

"I remember . . . knitting. . . ."

"That's right," Argyle said. "You were hiding in the pantry. You said you weren't coming out until Dad had gotten rid of the creature. Do you remember where Dad was?"

Mrs. Pennywhisker rubbed her temples. "I'm sorry. It's all still a bit blurry."

Argyle paced around the room. The gears in his brain turned rapidly. "This afternoon, everyone was drawn to the ski lift by the sound of Paisley playing her sousaphone. Even Mrs. Pennywhisker, who had been inside the house. But Dad wasn't there." He looked around at the others. "Neither were Mrs. Danworth and Mrs.

Paprikash, because they were trapped in the humidor. And Brewster was stuck in the car in his seat belt. So that means Dad must be stuck or trapped somewhere."

"Good point," Paisley said. "But where? We've been all over the first floor. Could he have gone upstairs?"

"Not likely," Mrs. Danworth said. "He had a lot of trouble with the entrance steps. I had to practically pull him up and into the house. I didn't think much about it at the time because I was sneezing so much."

"So he's got to be trapped somewhere."

"Or else he's hurt and can't move!" Paisley wrung her hands. "Mrs. Pennywhisker, please try to remember where you saw him."

"Hang on," Argyle said. "The tredecapus follows sounds, but remember when Mrs. Danworth and Mrs. Paprikash followed us into the card-playing room? They weren't following our sounds; they were following our scent. And then the smell of the cigars was so overwhelming, they went straight for the humidor instead of us. Are there cigars anywhere else in the house? Maybe that's where he is."

"Absolutely not," said Mrs. Danworth. "Cigars are allowed only in the card-playing room. There are ventilation fans built into the ceiling there expressly for that purpose."

"Then where else could he be?" Paisley cried.

"There must be some other place," said Argyle. "Somewhere that's either really noisy or really stinky."

Mrs. Pennywhisker gasped. "I remember everything!" She put her hand over her mouth.

"Well, tell us!" Paisley cried.

"I was going to take the cake carrier down to the cheese cellar to get the twins' annual smelly cheese birthday cake—though why we keep up that tradition I have no idea."

"Mr. Price likes the smelly cheese birthday cake," Mrs. Paprikash replied.

"Well, no one else does. It was funny the first time but every year?"

"Tell us what happened!" Argyle pleaded.

"Right. Well, I was going to take the cake carrier down, but just when I was opening the cellar door, Mr. Price came up behind me. When I turned around, he was wearing that thing on his face and I screamed. Then I realized he was just trying to scare me—that's what I figured, anyway—so I yelled at him and pushed him through the cellar door. I shut the door and told him not to come out until he'd gotten rid of that thing. Of course, I knew he wouldn't listen, so I got my knitting and went and hid in the pantry. Heavens, he must still be in the cheese cellar. I hope I didn't hurt him!"

"The cheese cellar's not very big," Mrs. Paprikash said. "It's just a little room under the kitchen that's chilled. I'm more concerned about him getting too cold."

"Oh no." Argyle turned to Paisley and Halle with his eyes wide. "The tredecapus prefers the cold. Remember when I was playing with the controls of the humidor? And it replicates like crazy when there's something smelly around."

Paisley jumped up. "He's been in there for hours! He could be suffocating in tredecapuses by now! We've got to save him!"

"We don't even know where the cheese cellar is," Argyle said.

"I'll take you there straightaway," Mrs. Paprikash said, hurrying to the doorway.

45

RAVEN ROSE

Halle got that old feeling in her stomach again—the one she always got when she thought about meeting Maximus Price—the feeling like worms were wriggling around in there. But instead of having a party, these worms were rioting. Her stomach twisted and churned at the thought of opening the cheese cellar and unleashing a horde of slimy creatures. The thing was, she knew she couldn't help this time. Her throat had turned raw from all that screaming. She wasn't sure she could even speak.

Instead of leaping into action like the twins, she was rooted to the floor. There was nothing she could do.

"Let's go," Argyle said, his eyes bright with determination.

Halle swallowed. "We need . . . helmets . . . or masks." Her voice was nothing more than a whisper.

"There's no time," Paisley said. "Our dad's been trapped in there all day. All you have to do is scream."

Her legs trembled as if filled with tredecapus slime. Everyone was counting on her, but she was about to let them all down. She really was nothing but a Damsel in Distress. Just a victim like those screaming horror movie Scream Queens—and she couldn't even do that anymore.

Argyle and Paisley doubled back and grabbed her hands. Before she knew it, they were pulling her through the corridors, past towering suits of armor, beastly claw-footed furniture, and demonic marble busts.

As they weaved through the west wing, her head filled with warnings:

Peril.

Dire Circumstances.

Run.

Hide.

They'd open the cellar door, and she wouldn't be able to scream. The creatures would get out, and the whole thing would start all over again. And this time, no one would be able to stop it.

"Come on!" Argyle urged her. "There's no time to lose."

"I . . . can't . . . ," she rasped as she pulled back against their grips. She came to a stop. "I can't do it."

"What are you talking about?" Argyle said. "You're the only one who can. No one else can scream like that. Not even Mrs. Pennywhisker."

They had reached the entrance hall. Halle stood by the front door. Her bicycle was just outside on the path. She couldn't save Maximus Price. She couldn't save any of them. But she could ride back to town—back to her mom and Aunt Rhona—and they could lock themselves in their apartment above the flower shop and wait until everything was safe again.

Paisley and Argyle looked at her, the brightness fading from their eyes.

"What's wrong?" Paisley asked.

Halle pointed to her throat. "I can't do it. My voice is gone. I'm useless. You guys have to save your dad. You've done everything else today. The humidor. The tapestries. Argyle, you figured out how these creatures work. And

Paisley, you rounded them all up with your sousaphone playing. I know you guys can figure out another way to help him. I can't do it."

"But we couldn't have done any of that without you," Paisley said. "We didn't even believe the creature was real until you came along. We would have been goners!"

"That's right," said Argyle. "Plus, you taught us so much: how to protect ourselves, how to avoid danger, even how to make sandwiches! You're like a real superhero, and we're just your sidekicks."

Halle shook her head. "I'm no hero. I spent most of the day running away. I was only able to scream because I was too terrified to do anything else, and now I can't even do that."

The twins stared at her, speechless. Worried expressions filled their faces: eyes wide, eyebrows slanted toward the center of their foreheads, mouths turned down. Then the worry shifted to something else. Their eyes dimmed, and they pressed their lips in a thin straight line. It was an expression she knew. She wore it on her own face every time her mother dismissed the movies Halle loved or made her stop practicing her scream. Every time Georgia and Dakota laughed at her or left her out. It was

Disappointment. She had failed Paisley and Argyle. And there was nothing she could do about it.

Mrs. Paprikash called to them from down the corridor. "This way, hurry!"

The twins held Halle's gaze for a moment longer, then turned and sped down the corridor after Mrs. Paprikash.

Halle was left alone in the entrance hall. Any moment now they'd open the cellar door to try and save their dad, only to unleash a slew of slimy parasites. If Aunt Rhona could see her now, she'd be screaming at her to leave. Halle reached for the front door. She had to get out of there before it was too late. She turned around for one last look at the entrance hall, and her eyes landed on the portrait of Raven Rose.

If only she were real. If only she were here right now. Aunt Rhona would never tell Raven Rose to get out. She would cheer her on. Not only could Raven Rose survive any ordeal, but she'd also save everyone else. Even when it looked like all hope was lost. They needed someone like that now. They needed a Raven Rose.

Halle's eyes fell on her flower arrangement under the portrait. It reminded her of something Aunt Rhona had said that morning. She'd said that Halle was the only one who could make that arrangement for Maximus Price. It was her job and her job alone. No one but Halle had the resources, the skill, and the intense admiration for Maximus Price and his work. Aunt Rhona had been right. But the twins were right too. At this point she might be the only one who could save their dad. She clenched her jaw. She couldn't leave now. She needed to finish her job. She needed to step up and be a Heroine, not a Damsel in Distress. She needed to be a Raven Rose.

And maybe, just maybe, instead of telling her to get out of there, Aunt Rhona would cheer her on like she always cheered on Raven Rose. After all, Aunt Rhona was the one person who always supported her. She encouraged her in all her interests. She didn't even mind when Halle practiced her scream.

But what could she do? Her voice was gone from all that screaming. It was exactly what Aunt Rhona had warned her about that morning.

A new thought suddenly popped into Halle's head. She raced down the corridor to catch up with the twins.

"Honey water! I need honey water!"

Halle, Paisley, and Argyle clustered around the narrow cellar door while they waited for Mrs. Paprikash to return with the honey water. Paisley put her ear to the door. "There are things moving around in there. I was hoping they'd already be dead from when you screamed earlier," she said to Halle. "From when you rescued Mrs. Paprikash and Mrs. Danworth in the humidor."

Argyle inspected the door. "It's got a pretty good seal around it. Probably to keep it cool inside. It must have muffled the sound."

"How many tredecapuses do you think are in there?" Paisley asked.

He looked at his watch. "Dad's been in there since about eight o'clock this morning. It's nearly five o'clock now. If they're replicating as fast as they did outside with one smelly cigar, there could be hundreds by now."

Halle's stomach worms churned and twisted inside her. Mrs. Paprikash hurried back with the honey water and handed it to her. With a trembling hand, she brought the glass to her lips and took a sip. The sweet, syrupy liquid coated her throat as she swallowed it.

"How big is the cheese cellar?" Paisley asked Mrs. Paprikash.

"Bigger than the humidor but not by much. When you open the door, there are six or seven steps down. The ceiling is so low you have to crouch, and the shelves are just loaded with cheese."

Paisley turned to Halle. "Are you ready?"

Halle imagined Aunt Rhona cheering her on as she took another big gulp of the honey water.

Paisley wrung her hands together, and Argyle chewed on his bottom lip as they waited for her to reply.

Finally she cleared her throat. "Listen," she told them, her voice only slightly less hoarse. "I want you to stand way back. If I can't scream, those things are going to jump out and get me. You'll need to run as fast as you can. Remember the rules. Take everyone else and go. Lock up the house and head into town. Promise me, okay?"

"We can't leave you," Paisley said.

"You'll have to. You're the only ones who know anything about these creatures. You can figure out some other way to stop them after you get to safety."

Argyle shook his head. "We won't have to. You can do this."

Halle paused. "Just promise me, okay?"

The twins looked at her and then at each other. They nodded and backed away.

Halle brought the glass to her lips once more and drank down the remaining liquid. She closed her eyes and breathed in and out, imagining the honey water filling her stomach, drowning the wriggling worms, and then seeping into her veins, pumping warmth and strength through her whole body—to her muscles, her bones, her skin.

She opened her eyes and handed the glass to Mrs. Paprikash, who took it and scurried back to stand with the twins.

Reluctantly, Halle reached for the cellar door. The handle was cold in her warm hand. She glanced back at Paisley and Argyle, who nodded in encouragement. She straightened her back and squared her shoulders. Then, with a burst of purpose and determination, she yanked the door open.

A waft of cold, putrid air hit her first, smelling like a mix of dirty socks and barnyard waste. She fought the urge to gag. Then a wave of creatures sloshed through the opening. Halle jumped back. Slime and tentacles oozed around her, slapping at her legs as the wave expanded down the hallway. Hundreds of tredecapus bodies wriggled and writhed together. She pressed her back against the wall. There were so many. As the mass spread outward toward the twins and Mrs. Paprikash, Halle swallowed.

Her mouth was suddenly dry again. She swallowed over and over, but she couldn't do it. She couldn't scream.

"Run," she rasped. "Run!"

And Mrs. Paprikash did.

But the twins didn't. Their eyes searched the wave of creatures for signs of their father. They craned their necks, trying to see inside the cellar.

"Run!" she pleaded again.

But instead, the twins grasped each other's hands. They waded straight into the expanding wave, slipping and sliding amid the wiggling bodies.

Paisley's and Halle's eyes met, and then Paisley did something unexpected. She let out a scream. It was a timid scream but a scream nonetheless. And then Argyle joined in with a scream of his own, lower pitched and throaty.

As the twins held their screams, they reached their free hands toward Halle, and she reached her hands toward them. They stomped and trudged toward each other through the gooey mass until their fingers touched, and she clasped their hands in hers, feeling their energy. Halle swallowed again, drew in an enormous breath, and screamed with every bit of power and force she possessed.

As soon as she joined in, the floor seemed to quake as the wave of tredecapuses roiled around their ankles. Halle and the twins squeezed their hands tighter together as if the floor might split open and suck them under. Paisley's scream died out. She was out of breath. Then Argyle, too, had to stop, but Halle held hers as the mass grew more and more turbulent around them. When her scream started to quaver, Paisley picked it up again, louder and stronger this time, and Argyle joined in again as well. Halle took another humongous breath while the twins held the vocal assault, and then she let out another scream.

The wave slowed to a bubbling mass and then ripples and then stillness before their screams faded and all was quiet again. Halle surveyed the hallway. Tredecapus bodies were suspended in a sea of slime, and the entire mass seeped down the corridor like molasses.

A cough and a groan came from inside the cheese cellar. Paisley and Argyle released Halle's hands and sloshed through the ooze toward the open cellar door.

"Dad? Dad? Are you okay?" Paisley cried as she peered down into the cellar.

Argyle joined her at the doorway and flipped on a

light switch. Halle peeked in between them. A pool of tredecapus bodies mixed with scattered wedges of cheese filled the tiny room, and in the middle of it all stood Maximus Price, covered in slime and trembling all over.

"Paisley? Argyle?" he stammered.

"Come on up, Dad. Everything is fine," Paisley said.

Maximus Price took a few feeble steps before the twins ran down into the muck to help him. They launched into explanations of tredecapuses and humidors and sousaphones while their father looked from one to the other with his eyebrows furrowed.

"Are you speaking some kind of gibberish?"

They laughed with delight, elated to see their father again. Argyle kicked tredecapus bodies off the steps, while Paisley held her father's arm to steady him.

"No, Dad. It's all real," she said. "You'll feel better in a few minutes, and we'll explain everything."

They reached the top of the stairs, where Halle stood waiting for them.

"Who is this?" Maximus Price asked.

Halle stared at him. His clothes were sopping wet, and his hair and beard were dripping with slime. Dark circles rimmed his baggy eyes. He was much smaller than she'd imagined he would be.

She stuck her hand out and shook his cold, wet one. "I'm Halle Thompson, the flower delivery girl."

She expected the worms to start wriggling again, but they were gone.

46

TERROR IN THE TUNDRA

While Argyle went to the kitchen to get food for everyone and Halle borrowed a phone to call her mom, Paisley escorted her father into the drawing room. She'd never seen him move more slowly, and his eyes had a strange, faraway look to them.

"Are you okay? Are you hurt? What do you need?" she asked.

He took a seat among the other adults and gazed around at the strange scene before him: The usually immaculate staff, now covered in slime, were sipping hot cocoa along with two equally slimy police officers and the mailman, the three of whom had all squeezed themselves together on a narrow settee.

He turned back to Paisley. "I'm a bit . . . cold," he said, rubbing his arms.

"Stay here. I'll be right back."

Paisley tore out of the room, through the halls, and up the steps to her bedroom. She threw open the dressing room door. There, in the back of the room, were the fluffy terry cloth bathrobes—one in every pastel color. She hesitated for a moment. There were so many to choose from. Then she bundled the whole lot of them in her arms and raced back down the stairs. Once in the drawing room, she passed the robes around to everyone before wrapping the last one around her father's shoulders and giving him a squeeze. She embraced him for several moments, pushing her warmth into his cold body. Then she stepped back to look at him. His eyes finally met hers.

"Thank you, Paisley. I don't know what I'd do without you."

Paisley didn't have words. Her father was okay. A bit worn out and quite a bit slimy, but he was okay.

Just then Halle arrived with a tray full of microwave popcorn bags, followed by Argyle carrying the big bowl of fruit from the kitchen.

"How's this for dinner?" Argyle asked.

Paisley took one of the bags and offered it to their dad.

He grabbed a handful of popcorn. "Thank you for making this, Halle. I'm famished."

"Oh, I didn't make it. Argyle did. I was on the phone with my mom. She and my aunt Rhona are coming to pick me up soon."

"Argyle, you should have let Mrs. Paprikash do that. You could have burned yourself."

"Don't worry. I was super careful. I knew the popcorn would be hot, but there were some giant mittens on the counter, so I put them on when I took the popcorn out of the microwave. I didn't want to bother Mrs. Paprikash."

Their dad stared at the popcorn in his hand, deep in thought.

Mrs. Paprikash grabbed an apple from the fruit bowl. "This is better than anything I could have managed right now. Thank you, Argyle."

Relief filled Paisley as the adults gradually gained their energy back. She glanced around the room and took everything in—the warm robes, the food, the grateful expressions on everyone's faces. She and Argyle were the ones taking care of *them* for once. There was an empty seat next to Halle, and Paisley sat down. If it weren't for Halle, this wouldn't have even been possible.

After a few moments of joyful snacking, her father stood up. He glanced around the room and cleared his throat. "I owe you all an apology. More than an apology. I owe you all my life!"

Mrs. Pennywhisker cut in. "We didn't do anything. It was the children."

The others nodded in agreement.

"The children? But how?"

As they relayed the story to their father, he sank back down in his chair. "I never imagined bringing that thing home with me would be dangerous. I thought it was dead."

"Where did it come from?" Argyle asked.

"After we finished wrapping up *Planet of the Prawn People*, I decided to do a quick location scout for my next movie, *Terror in the Tundra*. I found the perfect spot in the remote islands of Svalbard, and we cleared everything with the Norwegian government. While I was there, I got to explore some of the glacial caves, and I found what you call a 'tredecapus' impacted in the ice. I thought it was pretty neat, and I knew I could get a rise out of Mrs. Pennywhisker"—Mrs. Pennywhisker scowled at this—"but I never imagined it would come to life and attack everyone!"

"Obviously it wasn't dead," Argyle said. "The ice must have been keeping it dormant. Some types of bacteria and microbes can survive thousands of years in ice. Maybe this thing can too. Hey, maybe we just discovered some kind of prehistoric life-form!"

"I put it in an insulated container because I didn't want it to decompose before I got it home. I guess it must have thawed out on the way." He dropped his head into his hands.

Paisley should have been angry at her dad, but she was just so relieved that he was okay. She patted him on the back. "It's a good thing Halle showed up when she did. She's the one who saved us all."

Halle's eyes dropped to the floor. "Only after I spent most of the day running away."

"That was all right, though," Argyle said. "If we

hadn't been running away, we would have been attacked along with everyone else."

Paisley laughed. "Yeah. We thought it was our birthday prank."

Their dad looked up. "Your birthday prank? Good heavens! Your birthday isn't until tomorrow!"

47

BIRTHDAY PLANS

There was a loud rapping from the brass door knocker at the front door. Halle jumped up. "That's probably my mom and my aunt. I guess it's time for me to go."

"Nonsense," Maximus Price said. "Invite them to come in. Argyle made plenty of popcorn."

Mr. Belloft excused himself to answer the door.

"Halle," Maximus Price continued, "if there's anything I can do for you, just say the word. You not only saved me, but my children and my household as well."

"And probably the whole town," said Argyle.

"Maybe even the world!" Paisley added, pulling out

her notepad and checking the final items off her list.

"We did it together," Halle said. "I couldn't have done it alone."

"Still. If there's anything I can do for you, I'll do it," Maximus Price said.

"Well, there is one little thing. . . ."

"Anything!"

Halle smoothed her hair and bit her lip. "You could maybe . . . give me a tiny part as a Scream Queen in one of your movies. . . . I'm really good at screaming . . . and running . . . and tripping."

Maximus Price frowned and rubbed his beard. "Oh, I don't know that you've got what it takes to be one of my Scream Queens."

Halle's eyes dropped to her lap. "Well, maybe not now. Maybe when I'm a little older. I'll keep practicing! And you wouldn't even need to give me any lines. I'll just run and trip and scream."

Maximus Price shook his head. "No, no. You're definitely not Scream Queen material."

"Oh," she said. Her heart seemed to shrivel inside her chest. She swallowed and blinked back tears.

Maximus Price leaned over and lifted her chin.

"That's because you are a Heroine, Halle."

She looked at him, not sure she had heard him correctly.

"You may be good at acting like a victim, but you're not one. Anyone can see that. Deep inside, you are mighty. You are brave. That's what I need. I haven't had a true heroine in my movies since . . ." His eyes grew misty for a moment and then he focused back on Halle. "Are you interested?"

At first, she couldn't speak. Was he really saying what she thought he was saying? He kept looking at her, and she realized he was actually waiting for an answer. "Yes! Yes! A thousand yesses!"

Someone squealed behind her, and she turned to see Aunt Rhona and her mom. Aunt Rhona had the biggest grin on her face. They'd clearly heard everything. Halle ran to them and wrapped her arms around them. Tears leaked from her eyes and seeped into her mother's shirt.

Aunt Rhona smoothed down Halle's hair. "I'm afraid we've just missed tonight's showing of *Raven Rose and the Isle of Empty Souls*, but there's a matinee tomorrow afternoon. Maybe you can invite your new friends."

Halle wiped her tears and turned to the twins, her

eyes glowing with excitement. New friends. That was exactly what they were. "That would be amazing! We could take you for your birthday!"

Maximus Price cleared his throat. "Of course they can see the movie. They'll be thirteen tomorrow. But we can have a private showing right here at Price Manor. You're all welcome to come. There's no need for us to go into town."

Halle turned to Argyle, and their eyes met. She gave him a nod of encouragement.

He stood up and stepped toward his father. "Dad? Can we see the movie in a real movie theater?"

"We have a full-sized screening room right here at the house. It's practically the same thing."

"But it's not the same, Dad. It doesn't smell like popcorn, and we don't get to see any previews. And there's no huge crowd in the audience to enjoy the movie with. Can we please go? It would just be for a few hours."

Maximus Price shook his head. "It's not safe, Argyle. Something terrible could happen, especially with your delicate condition. What if you passed out?"

"We'll get popcorn and snacks!"

"But you and Paisley have never been into town

before. You could be hit by a car while crossing the street."

"You could teach us how to cross safely. If other kids can do it, we can too. Right, Paisley?"

Paisley hesitated.

Maximus Price continued. "We have no control over what happens out there. There could be a fire at the movie theater. Or a flood. It's better if you stay here. This is The Safest Place on—" Maximus Price glanced around at the roomful of people, their matted, slime-covered hair and their lumpy bathrobes. "Oh. Right." He turned back to Argyle.

"You're keeping us alive, Dad, but we need to *live*."

Maximus Price bowed his head. "Well, I suppose if we're *very careful* . . ."

"Really? We can go?" Argyle asked. "And can we go to the ice cream parlor afterward? And the mini-golf place?"

Maximus Price squirmed and wiped sweat from his brow.

Officer Munrap sat up straight and puffed out his chest. "We'll be on duty, Mr. Price. No need to worry."

Halle wasn't sure how helpful that was coming from a police officer wearing a fuzzy bathrobe who'd been pretty much a zombie only a few hours ago, but Maximus Price sighed and nodded his head.

Argyle jumped out of his chair and did a happy dance. “Yes! Yes! Yes!”

“Paisley? What about you?” her father asked.

Paisley still looked a little unsure.

Halle took her hand. “You helped me roller-skate. Let me show you what it’s like outside of Price Manor. Maybe we could even go to the roller-skating rink sometime.”

Paisley’s eyes lit up. “Really?” She turned to her father. “Would that be okay?”

Maximus Price let out a breath and smiled. “As long as you’re with Halle, I’m sure you’ll be just fine.”

Halle, her mother, and Aunt Rhona stayed late into the night. She filled them in on the whole story. Aunt Rhona squealed and shrieked at all the good parts, just like she always did when they discussed movies at the ice cream parlor. Not only that but she convinced Maximus Price to have Argyle tested for anemia. A simple course of vitamin B12 might be all he needed to keep his energy up permanently. Meanwhile, Argyle replenished the popcorn several times, and Paisley performed a few solos on the sousaphone. Finally, when Halle could barely keep her eyes open any longer, she suggested they go home.

As they were leaving, the enormous grandfather clock bonged in the entrance hall. It was midnight. At the same time, an obnoxious beeping sound filled the air.

Everyone put their hands over their ears.

“What is that horrible noise?” Paisley asked.

“It’s our watches! The alarms must have been set for midnight,” Argyle said.

“Ugh. How do we turn them off?”

Their father hurried over. “Ah yes. That means it is officially your birthday! Just press that little button on the left side to shut off the alarm.”

The twins both pressed the buttons. The instant the alarms stopped, a stream of water shot out from the watches and doused their faces.

“Really, Dad? Squirting watches?” Argyle said.

Maximus Price doubled over with laughter. “Happy birthday, my dears!”

The twins wiped the water from their eyes.

Paisley glared at her father. “That’s it? That’s our birthday prank?”

Maximus Price winked. “You weren’t expecting that, were you?”

Mrs. Pennywhisker chuckled and patted the twins on their backs. “Happy birthday, sweethearts. Now, what on earth did I do with that cake carrier?”

ACKNOWLEDGMENTS

This book would not have been possible without my true love, my partner in this wonderful life, my husband—Angel. I thought being a published author was a distant dream, but you made me realize that dream could be a reality. Your support, encouragement, and motivation were essential to me on this journey, and I wouldn't have wanted to take it with anyone else.

Humongous thanks to my agent, Kaitlyn Katsoupis, for your enthusiasm for this wacky story, for encouraging me to develop these characters even further, and for championing this book to its home.

Deepest gratitude goes to my editor, Sarah Homer, for making my dream come true and for your guidance and hard work in making this book shine, and also to Tara Weikum for steering the ship. Thanks to all the amazing folks at HarperCollins who have transformed this story from words on my laptop to an actual physical book.

Tremendous appreciation goes to my incredible and talented critique partners, Hilda Burgos, Nicole Wolverton, Amy Sisson, and Susan North, for years of useful feedback on numerous drafts and for all the celebrating and commiserating along the way.

An abundance of gratitude to my beta readers who read early versions or sections of this book and gave initial feedback and advice: Chris Higgins, Terrilyn McCormick, Brian Poucher, and Nicole Valentine. A mountain of thanks to Sharon Nicolary and Karen Jones for weekly feedback and in-person discussions at the Green Line Cafe. A cornucopia of appreciation goes to Kelly Chancey for enthusiastically providing suggestions on *both* versions of this story and believing in and encouraging me on this journey. To the fierce and fabulous Kim Briggs, you get me as much as I get you.

Thank you for the manuscript critique swap and for your friendship and collaboration on so many other projects. And a very special shout-out goes to my young beta readers, Millie, Alfie, Flare, Griffin, and Jove, for your early thoughts and impressions.

To my SCBWI family: Rona Shirdan, Alison Green Myers, Lindsay Bandy, Kristen Strocchia, Berrie Torgan-Randall, Heather Stigall, Joanne Roberts, and Virginia Manning. I'm so grateful for the support and friendship of our wonderful team.

I could never have gotten anywhere without my best friend and muse, Hannah Dalton. Thank you for sharing and cultivating my appreciation for the absurd, the weird, and the wonderful.

To my kids, Millie and Owen, thank you for continuing to endow our home with creativity, music, imagination, and humor.

I'm incredibly grateful for the support of my family. To my mom, Edie Parnum, who read this story and thought it was publishable (and who wasn't just saying that because she was my mom), thank you for believing in me. Those overly eager bird-watchers were written for you. And to Sarah, Ben, Sam, Flare, Greg, Brian, Lois,

Scott, Morgia, Eddie, Chris, Jen, Tammy, Luis, Sarina, Azire, Beki, Tres, Cristian, Emma, Shelly, Tom, Eli, Max, Mimi, and Papá, thanks for cheering me on from the sidelines and for always asking, “How’s the book coming along?” It’s finally here.

And last but not least, thank you to you, dear readers, for going on this journey with Halle, Paisley, and Argyle. You are all heroes and heroines in your own lives.